Tune in the World with Ham Radio

Eighth Edition

The American Radio Relay League, Inc.
Newington, CT 06111

EDITED BY

Larry D. Wolfgang, WA3VIL
Kirk Kleinschmidt, NTØZ
Joel P. Kleinman, N1BKE
Gerald L. Hall, K1TD
Bruce S. Hale, KB1MW

PRODUCTION STAFF

Mark Wilson, AA2Z
David Pingree
Steffie Nelson, KA1IFB
Sue Fagan
Michelle Chrisjohn, WB1ENT
Leslie K. Bartoloth, KA1MJP

COVER ARTWORK

Bev Rich

CODE TAPE NARRATION

Loraine McCarthy, N6CIO

This work is publication No. 28 of the Radio Amateur's Library, published by the League.

Printed in USA

Library of Congress Catalog Card Number: 76-13248

Book only ISBN: 0-87259-246-4
Kit ISBN: 0-87259-247-2

Eighth Edition
Second Printing

Foreword

Whatever your interests have been, Amateur Radio will add new dimensions and new horizons to your world.

I know. It literally took me out of this world. When I became a radio amateur in my early teens I had no idea where it would lead. But it turned out to be the first step into our NASA space program, into my becoming an astronaut and a member of the Skylab 3 team. In December 1983, I was privileged to be the first radio amateur to operate from space, while on board the space shuttle "Columbia." Tens of thousands of amateurs around the world heard my signals, and together we were able to perform some meaningful experiments.

Amateur Radio today is many different things. What excites you may be a new friend on the other side of the world or across town, finding yourself acting as a vital communications link in a very real emergency or experimenting with new skills and new techniques. Best of all, ham radio lets you do what you want in a variety of challenging directions without too many bothersome restrictions and limitations.

ARRL helped me and countless thousands of hams to get started over the past 75 years. Now they have made obtaining a Novice license easier and more fun than ever.

Let me wish you the best of everything in joining our special fraternity. Perhaps we'll meet on the air one of these days.

73,

Owen K. Garriott, W5LFL

Preface

Welcome! If you are reading these lines, chances are that you would like to join the nearly two million people in every corner of the globe who call themselves radio amateurs, or "hams." They have earned the very special privilege of being able to communicate directly with one another, by radio, without regard to the barriers that so often interfere with our understanding of the world. Whether across town or across the sea, hams are always looking for new friends; so, wherever you may happen to be, you are probably near someone—perhaps a whole club—who would be glad to help you get started.

Most of the active radio amateurs in the United States are members of the American Radio Relay League. ARRL has been the hams' own organization for more than 75 years, providing training materials and other services and representing its members nationally and internationally. *Tune in the World with Ham Radio* is just one of dozens of ARRL publications for all levels and interests in Amateur Radio. You don't need a ham license to join. If you're interested in ham radio, we're interested in you. It's as simple as that!

Most of you purchased this book as part of a package that includes two cassette tapes to help you learn the Morse code. Or, you may have purchased just the book by itself; if you find that you want the cassette tapes, they are available from ARRL Headquarters. Others of you received your copy along with either the AEA Morse University™ or the GGTE Morse Tutor computer program. If so, you have an advanced Morse-code-teaching program at your fingertips and probably will not need the tapes.

This edition of *Tune in the World* is based on the pool of Novice examination questions prepared for use beginning November 1, 1989. You should also be aware that the Federal Communications Commission reorganized its rules governing Amateur Radio effective September 1, 1989. The new "Part 97" of the FCC Rules, with additional explanatory material, is contained in the Eighth (or later) Edition of another ARRL publication, *The FCC Rule Book*; you will want to obtain a copy for reference while you're studying, and also to have in your "ham shack" when you're licensed to operate. The new question pool was developed before the new version of the rules was released. These new rules may affect the answers to some questions. To avoid confusion, Volunteer Examiners have the discretion to not include such questions on the examinations they administer.

The Volunteer Examiner Coordinators' Question Pool Committee is expected to release a supplement to the Element 2 (Novice) Question Pool on or before March 1, 1990. This supplement will include revisions to questions and answers affected by the new rules. The supplement is expected to be put into use on July 1, 1990.

Now—on to your studies!

David Sumner, K1ZZ
Executive Vice President

Newington, Connecticut
August 1989

Contents

Important Note: Please see the information on page 12-1 concerning the Element 2 question pool and the Element 2 supplement before you begin this book. The Element 2 Supplement follows page 12-35, at the end of the Novice Question Pool Answer Key.

HOW TO USE THIS PACKAGE

You're about to begin an exciting adventure: a fun-filled journey into the world of Amateur Radio. *Tune in the World With Ham Radio* is *the* study guide to help you reach your goal. This book introduces you to basic radio theory, which is far less frightening than it sounds. You'll also learn Federal Communications Commission (FCC) rules and regulations and the international Morse code—just as the FCC requires. You will learn enough to pass your Novice exam with ease. But we in the American Radio Relay League care much more about you and your future in Amateur Radio than to stop there! We won't abandon you once your Novice "ticket" is in hand.

Tune in the World is also a pathway to a successful beginning after you've passed the Novice license exam. In this book you'll find the practical knowledge needed to become an effective, competent communicator. We all hope you will take enough pride in the achievement of earning your license to be a considerate communicator as well.

But that's all a few weeks and a few pages down the line. To ensure your success and head you in the right direction toward your first on-the-air contact, here's how to use this package to your best advantage.

Tune in the World has been designed and written by a staff with a great deal of experience, backed by decades of tradition in Amateur Radio and its teachings. The cassette tapes with this package give you everything you'll need to learn the Morse code at 5 words per minute. The package also provides everything you'll need to learn—and understand—the theory you should know to operate an Amateur Radio station. The book explains all the rules and regulations required of a Novice candidate, not only to pass the test, but to operate properly (legally) once your license is hanging proudly in your shack.

SELF STUDY OR CLASSROOM USE?

We designed *Tune in the World With Ham Radio* both for self study and for classroom use. An interested student will find book complete, readable and easy to understand. Read carefully, and test yourself often as you study. Before you know it, you'll be ready to pass that exam!

Why deprive yourself of the company of fellow Novice candidates and the expertise of those "old-timers" in your hometown, though? *Tune in the World* goes hand-in-hand with a very effective ARRL-sponsored training program run by over 6000 volunteer instructors throughout the United States. If you would like to get in touch with a local class, just contact the Educational Activities Branch at ARRL Headquarters—we'll be happy to assist.

Amateur Radio is not necessarily a loner's hobby—as you'll soon discover. Hams are very social animals who derive a great deal of pleasure from helping a newcomer along the way. The most effective learning situation is often the one you share with others. There are knowledgeable people to turn to when you have a question or problem. You can practice the Morse code with fellow students and quiz one another on the basic electronics concepts.

It doesn't matter if you're studying on your own or joining a class. Use *Tune in the World* to study for your Novice exam and you'll be on the air in no time at all.

USING THE PACKAGE

Tune in the World leads you from one subject to the next in a logical sequence that builds on the knowledge learned in earlier sections. It presents the material in easily digested and well-defined "bite-sized" sections. You will be directed to turn to the question pool in Chapter 12 as you complete a section of the material. This review will help you determine if you're ready to move on. It will also highlight those areas where you need a little more study. In addition, this approach takes you through the entire question pool. By the time you complete the book, you will be familiar with all the questions used to make up your test. Please take the time to follow these instructions. Believe us, it's better to learn the material correctly the first time than to rush ahead, ignoring weak areas and unresolved questions.

Every page of this package presents information you'll need to pass the exam and become an effective operator. Pay attention to diagrams, photographs, sketches and captions; they contain a wealth of information that you should know. You'll also find a few anecdotes and "mini-articles" (called Sidebars) that will help put the tradition of Amateur Radio in perspective. Our roots go back to the beginning of the 20th Century, and our community service continues even as you read this.

Start at the beginning. Chapter 1 summarizes the fun you will have when you earn your license and join the thousands of other active radio amateurs. You should read it casually, with the goal of learning a little bit about the group you'll soon be joining. Chapter 2 explains the need for international and national regulation. It defines who the regulating bodies are and describes Part 97 of the FCC Rules and Regulations as they apply to the Novice Amateur Radio license.

Chapter 3 lets you in on a little secret. The Morse code is not only easy to learn; it can be so enjoyable that you may become "addicted." You should read this chapter before beginning any serious study of the Morse code. (You've undoubtedly had one of the code tapes in a cassette player already!) People try to learn the code by many different methods. The code cassettes included with this package teach you the Morse code at your own pace. You should rewind the tape to review any characters that give you difficulty. The technique used with these tapes also makes it easy to increase your speed once you've learned the code. Trust us, stay with our system, and master each section as it is presented before going on. Be sure to set aside the time for a couple of short Morse code practice sessions *every* day. Marathon, last-minute cramming is simply not the way to learn the code—or to learn any acquired skill.

The code-teaching tape presents the letters, numbers and punctuation marks one at a time. Take your time; we've found that frequent, short practice sessions are more effective than long sessions spaced further apart. Twenty minutes of practice, twice a day, is ideal. After reading Chaper 3, begin studying the code and theory in "parallel"; that is, split your study time between daily sessions with the code and study sessions with the text. Move ahead at a pace that allows you to master the material. You do have to keep moving, however. Don't allow yourself to stay on one section too long. Move ahead, even if you have to come back and review later.

Chapter 4 does for the theory what the cassettes do for the code. Basic radio theory is broken into well-defined sections. Each section builds on the previous material to explain the theory you'll need to answer the questions in Chapter 12.

Study these sections one at a time. Be sure to follow the instructions at specific places in the text that direct you to study the actual test questions. This is the best way to determine how well you understand the most important points. The text explains the theory in fairly simple terms, so you shouldn't have any problems. Don't be afraid to ask for help if you don't understand something, though.

If you're participating in an official ARRL-registered class, you'll have the chance to ask the experts for help. Ask your instructor about anything you are having difficulty with. You may also find it helpful to discuss the material with your fellow students. If you're not in a class and run into snags, don't despair! The Educational Activities Branch at ARRL Headquarters will be happy to put you in touch with an Amateur Radio operator in your area who can help answer your questions.

Chapters 5 and 6 cover some basic circuit components and describe how to put them together to make a working electronics circuit. You will also learn how to connect various pieces of equipment to form an Amateur Radio station.

Chapter 7 offers some guidelines to help you select the equipment you'll need to set up your own radio station. Chapter 8 gives you all the information about antennas that you'll need to pass your exam and get started in Amateur Radio. Full construction details are provided for several simple antennas suitable for Novice operation.

Chapter 9 will help you assemble an effective Amateur Radio station. After your license arrives from the FCC and your station is set up, Chapter 10 leads you through those first few contacts. How do you know when to operate on what band? Where do you turn to make contact with a specific foreign country or faraway state? How do you establish contact in the first place and what do you say? What sorts of activities will you want to become involved with on the air? Chapter 10 has the operating information to make you feel at ease on the air. Chapter 11 will help you identify and solve the problems you are most likely to encounter as a new Novice.

All these chapters cover information necessary to pass the exam and the practical information you'll need to be a good operator. Study the material presented in this book and follow the instructions to review the exam questions. You'll cover small, bite-sized pieces of the text and a few questions at a time. Review the related sections if you have any difficulty, and then go over the questions again. In this way you will soon be ready for your Novice exam. Before you know it, you'll be on the air talking to hams around the world!

Most people learn more when they are actively involved in the learning process. Turning to the questions and answers when directed in the text helps you be actively involved. If you would like to mark the correct answers in the question pool, you can mark them as you study. This will reinforce the material in your mind. Make an asterisk or check mark in the left margin next to the correct answer. Then you can cover those marks with a slip of paper when you want to review later. Paper clips make excellent place markers to help you find your spot in the text, the question pool and the answer key.

THE TEST

The FCC requires two licensed hams to administer a Novice exam. If you're taking a class, you'll have no problem in locating two experienced hams to administer the code test and written exam. Who may give the Novice test? Sections 97.513 and 97.515 of the new FCC rules (effective September 1, 1989) list the qualifications for the two people who administer a Novice test. Both Novice examiners must:

1) hold a current General, Advanced or Amateur Extra class operator license issued by the FCC;

2) be at least 18 years of age;

3) not be related to the applicant;

4) not own a significant interest in, or be an employee of, any company or other organization engaged in the manufacture or distribution of equipment used in connection with Amateur Radio transmissions, or in preparation or distribution of any publication used in preparation for obtaining an Amateur Radio license. An employee who can demonstrate that he or she does not normally communicate with that part of an organization engaged in such manufacture or publishing is eligible to be a volunteer examiner, however;

5) never had their Amateur Radio station or operator's license revoked or suspended.

The code test requires you to show your ability to send and receive the Morse code at 5 words per minute. Here is a simple way to estimate code speed: Five characters make one word, with punctuation and numerals counting as two characters each. You must know the 26 letters, 10 numerals and the basic punctuation marks: comma, period, question mark, double dash (= also called $\overline{\text{BT}}$) and fraction bar (/ also called $\overline{\text{DN}}$). You must also know some common procedural signals (prosigns): $\overline{\text{AR}}$ (the + sign, also used to mean "over" or "end of message") and $\overline{\text{SK}}$ ("clear; end of contact"). For further details on the code test, please refer to Chapter 3.

For the written exam, you must answer 30 questions about Amateur Radio rules, theory and practice. Your test questions will be drawn from a pool of 372 questions. Most VECs (including the ARRL VEC) agreed to use the question pool printed in Chapter 12, beginning on November 1, 1989. These questions include multiple-choice answers that most VECs have agreed to use.

The question pool is divided into nine subelements. Each subelement is further divided into blocks, with one question to come from each block. Your test must include one question from each of the 30 blocks. Your examiners may select the questions, or they may contact the Educational Activities Branch at ARRL Headquarters for a printed exam. Your examiners could simply go through the question pool printed in Chapter 12 and have you answer the chosen questions right from the book. (Provided you have not marked the correct answers beforehand!) The passing grade for an amateur license is 74%, so you will need 22 or more correct answers out of the 30 questions on your exam to pass the test. (Another way of putting this is that you may have as many as eight incorrect answers and still pass!)

The questions must be used exactly as printed (and as reproduced in Chapter 12 of this book). Novice examiners may use their discretion about the examination format. This opens several possibilities for the examiners. The exam could be multiple-choice, true-false, or essay type. They may even conduct your test on an oral "interview" basis, with the examiners asking the questions and the applicant responding. The examiners aren't *required* to use the multiple-choice answers printed with the questions in Chapter 12, but we highly *recommend* it.

Before you go to take the Novice exam, you should fill out an FCC Form 610. (Chapter 2 has full details about how to fill out this form.) After you complete the test, your examiners will grade it. When you pass both the code and written tests, the examiners will complete the Administering VE's Report on the front of the form and the Certification section on the back, indicating that you passed both the code and written elements.

The examiners send the completed application to the FCC in Gettysburg, PA. Now comes the hard part! It may take as long as six weeks for your license to arrive in the mail. The FCC staff processes your application along with the thousands of other 610 forms they receive each month.

If you need help in locating someone to administer the test, drop us a note at the Educational Activities Branch at ARRL Headquarters. We can put you in touch with examiners and clubs in your area.

Give *Tune in the World* a chance to guide you the way it was intended—by following these instructions. You'll soon be joining us on the air. Each of us at the American Radio Relay League Headquarters and the entire ARRL membership wishes you the very best of success. We are all looking forward to that day in the not-too distant future when we hear your signal on the Novice bands. 73 (best regards) and good luck!

Chapter 1

Exploring Ham Radio

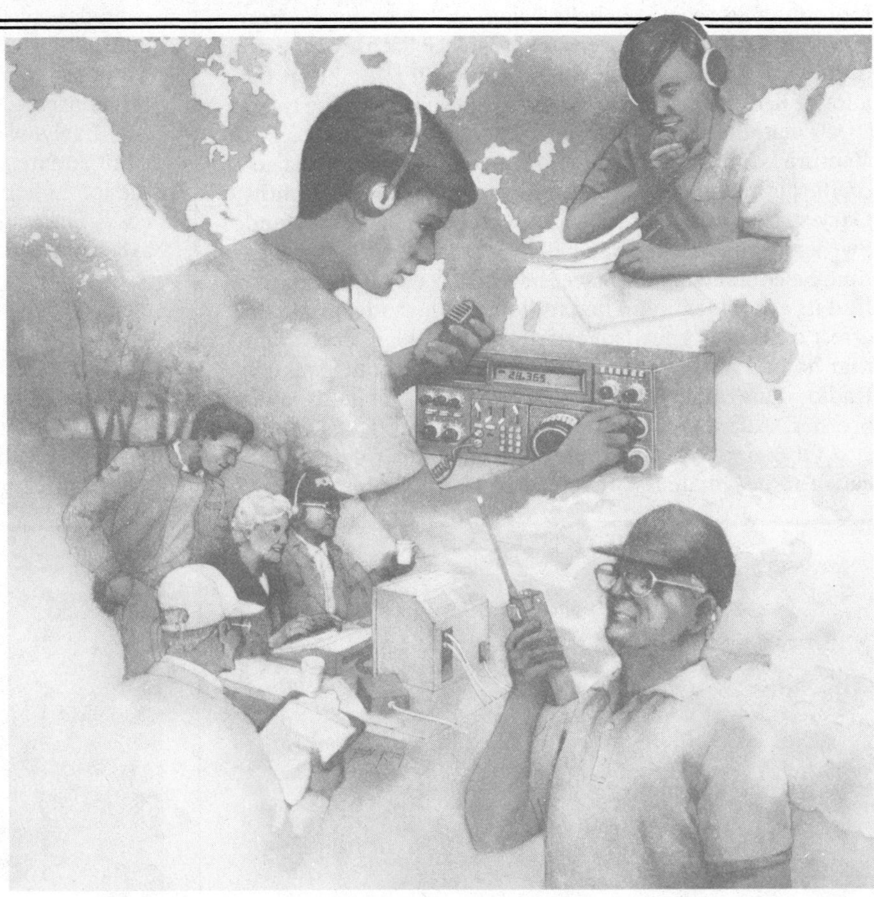

Hobby of a Lifetime: Ham Radio

Amateur Radio . . . ham radio. What does it mean to you? What images come to mind when you hear the term? Morse code? That evening news report about the Amateur Radio operators who set up emergency transmitters after an earthquake? The funny-looking antenna in your neighbor's yard? A birthday greeting via Amateur Radio from your uncle who lives clear across the country? No matter what images come into your mind, ham radio is FUN!

Obviously, you know something about ham radio; you have this book. But you want to know more. *Tune in the World with Ham Radio* will introduce you to the wonderful hobby of Amateur Radio. It will answer your many questions on the subject, and lead you to your first license, the Novice license.

WE COME FROM ALL WALKS OF LIFE

Communicating and experimenting. That's what Amateur Radio is all about. That's why people from all walks of life become hams. Young or old, we all enjoy the thrill of meeting and exchanging ideas with people from across town or from the other side of the earth. The excitement of building a new project or getting a circuit to work properly is almost beyond description.

Carm Prestia is a diamond-in-the-rough police sergeant, patrolling a bustling university town tucked away in the mountains of Pennsylvania. By night, he packs a .38-caliber Police Special to protect thousands of his fellow townspeople. By day, he wields a soldering iron in pursuit of the world's greatest hobby: Amateur Radio.

"I love to talk to people on the radio," Carm explains as he unstraps his portable radio from his uniform belt. "I talk all night at work, but I still go home and fire up the ham gear."

Carm's shack (hams all over the world affectionately call this room their "shack") is in a corner of his basement. His equipment table holds a transmitter for sending and a receiver for listening. His radio gear works with an antenna outside, above his backyard. Carm can talk with a friend in the next town one minute and with a ham halfway around the world in Australia the next.

Each Amateur Radio station has its own distinctive call sign. The Federal Communications Commission (FCC) issued Carm his call sign, WB3ADI. The FCC is the regulatory agency responsible for issuing licenses in the United States.

Ham radio operators are so proud of their call signs that the two often become inseparable in the minds of friends. Barry, K7UGA, of Arizona has worked (talked with) thousands of hams on the air. Many of them didn't know that his last name is Goldwater or that he was a United States senator. King Hussein of Jordan is also a ham, known simply as JY1 to all his on-the-air friends.

AGE IS NO BARRIER

Age is no barrier to getting a ham license and joining in the fun. There are hams of all ages, from five years to more than 80 years. Michelle Allen, an "A" student from New Haven, Indiana, received her Novice license at the age of 12. Known to her ham friends as KA9FUL, she spends a lot of her time (aside from school and sports) on the radio.

Then there's 10-year-old Gary Lieb, KA6DLE, of Ventura, California. Every morning before school, he gets on the air and makes several contacts. In just a few months, Gary spoke (using Morse code) with hams in Switzerland, Sweden, Canada, Japan and all over the United States. Each time he contacts a new place, he looks it up on the map to find its exact location. Ham radio has thus had a profound effect on Gary's knowledge of geography. Gary does admit that he had a small advantage in learning about Amateur Radio. His father, Jerome, is WA6GSA, and his older brother, Adam, is WA6JGK.

Of course, young people aren't the only ones who are active radio amateurs. Involved in many hobbies, Evelyn Fox of Merrimac, Wisconsin, plays contract bridge with her AARP (American Association of Retired Persons) group on the 40-meter band. Electronics may seem a little mysterious to many people even though it plays a major part in their everyday lives. A stranger to the field may at first be confused by the volt, ohm and ampere. After someone explains the basic concepts in simple terms, however, these terms aren't as frightening.

Evelyn was over 75 years old when she became interested in Amateur Radio. That didn't stop her from taking on the job of learning radio theory and the international Morse code. She found the theory a pleasant challenge. She joined a club, attended its classes, and now holds the call sign WB9QZA. Not bad for someone who knew nothing about electronics when she first got started.

"BREWING" IT AT HOME

Ham radio operators pop up in some of the least expected places. Dr. Peter Pehem, 5Z4JJ, is one of Africa's flying doctors. He works out of a small village on the north slope of Mount Kilimanjaro in Kenya. Pete has been bitten by an

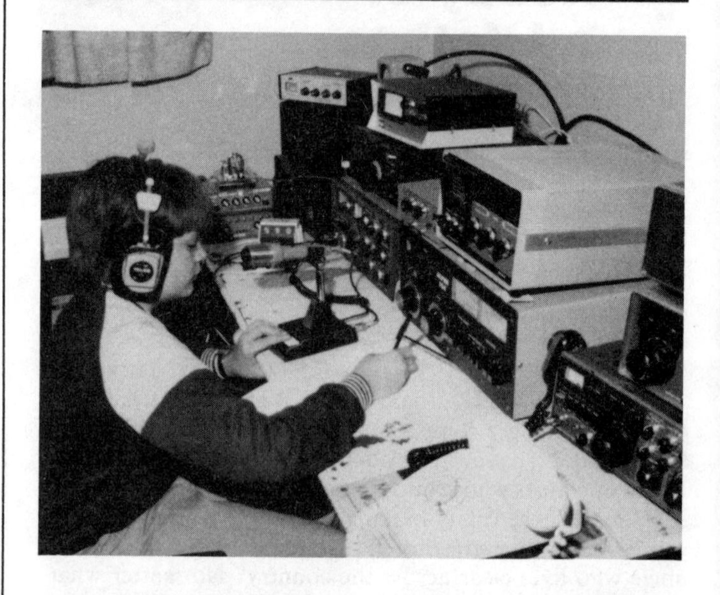

OSCAR bug, but can't do anything about it while on medical duty. But he can and does attack it with great pleasure when he's off duty.

Somebody gave Pete an old radiotelephone, a vacuum tube and some coaxial cable. The doctor added empty aspirin tins and a quartz crystal from his airplane radio. Right out there in the African bush, he fired up a homemade transmitter, built on the aspirin tins. Then he talked to the world through OSCAR, the Orbiting Satellite Carrying Amateur Radio.

Pete proved something with his homemade gear; you don't need a shack full of the latest commercial equipment to have fun on the air. New hams find this out every day. Tom Giugliano, WA2GOQ, of Brooklyn, New York, contacted 26 of the 50 states using a pre-World War II transmitter and receiver. He used a simple homemade wire antenna. Other hams have been successful in bridging the oceans to contact hams in Europe and Japan using simple equipment running less than 1 watt of power. As a Novice, you'll be permitted 200-watts peak envelope power (PEP) output. That's more than enough to contact other hams around the world.

There was a time, many years ago, when no commercial equipment was available. The earliest hams, beginning more than 75 years ago, tried to find more efficient ways of communicating with each other. All early radio sets were home-brew (homebuilt) and were capable only of communication over several miles. Some transmitters were nothing more than a length of copper wire wrapped around an oatmeal box, attached to a few other basic parts and a wire antenna. Often the transmissions were one-way, with one transmitting station broadcasting to several receive-only stations. Over the years, hams have continually looked for ways to transmit farther and better. They are constantly developing and advancing the state of the art in their quest for more effective ways of communicating.

PEERING BACK THROUGH TIME

It all started on a raw December day in 1901. Italian inventor and experimenter Guglielmo Marconi launched the Age of Wireless from an abandoned barracks at St John's, Newfoundland. He listened intently for a crackling series of buzzes, the letter S in international Morse code, traversing the 2000 miles from Cornwall, England. That signal was the culmination of years of experimentation.

Later, Marconi set up a huge station at Cape Cod that was unlike anything today's ham has experienced. Marconi's 3-foot-diameter spark-gap rotor fed 30,000 watts of power to a huge antenna array suspended from four 200-foot towers on the dunes at South Wellfleet, Massachusetts.

By 1914, Marconi had set up a station and antennas for daily transmission across the Atlantic. And Amateur Radio operators all over America were firing up their own homebuilt transmitters. Soon, several hundred amateurs across the country joined Hiram Percy Maxim in forming the American Radio Relay League (ARRL), based in Hartford, Connec-

The state-of-the-art in amateur gear has come a long way since the days of "Old Betsy," Hiram Percy Maxim's own spark-gap transmitter shown below. Spark-gap transmitters were the very first type of radio transmitters, used in the early 1900s. The modern equipment shown above has been part of the W1AW visitors' operating position. This station is capable of operating on the 1.8- to 435-megahertz (MHz) amateur bands and can be used for radioteletype and the OSCAR satellites.

ticut. These amateurs set up a series of "airborne trunk lines" through which they could relay messages from coast to coast. If you're interested in learning more about the history of Amateur Radio, you'll enjoy *200 Meters and Down* by Clinton B. DeSoto. *Fifty Years of ARRL* is also an interesting account of the ARRL's first half century. Both books are available from your local ham radio dealer or directly from ARRL.

There were more hams experimenting all the time. Commercial broadcasting stations began to spring up after World War I. This brought a great deal of confusion to the airwaves. Congress created the Federal Radio Commission in 1927 to unravel the confusion and assign specific frequencies for specific uses. Soon amateurs found themselves with their very own frequency bands.

Continued experimentation over the years has brought us tubes and transistors. Equipment has grown smaller and more sophisticated. In the early days of radio communications, equipment was large and heavy. Sometimes it would take up a whole room for what now can be done with the circuitry in a tiny box.

WE PITCH IN WHEN NEEDED

Traditionally, amateurs have served their countries in times of need. During wartime, amateurs have patriotically taken their communications skills and technical ability into the field. During natural disasters, when normal channels of communication are interrupted, hams provide an emergency communications system. Practically all radio transmitters operating legally in the United States provide some public service at one time or another. Hams don't wait for

Public service has been a ham tradition since the very beginning. Whether it's a walk-a-thon, the Olympic Torch Run or the aftermath of a tornado, hams are always there to help with communications, which they provide at no cost whatsoever to the group involved.

someone to ask for their help; they pitch in when needed. They provide communications on March of Dimes walk-a-thons, help plug the dikes when floods threaten and warn of approaching hurricanes. Hams bring assistance to sinking ships, direct medical supplies into earthquake zones, and search for downed aircraft and lost children.

Amateur Radio operators recognize their responsibility to provide public-service communications when necessary. They train in various ways to be effective communicators in times of trouble. Every day, amateurs relay thousands of routine messages across the country. They send many of these messages through "traffic nets" devoted to developing the skill of sending and receiving messages efficiently. (A net is a gathering of hams on a single frequency for some specific purpose. In this case the net's purpose is to "pass traffic"—relay messages.) This daily operation helps prepare hams for real emergencies. Also, each June thousands of hams across the country participate in the ARRL Field Day. They set up portable stations, including antennas, and use emergency power. These operating events help hams to test their emergency communications capabilities. Such events also help them identify problems that could arise in the event of a disaster.

MAYDAY—WE'RE GOING DOWN

Hams often provide relays for ships that run into trouble on the high seas. A huge, destructive wave struck the 35-foot yacht *Gambit* while the yacht was sailing in the South China Sea. The force of the wave smashed the porthole glass and broke the main boom. The crew of the yacht suddenly found their ship adrift and taking on water.

Fortunately, the yacht's captain, Dean Pregerson, WH2ABD, is a ham, and he had a radio on board. He quickly put out a distress call on the 20-meter amateur band. Hams in Hong Kong, Japan, Indonesia and Guam heard his call. Tony Armstrong, VS6AG, notified Hong Kong Marine Search and Rescue Center. The hams on Guam alerted the US Air Force base there. Hams in Hong Kong kept in constant touch with the sinking yacht while search planes and ships combed the area. After more than 18 hours of continuous searching, the rescuers found the ship and saved the crew.

IN QUAKES AND FLOODS: HAMS ARE THERE

In September 1985 a massive earthquake rocked Mexico City, and a smaller but just-as-terrifying aftershock hit two days later. Amateur Radio operators across North America sprang into action. Thousands of inquiries came in to ARRL Headquarters and to individual hams across the country. Friends, relatives and business associates worried about people in the affected area. They wanted to learn about the safety of those in Mexico City.

Why did they choose to ask Amateur Radio operators to help them? Over the decades, hams have volunteered their services in times of emergency to relay vital information to and from stricken areas. The September 1985 Mexican quakes knocked out most means of communications. Especially in rural areas, Amateur Radio was the only way news of the disaster could reach the rest of the world. Hundreds of Amateur Radio operators spent days and nights seeking news of individuals and conditions in certain areas. They went back to their regular routine of working and spending time with their families only after workers restored regular communications channels.

The small college town of Rexburg, Idaho, was suddenly smashed by a 5-foot wave of onrushing water. A tremendous 12-foot wall of water followed immediately. Everyone was in a panic. The Teton Dam had burst, pouring millions of gallons of water in a raging torrent down the Snake River Valley. The residents of Rexburg were right in the water's path. While the residents scrambled to high ground, hams waded through the murky water to fire up their transmitters and receivers.

The tremendous impact of the water had uprooted telephone poles and knocked out commercial transmitters. Hams were able to keep civil-preparedness workers and town officials in communication with each other, however. Commercial radio stations in nearby towns let amateurs set up equipment inside their studios to handle emergency messages from the disaster. Hams remained calm and helped out wherever they could, providing communications and additional manpower.

Hams immediately jump into service, even when the disaster is halfway around the world. Italy was shaken to its roots by earthquakes late in 1980. Hams worked for days manning their stations to receive and transmit disaster information. North American and Italian amateur stations relayed thousands of messages. Concerned individuals asked the hams to send messages about the health and welfare of friends and relatives living in stricken areas. WB2JSM, the Hall of Science Radio Club station in New York City, received nationwide publicity on network news broadcasts for its role in handling messages for earthquake victims.

LEND A HELPING HAND

Amateur Radio holds no roadblocks for handicapped individuals. Many people who are unable to walk, see or talk are still able to earn an Amateur Radio license. Then they can converse with friends in their home town or across the world using Amateur Radio. Many local ham clubs even take classes to a home to help a person with a disability discover a rewarding hobby.

Otho Jarman is a paraplegic unable to leave his bed. Bill Haney, WA6CMZ, helped Otho learn the code and pass his Novice exam through such a club program. Bill spent one hour each week teaching Otho the code, and in seven weeks he passed his Novice exam. Other members of the Barstow (California) Amateur Radio Club helped Otho put his station together. They obtained equipment for him and constructed antennas.

Soon, Otho was on the air talking with hams in Mozambique, Nicaragua and Puerto Rico. He could communicate with the world from his bed. Sixteen years earlier, at age 22, Otho had broken his spine when he dove into a reservoir to rescue a drowning child. Although interested in Amateur Radio for years, he had not had an opportunity to learn about it until the club came along. Now he monitors local frequencies from 7 AM to 10 PM, often just chatting or giving directions to motorists.

"Amateur Radio can take a disabled person out of his living room, out of his bed or out of his wheelchair, and put him in the real world," Otho says.

The Courage Center, of Golden Valley, Minnesota, sponsors the HANDI-HAM system to help people with physical handicaps obtain amateur licenses. The system provides materials and instruction to handicapped persons interested in obtaining ham licenses. They also provide information to nonhandicapped hams, "verticals," who wish to help handicapped people earn a license.

The Chinese saying, "One picture is worth a thousand words," is the spirit embodied in the Courage Center symbol. It is the letter "C," inside of which a figure is represented. The figure's outstretched arms hold a crutch in the left hand and use the C for support on the right side. There can be no better way to convey the message of HANDI-HAMS, a part of the Courage Center, a rehabilitation facility that uses Amateur Radio to help persons with disabilities. "HANDI-HAM Radio knows no barriers."

THE NOVICE LICENSE

Once you've earned your Novice license, you can join the thousands of other hams on the air. Then you'll begin to experience the thrill of Amateur Radio firsthand. As a Novice, you can contact other hams in your town and around the world. You can talk into a microphone to use either single sideband (SSB) or frequency modulation (FM) to communicate. You can experience the thrill of using the international Morse code (called "CW" by hams). You can even communicate by packet radio or radioteletype using a computer. You can do all this on parts of special frequency bands set aside by the FCC for Amateur Radio operators.

For your first contact, you'll probably tune around looking for a station calling "CQ" (calling for any station to make contact). Perhaps you'll even try calling a CQ on your own. Suddenly you'll hear your own call coming back! It's hard to describe the excitement. Someone else is sending your call sign back to let you know that they hear you and want to make contact.

Each time you send a CQ, you'll wonder who will answer. It could be a ham in the next town, the next state or clear across the country. The whole world is full of hams to talk to.

There are many wonderful things you can do as a Novice. After you have been on the air for a while, you'll be known as one of the regulars on the band. It's surprising how many people from all over you'll recognize and who will recognize you. Many a fast friendship has developed through repeated on-the-air contacts.

Soon you'll be collecting contacts with different states and exchanging QSL cards (postcards) with other hams you've talked to. These special cards commemorate each contact. They also serve as proof of the contact as you begin working toward some of the awards issued by the American Radio Relay League. [The Worked All States (WAS) award

is one popular example.] (You'll learn more about the special language and abbreviations that hams use as you study. Chapter 10 has a list of common "Q signals" used by most hams.)

There is a certain intrigue about DX (long-distance communication) that catches many hams. Talking to hams from other lands can be quite an experience. After all, foreign hams are people just like you who enjoy finding out about other people and places! Also, hams in other countries often speak enough English to carry on a limited conver-

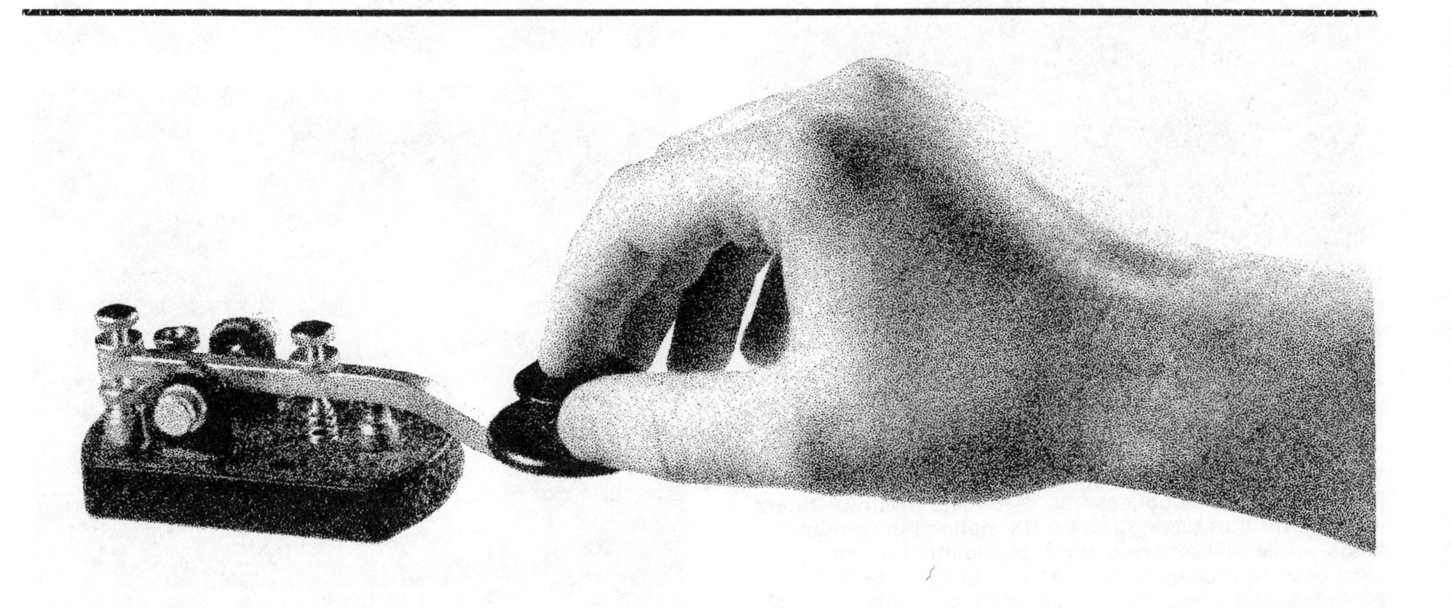

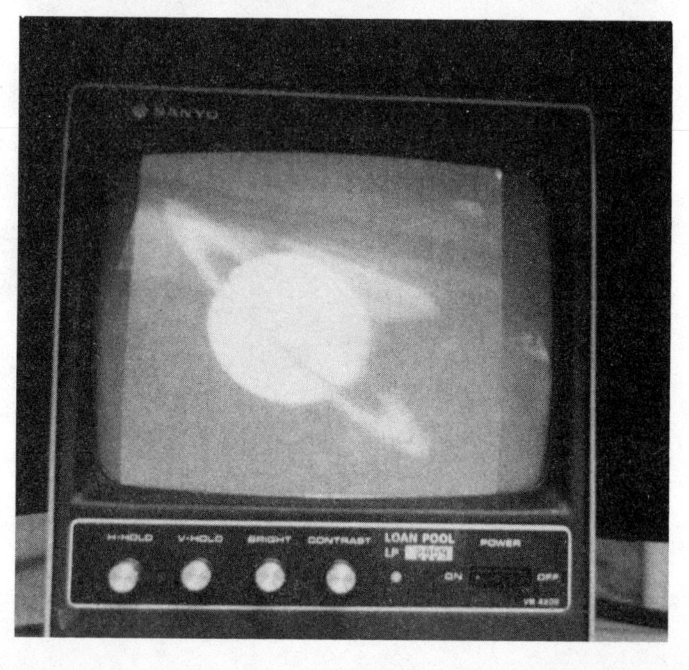

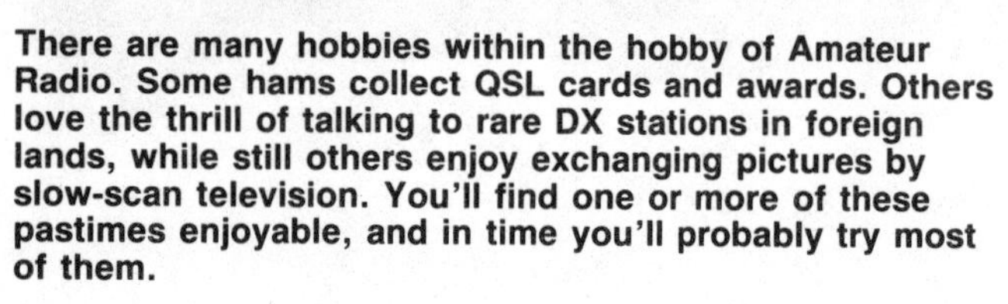

There are many hobbies within the hobby of Amateur Radio. Some hams collect QSL cards and awards. Others love the thrill of talking to rare DX stations in foreign lands, while still others enjoy exchanging pictures by slow-scan television. You'll find one or more of these pastimes enjoyable, and in time you'll probably try most of them.

sation, so you'll have little problem there. The Morse code, Q signals and many abbreviations are understood internationally. This makes CW ideal for communicating with hams in foreign countries. Plenty of DX stations operate in the Novice bands. Quite a few Novices have even contacted hams in more than 100 different countries.

If you enjoy a little competition, perhaps you'll like on-the-air contesting. The object of a contest is to work as many people in as many different areas as possible in a certain time. Each year, the American Radio Relay League sponsors a "Novice Roundup" contest. In the Novice Roundup, you will have the chance to contact old friends and make new ones. You might work some new states or countries, and if you use Morse code you will increase your code speed. You are certain to improve your general operating skills and ability. Most of all, though, you will have fun.

OTHER MODES

Novice class operators may use just about every operating mode available to Amateur Radio operators. You should become familiar with some of these other modes. In addition to voice and Morse code privileges, you may want to investigate some of the more exotic modes. Here is a brief description of some of your options:

With slow-scan television (SSTV), hams send still photos to each other, one frame at a time. It takes about 8 seconds for the bright band of light to creep down the screen to make a complete picture. (Your home TV makes 30 complete pictures per second.) SSTV pictures are more like those shots of the moon or Saturn that you may have seen transmitted from space. SSTV pictures may be transmitted around the world via shortwave ham transmitters. Amateur Radio operators were the first to flash TV snapshots of Mars to foreign countries.

Facsimile (FAX) is a means of sending drawings, charts, maps and graphs. You can even play games over the air by transmitting FAX pictures of each move. News services send photos from around the world using FAX.

With radioteletype transmission (RTTY), a ham can type out a message and send it over the air to a friend's station. Even if the friend is away, his or her radioteletype system can receive and hold the message until he or she returns. Early RTTY systems used mechanical machines cast off by news services. Today, many hams use personal computer systems. These display the message silently on a TV screen rather than using roll after roll of paper with noisy, clacking typewriter keys. Packet radio is a modern computer-controlled system capable of relaying messages and storing them for later reception by the intended ham.

HAM SATELLITES

Hams even have their own satellites, which they can talk through to points around the globe. The Orbiting Satellites Carrying Amateur Radio (OSCARs) have been hurtling through near-Earth space since 1961. Hams use these satellites to communicate in voice, code, radioteletype and packet radio around the world.

Some schools use the OSCARs to instruct students in science and math. No license is needed to listen in. So many students across the country have become eavesdroppers on Amateur Radio transmissions through an OSCAR. All it takes is a receiver and an antenna to introduce students to the exciting world of space technology.

Hams from many countries worked in a joint effort to build several OSCARs. Weighing less than your living-room TV and powered by sun-charged batteries, OSCARs retransmit hams' "uplink" signals down to other earthbound stations.

You can operate in any of these exciting modes when you become an Amateur Radio operator. Take the time to explore the adventure of ham radio!

ELEMENT 2 (NOVICE) SYLLABUS

(Required for all operator licenses)

The complete syllabus contains 30 blocks, with one question to come from each block.

Subelement 2A Commission's Rules (10 Exam Questions)

One question must be from the following:

2A-1 Basis and purpose of the amateur service {97.1}
2A-2 Amateur service, definition {97.3(a)(4)}
2A-3 Amateur communication, definition {97.3(a)(4)}
2A-4 Amateur operator, definition {97.3(a)(1)}

One question must be from the following:

2A-5 Operator license, definition {97.5(d)(1)}
2A-6 Primary station license, definition {97.5(a), (d)(1)}
2A-7 Amateur station, definition {97.3(a)(5)}
2A-8 Control operator, definition {97.3(a)(11)}
2A-9 Amateur operator license classes {97.9(a)}

One question must be from the following:

2A-10 Novice control operator frequency privileges {97.301(e), (f)}

One question must be from the following:

2A-11 Amateur operator license eligibility {97.5(d)(1)}
2A-12 Novice license exam elements {97.501(e); 97.503(a)(1), (b)(1)}
2A-13 Primary station license eligibility {97.5(d)(1)}
2A-14 Mailing address furnished by licensee {97.21}
2A-15 Valid US call sign assignment {97.17(f)}
2A-16 Length of time license is valid {97.23(a)}

One question must be from the following:

2A-17 Novice control operator emission privileges {97.3(c); 97.305(c); 97.307(f)(9), (10)}

One question must be from the following:

2A-18 Transmitter power on 3700-3750 kHz, 7100-7150 kHz, 21100-21200 kHz {97.3(b)(6); 97.313(a), (c)(1)}
2A-19 Novice transmitter power on 28.1-28.5 MHz, 220-225 MHz, 1270-1295 MHz {97.3(b)(6); 97.313(c)(2), (d), (e)}
2A-20 Digital communications (Limited to concepts only) {97.3(c)(2); 97.305(c)}

One question must be from the following:

2A-21 Station licensee responsibility {97.103(a), (b)}
2A-22 When control operator is required {97.7; 97.103}
2A-23 Who may be control operator {97.7}
2A-24 Control operator location {97.109(a), (b)}
2A-25 Amateur operator license availability {97.9(a)}
2A-26 Primary station license availability {97.5(e)}

One question must be from the following:

2A-27 Station identification {97.119(a)}
2A-28 Points of communication {97.111(a)(1), (2), (3), (4)}
2A-29 Operation away from fixed-station location {97.5(e)}
2A-30 Business communication {97.113(a)}

One question must be from the following:

2A-31 International communications {97.111(a)(1)}
2A-32 Messages for hire {97.113(b)}
2A-33 Broadcasting {97.3(a)(10); 97.113(b)}
2A-34 Third-party traffic (Definition) {97.3(a)(38); 97.115(a)}
2A-35 Transmission of music {97.113(d)}
2A-36 Codes and ciphers {97.113(d)}

One question must be from the following:

2A-37 False signals {97.113(d)}
2A-38 Unidentified communications {97.119(a)}
2A-39 Malicious interference {97.101(d)}
2A-40 Notices of violation {This topic is not covered in the new Part 97 Rules.}

Subelement 2B Operating Procedures (2 Exam Questions)

One question must be from the following:

2B-1 Choosing a frequency
  2B-1-1 Tune-up and operation
  2B-1-2 For emergency operation
2B-2 Morse Radiotelegraphy
  2B-2-1 Establishing a contact
  2B-2-2 Choosing a sending speed
  2B-2-3 Proper use of $\overline{AR}$, $\overline{SK}$, $\overline{BT}$, $\overline{DN}$ and $\overline{KN}$
  2B-2-4 Proper use of CQ, DE and K
  2B-2-5 RST signal reporting system
  2B-2-6 Proper use of QRS, QRT, QTH, QRZ and QSL
2B-3 Radiotelephony
  2B-3-1 Establishing a contact
  2B-3-2 Standard International Telecommunication Union phonetics

One question must be from the following:

2B-4 Radioteleprinting
  2B-4-1 Establishing a contact
  2B-4-2 Sending speed (Limited to the concept that both stations must be using the same speed)

2B-5 Packet radio
  2B-5-1 Connecting to and monitoring other packet radio stations
  2B-5-2 Digipeaters and networking (Limited to concepts only)
  2B-5-3 Sending speed (Limited to the concept that both stations must be using the same speed)

2B-6 Repeater use
  2B-6-1 Establishing a contact through a repeater
  2B-6-2 Purpose of repeater operations
  2B-6-3 Input/output frequency separation (Limited to concept only; specific frequency splits not required)
  2B-6-4 Simplex versus repeater operation
  2B-6-5 Special features of repeaters (Autopatch, time-out timers)

Subelement 2C Radio-Wave Propagation (1 Exam Question)

One question must be from the following:

2C-1 Line of sight
2C-2 Ground wave
2C-3 Sky wave
2C-4 Sunspot cycle
2C-5 Sunspots and their influence on the ionosphere
2C-6 Reflecting VHF/UHF radio waves

Subelement 2D Amateur Radio Practices (4 Exam Questions)

One question must be from the following:

2D-1 How to prevent use of amateur station by unauthorized persons
2D-2 Station lightning protection
2D-3 Ground system

One question must be from the following:

2D-4 VHF/UHF RF safety precautions
2D-5 Purpose of safety interlocks and other safety devices
2D-6 Antenna installation safety procedures

One question must be from the following:

2D-7 Standing wave ratio
  2D-7-1 SWR Meter
  2D-7-2 Acceptable SWR readings
  2D-7-3 Common causes of high SWR readings

One question must be from the following:

2D-8 Radio frequency interference
  2D-8-1 RF overload of consumer electronic products
  2D-8-2 Harmonic radiation interference to consumer electronic products
  2D-8-3 Handling RFI complaints

Subelement 2E Electrical Principles (4 Exam Questions)

One question must be from the following:

2E-1 Metric prefixes
  2E-1-1 giga (G)
  2E-1-2 mega (M)
  2E-1-3 kilo (k)
  2E-1-4 centi (c)
  2E-1-5 milli (m)
  2E-1-6 micro (μ)
  2E-1-7 pico (p)

One question must be from the following:

2E-2 Concept of current
  2E-2-1 Electron movement
  2E-2-2 Current units

2E-3 Concept of voltage
  2E-3-1 Electrical pressure
  2E-3-2 Voltage units

2E-4 Concept of conductor

2E-5 Concept of insulator

2E-6 Concept of resistance
  2E-6-1 Opposition to electron movement
  2E-6-2 Resistance units

One question must be from the following:

2E-7 Ohm's Law (Any calculations will be kept to a very low level; integer math only—NO fractions or decimal math will be included)

2E-8 Concept of energy

2E-9 Concept of power
  2E-9-1 Rate of using energy
  2E-9-2 Power units

2E-10 Concept of open circuit

2E-11 Concept of short circuit

One question must be from the following:

2E-12 Concept of frequency
  2E-12-1 Concept of dc
  2E-12-2 Concept of ac
  2E-12-3 Frequency units
  2E-12-4 Concept of AF
  2E-12-5 Concept of RF

2E-13 Concept of wavelength

Subelement 2F Circuit Components (2 Exam Questions)

One question must be from the following:

2F-1 Schematic representation of a resistor
2F-2 Schematic representation of a switch
2F-3 Schematic representation of a fuse
2F-4 Schematic representation of a battery

One question must be from the following:

2F-5 Schematic representation of a ground
2F-6 Schematic representation of an antenna
2F-7 Schematic representation of a bipolar transistor
2F-8 Schematic representation of a triode vacuum tube

Subelement 2G Practical Circuits (2 Exam Questions)

One question must be from the following:

2G-1 Functional layout of Novice station equipment
- 2G-1-1 Transmitter, receiver, (or transceiver), power supply, antenna switch, antenna feed line, antenna
- 2G-1-2 Transceiver, antenna switch, SWR meter, impedance matching device, antenna

One question must be from the following:

2G-2 Morse telegraphy station equipment layout (Block diagram)

2G-3 Radiotelephone station equipment layout (Block diagram)

2G-4 Radioteleprinter station equipment layout (Block diagram)

2G-5 Packet-radio station equipment layout (Block diagram)

Subelement 2H Signals and Emissions (2 Exam Questions)

One question must be from the following:

2H-1 Emission types (Definition)
- 2H-1-1 A1A
- 2H-1-2 F1B
- 2H-1-3 F3E
- 2H-1-4 J3E

2H-2 Key clicks

2H-3 Chirp

2H-4 Superimposed hum

One question must be from the following:

2H-5 Undesirable harmonic radiation and other spurious emissions

2H-6 Electromagnetic radiation and safety awareness

2H-7 Adjacent channel interference (Proper settings for drive, mic gain, use of speech compression, concept of deviation in FM modulation)

Subelement 2I Antennas and Feed Lines (3 Exam Questions)

One question must be from the following:

2I-1 1/2-wavelength dipole (Approximate lengths)
2I-2 1/4-wavelength vertical (Approximate lengths)
2I-3 Advantages of 5/8-wavelength vertical antennas

One question must be from the following:

2I-4 Yagi antennas
- 2I-4.1 Concept of a directional antenna
- 2I-4.2 Names of Yagi elements (driven element, reflector and director)

2I-5 RF safety near antennas

One question must be from the following:

2I-6 Coaxial cable
2I-7 Parallel-conductor feed line
2I-8 Antenna matching device
2I-9 Balun
2I-10 Horizontal antenna polarization (Element orientation)
2I-11 Vertical antenna polarization (Element orientation)

KEY WORDS

Amateur Radio communication—Noncommercial radio communication by or among Amateur Radio stations solely with a personal aim and without pecuniary or business interest. (*Pecuniary* means payment of any type, whether money or other goods.)

Amateur Radio operator—A person holding a valid license to operate an Amateur Radio station. In the US this license is issued by the Federal Communications Commission.

Amateur Radio Service—A radio communication service of self-training, intercommunication, and technical investigation carried on by radio amateurs.

Amateur Radio station—A station licensed in the Amateur Radio Service, including necessary equipment at a particular location, used for Amateur Radio communication.

Control operator—A licensed amateur designated to be responsible for the transmissions of an Amateur Radio station.

Digital communications—Amateur communications that are received and printed automatically, or used to transfer information directly from one computer to another.

Emission—The transmitted signal from an Amateur Radio station.

Emission privilege—Permission to use a particular emission type (such as Morse code or voice).

False or deceptive signals—Transmissions that are intended to mislead or confuse those who may receive the transmissions. For example, distress calls transmitted when there is no actual emergency are false or deceptive signals.

Frequency bands—A group of frequencies where amateur communications are authorized.

Frequency privilege—Permission to use a particular group of frequencies.

Malicious interference—Intentional, deliberate obstruction of radio transmissions.

Mobile operation—Amateur Radio operation conducted while in motion or at temporary stops at different locations.

Operator license—The portion of an Amateur Radio license that gives permission to operate an Amateur Radio station.

Peak envelope power (PEP)—The average power of a signal at its largest amplitude peak.

Portable operation—Amateur Radio operation conducted away from the location shown on the station license.

Station license—The portion of an Amateur Radio license that authorizes an amateur station at a specific location. The station license also lists the call sign of that station.

Third-party participation (or communication)—The way an unlicensed person can participate in Amateur Radio communications. A control operator must ensure compliance with FCC rules.

Third-party traffic—Messages passed from one amateur to another on behalf of a third person.

Unidentified communications or signals—Signals or radio communications in which the transmitting station's call sign is not transmitted.

Chapter 2

The Radio Spectrum: A Limited Resource

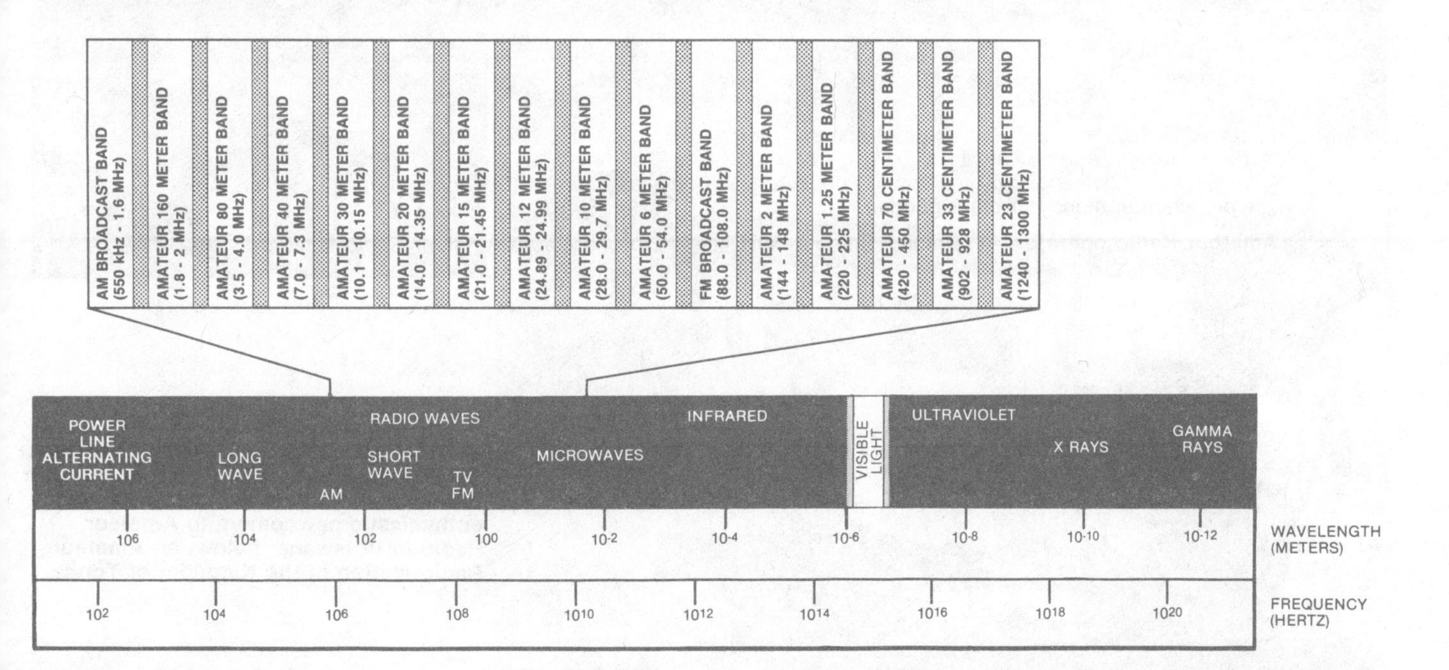

Figure 2-1— The electromagnetic spectrum, showing the range from the 60-hertz ac power line in your home, through the radio frequencies, X-rays and gamma rays. Amateur Radio operations use a small, but significant, portion of these frequency bands.

Speed of light = C = 10^8 meter/sec

freq × WL = C = speed light

$f \times \lambda = c$

When you tune an AM or FM broadcast radio to your favorite station, you select a specific spot on the tuning dial. There are many stations spread across that dial. Each radio station occupies a small part of the entire range of "electromagnetic waves." Other parts of this range, or spectrum include microwaves, X-rays and even infrared, ultraviolet and visible light waves.

Figure 2-1 shows the electromagnetic spectrum from below the radio range all the way through the X-rays. Amateur Radio occupies only a small part of the total available space; countless users must share the electromagnetic spectrum.

You may be thinking: "Who decides where Amateur Radio frequencies will be, and where my favorite FM broadcast station will be?" That's a good question, and the answer has several parts.

Radio signals travel to distant corners of the globe, so there must be a way to prevent total chaos on the bands. The International Telecommunication Union (ITU) has the important role of dividing the entire range of communications frequencies between those who use them. There are many radio services that have a need for communications frequencies. These services include commercial broadcast, land mobile and private radio (including Amateur Radio). ITU member nations decide which radio services will be given certain bands of frequencies, based on the needs of the different services. This process takes place at ITU-sponsored World Administrative Radio Conferences (WARCs).

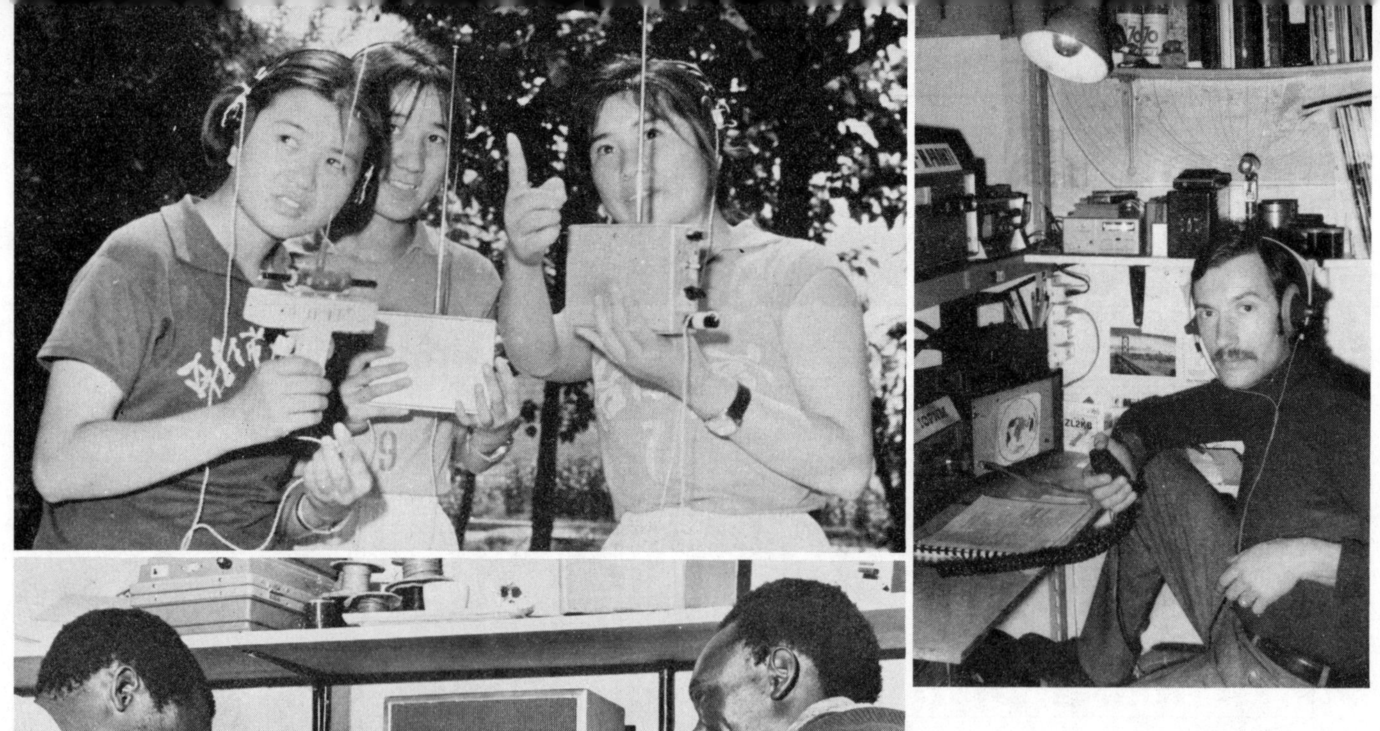

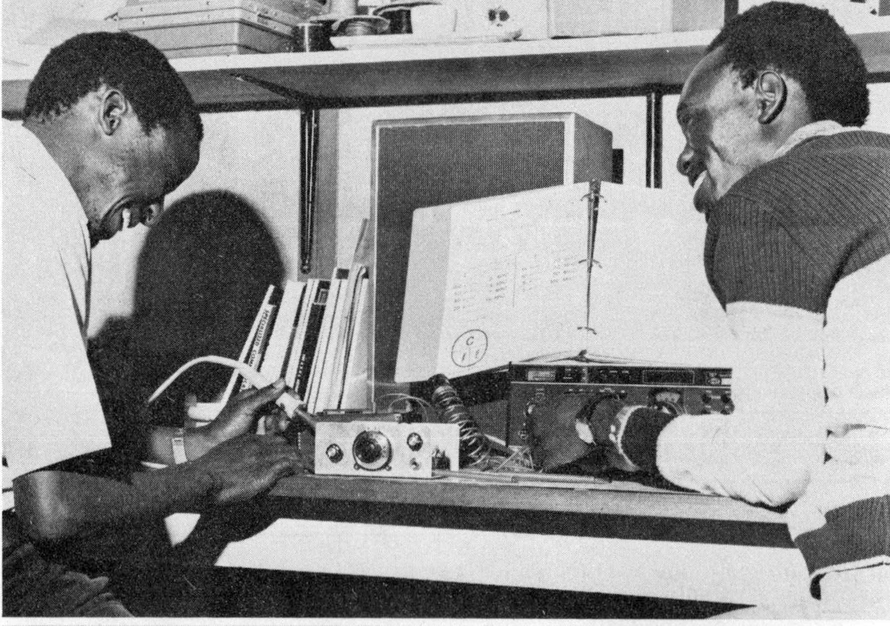

Nearly every country on earth has its share of active Amateur Radio operators. Above left, a direction-finding exercise in China; above a ham in Rome; left, two enthusiastic newcomers to Amateur Radio in Botswana; below, an Amateur Radio station in the Kingdom of Tonga.

WHY IS THERE AMATEUR RADIO?

In the case of Amateur Radio, the ITU has long recognized that hams are invaluable in times of emergency or disaster. At an ITU WARC held in 1979, Amateur Radio gained several new bands of frequencies. Mighty praise indeed!

The ITU makes these allocations on an international basis. The Federal Communications Commission (FCC) decides the best way to allocate **frequency bands** to those services using them in the US. The FCC is the United States governing body when it comes to Amateur Radio. In fact, an entire part of the FCC Rules is devoted to the Amateur Radio Service—Part 97. As your amateur career develops, you'll become increasingly familiar with Part 97.

To gain the privilege of sending a radio signal over the airwaves in the US, you must pass a license exam. The exam for a Novice license covers basic radio theory, FCC regulations and the Morse code. This book, and the code material that comes with it, will prepare you to earn your first Amateur Radio license.

THE FIVE PRINCIPLES

In Section 97.1 of the Amateur Radio Service rules, the FCC describes the fundamental purposes of the Amateur Service. The basis and purpose of the Amateur Radio Service includes five principles:

(a) Recognition and enhancement of the value of the amateur service to the public as a voluntary noncommercial communication service, particularly with respect to providing emergency communications.

Probably the best-known aspect of Amateur Radio is our ability to provide life-saving emergency communications. Normal communications channels often break down during hurricanes, earthquakes, tornadoes, airplane crashes and other disasters.. Amateur Radio is frequently the first available means of contact with the outside world from the affected area. Red Cross and other civil-defense agencies rely heavily on the services of volunteer radio amateurs.

One of the more noteworthy aspects of Amateur Radio is its noncommercial nature. In fact, amateurs may not accept any form of payment for operating their radio stations. (There is one limited exception to this rule, which we will explain later.) This means that hams make their services available free of charge. This is true whether they are assisting a search-and-rescue operation in the Sierra Nevadas, relaying health-and-welfare messages from a disaster-stricken Caribbean island or providing communi-

In the days before the giant eruption that demolished vast areas around the Mount St. Helens volcano in Washington, volunteer amateurs monitored its behavior. Hams are ready to do whatever they can when someone needs their communications services. Because of the suddenness and extent of that eruption, two hams lost their lives while helping to monitor the volcano.

cations assistance at the New York City Marathon. Talk about value!

Why do hams work so hard if they can't be paid? It gives them an immense feeling of personal satisfaction! It's like the good feeling you get when you lend a hand to an elderly neighbor, only on a much grander scale. Hams operate their stations only for personal satisfaction and enjoyment. They don't talk about business matters on the air. (FCC rules don't allow business communications over Amateur Radio.)

(b) Continuation and extension of the amateur's proven ability to contribute to the advancement of the radio art.

In the early days of radio there were no rules, but in 1912 Congress passed a law to regulate the airwaves. Amateurs had to keep to a small range of frequencies (known as "short waves"). There they would remain "out of the way"—everyone knew that radio waves couldn't travel very far at those frequencies. Ha! The amateurs soon overcame the restrictions. They were among the first to experiment with radio propagation, the study of how radio waves travel through the atmosphere. When vacuum tubes became available, amateurs began to develop much-improved radio communication circuits.

Today, the traditions and spirit in Amateur Radio remain. Amateurs continue to experiment with state-of-the-art technologies. Advancement of the radio art takes a major portion of an amateur's energies. The FCC promotes this amateur experimentation and technical development by establishing rules that are consistent with amateur techniques.

Section 97.1 of the amateur rules continues:

(c) Encouragement and improvement of the amateur radio service though rules which provide for advancing skills in both the communications and technical phases of the art.

Along with some of the technical aspects of the service, amateurs also hold special training exercises in preparation for communications emergencies. Simulated Emergency Tests and Field Days, where amateurs practice communicating under emergency conditions, are just two ways amateurs sharpen their operating skills.

The Commission's rules even specify a "service within a service" called the Radio Amateur Civil Emergency Service, or RACES. RACES provides amateur communications assistance to federal, state and local civil defense in times of need.

(d) Expansion of the existing reservoir within the amateur radio service of trained operators, technicians and electronics experts.

Self-training, intercommunication and technical investigation are all important parts of the Amateur Radio Service. We need more amateurs who are experienced in communications methods, because they are a national resource to the public.

(e) Continuation and extension of the amateur's unique ability to enhance international goodwill.

Hams are unique, even in this time of worldwide jet travel. They journey to the far reaches of the earth and talk with amateurs in other countries every day. They do this simply by walking into their ham shacks. International peace and coexistence are very important today. Amateurs represent their countries as ambassadors of goodwill. Amateur-to-amateur communications often transcend the cultural boundaries between societies. Amateur Radio is a teacher in Lincoln, Nebraska, trading stories with the headmaster of a boarding school in a London suburb. It is a tropical-fish hobbyist learning about fish in the Amazon from a missionary stationed in Brazil. Amateur Radio is a way to make friends with other people everywhere.

The five principles just described provide the basis and purpose for the Amateur Radio Service, as set down by the FCC. These principles place a large responsibility on the amateur community—a responsibility that you will share. It is the Commission's duty to ensure that amateurs are able to operate their stations properly, without interfering with other radio services. *All* amateurs must pass an examination before the FCC will issue a license authorizing amateur station operation. It's a very serious matter.

[At this point you should turn to Chapter 12 and study the questions that cover this material in the Novice question pool. Before you turn to Chapter 12, though, it may be helpful to understand a bit about the numbering system used for these questions. The numbers match the study guide or syllabus printed at the end of Chapter 1. The syllabus forms a type of outline, and the numbering system follows this outline format. There are nine subelements, labeled A through I. This chapter covers subelement A, so all the question references will be to questions with numbers that begin "2A." There are 40 syllabus points in subelement 2A. You will be told to study questions 2A-1, 2A-2 and so on. Finally, each individual question about a particular syllabus point has its own number, so the complete question numbers take the form, "2A-1.1," "2A-1.2" and so on.

You should study questions 2A-1.1 through 2A-1.4 in Chapter 12 now. If you have difficulty with any of these questions, review the material in this section.]

THE AMATEUR RADIO SERVICE

The FCC defines some important terms in Section 97.3 of the amateur rules. The **Amateur Radio Service** is "A radio communication service of self-training, intercommunication, and technical investigation carried on by amateur radio operators." Amateurs learn various communications skills on their own, and carry out technical experiments in electronics and radio principles.

How does the FCC define **Amateur Radio communication**? It is "noncommercial radio communication by or among Amateur Radio stations solely with a personal aim and without pecuniary or business interest." *Pecuniary* means related to money or other payment. In other words, you can't be paid for operating your station or providing a communications service for some individual or group.

An **Amateur Radio operator** is "a person holding a valid license to operate an Amateur Radio station." In the United States, the Federal Communications Commission issues these licenses. An Amateur Radio operator performs communications in the Amateur Radio Service.

How does the FCC define an **Amateur Radio station**? "A station licensed in the Amateur Radio Service, embracing necessary apparatus at a particular location, used for Amateur Radio communications." The person operating an Amateur Radio station has an interest in self-training, intercommunication and technical investigations or experiments.

An Amateur Radio license is really two licenses in one—an **operator license** and a **station license**. The operator license is the one that lets you operate a station within your Novice privileges. The station license authorizes you to have an Amateur Radio station. It also lists the call sign that identifies that station. Figure 2-2 shows an actual Amateur Radio license.

One piece of paper includes both licenses. You won't even be able to tell which part is which license. It is important to realize that your license includes both of these parts, however.

The operator license lists your license class and gives you the authority to operate an Amateur Radio station. The station license includes the address of your primary, or main, Amateur Radio station. The station license also lists the call sign of your station.

You must have your original operator license, or a photocopy, in your possession any time you are operating an Amateur Radio station. You are using the *operator's license* portion when you serve as the control operator of a station. It is a good idea to carry a copy of your license in your wallet or purse at all times. That way you are sure to have it if a chance to operate arises.

You must also have your original station license, or a photocopy of it, in your station any time someone operates the station. The station license can be in the control operator's possession or you can have it on display in the station.

Remember that one piece of paper is both the station license and the operator's license. You can't post the original at your station and carry it with you at all times. Most people make a photocopy of the license to carry with them. You can post the original license in your station. Many hams make another photocopy to put in their station, and put the original license in a safe place. Your original license must be available for inspection by any Government official or representative of the FCC, however. Don't lose the original license!

The FCC issues all licenses for a 10-year term. You should always renew your license for another 10 years before the present one expires. If your Novice license was issued in November 1989, it will expire in November 1999. It's a good idea to form the habit of looking at the expiration date on your license every now and then. You'll be less likely to forget to renew the license in time, then.

If you do forget to renew your license, you have up to 2 years to apply for a new license. After the 2-year grace period, you will have to take the exam again. Your license is not valid during this 2-year grace period, however. You may not operate an Amateur Radio station with an expired license. All the grace period means is that the FCC will renew the license if you apply during that time.

[Turn to Chapter 12 and study those questions with numbers that begin 2A-2, 2A-3, 2A-4, 2A-5, 2A-6 and 2A-7. Also study those questions with numbers that begin 2A-16, 2A-25 and 2A-26. Review this section if you have difficulty with any of these questions.]

THE CONTROL OPERATOR

A **control operator** is the licensed Amateur Radio operator who is operating an Amateur Radio station. Only a licensed ham may be the control operator of an Amateur

AMATEUR RADIO LICENSE		NOT TRANSFERABLE		
EFFECTIVE DATE	EXPIRATION DATE	CALL SIGN	OPERATOR PRIVILEGES	STATION PRIVILEGES
12/07/84	12/07/94	KA1MJP	NOVICE	PRIMARY

NAME AND ADDRESS
LESLIE K BARTOLOTH
168 STILLWELL DR
PLAINVILLE CT 06062

FIXED STATION OPERATION LOCATION
SAME AS MAILING ADDRESS

FOLD HERE

THIS LICENSE IS SUBJECT TO CONDITIONS OF GRANT ON REVERSE SIDE

UNITED STATES OF AMERICA
FEDERAL COMMUNICATIONS COMMISSION
GETTYSBURG, PA 17325

Leslie K. Bartoloth
(LICENSEE'S SIGNATURE)

FCC FORM 660
JULY 1984

FEDERAL
COMMUNICATIONS
COMMISSION

Figure 2-2—This is what an FCC Amateur Radio license looks like.

Radio station. If another licensed radio amateur operates your station with your permission, he or she assumes the role of control operator.

Any Amateur Radio operator may designate another licensed operator as the control operator, to share the responsibility of station operation. The FCC holds both the control operator and the station licensee responsible for proper operation of the station. If you are operating your own Amateur Radio station, then you are the control operator at that time.

A control operator must be present at the station control point whenever the transmitter is operating. This means that you may not allow an unlicensed person to operate your radio transmitter while you are not present. There is one time when a transmitter may be operating without a control operator being present, however. Some types of stations, such as repeater stations, may be operated by automatic control. In this case there is no control operator at the transmitter control point.

Your Novice license authorizes you to be the control operator of an Amateur Radio station in the Novice frequency bands. This means you can be the control operator of your own station or someone else's station. In either case, you are responsible to the FCC for the proper operation of the station.

If you allow another licensed ham to operate your station, you are still responsible for its proper operation. You are always responsible for the proper operation of your station. Your primary responsibility as the station licensee is to ensure the proper operation of your station.

[Before you go on to the next section, you should turn to Chapter 12 and study the questions with numbers that begin 2A-8, 2A-21, 2A-22, 2A-23 and 2A-24. If you have difficulty with any of these questions, review the material in this section.]

AMATEUR LICENSE CLASSES

There are five kinds, or levels, of amateur license. They vary in degree of knowledge required and **frequency privileges** granted. Higher class licenses have more comprehensive examinations. In return for passing a more difficult exam you earn more frequency privileges (frequency space and modes of operation).

The first step is the Novice license. The FCC issues this "beginner's" license to those who demonstrate the ability to adjust properly and operate safely an Amateur Radio transmitter. An applicant must show a basic proficiency in Morse code by passing a test at 5 words per minute (WPM). The exam also covers some very basic radio fundamentals and knowledge of a few rules and regulations. With a little study and some common sense, you'll soon be ready to pass the Novice exam.

Anyone (except an agent of a foreign government) is eligible to qualify for a Novice or higher class Amateur Radio operator's license. There is no age requirement. To hold an Amateur Radio station license, you must have a valid operator's license. (Remember, both licenses are printed on the same piece of paper.)

A Novice license gives you the freedom to develop operating and technical skills through on-the-air experience. These skills will help you upgrade to a higher class of license, with additional privileges.

As a Novice, you will be able to communicate with other amateur stations in the exotic reaches of the world. Novices provide public service through emergency communications and message handling. They enhance international goodwill just like operators with higher license classes.

The Novice population is the reservoir from which tomorrow's top-notch operators will be drawn. Ultimately, most Novices want to earn greater operating privileges. As they gain operating experience, they prepare for a higher-class license—the gateway to more frequencies and modes.

The next step up is the Technician license. To obtain the Technician license, a Novice need only pass a more comprehensive theory exam with a few additional rules and regulations. There is no additional code exam to upgrade from Novice to Technician. The Technician license carries more frequency privileges in the VHF (very high frequency) region.

There is a certain thrill to having more frequencies available to talk with other amateurs on the far side of the globe. That is a powerful incentive to upgrade to the General class license (the next step up from Technician). You will have to pass a 13-WPM Morse code exam and another theory exam to earn the General class license. The General class license gives voice privileges on eight high-frequency (HF) bands. These bands typically carry signals over great distances. The General class license is by far the most popular license of the bunch.

As you progress and mature in Amateur Radio, you will develop specialized interests in exciting areas like amateur television and satellite communication. You'll also want even more privileges, and that is where the Advanced and Amateur Extra licenses come in. To obtain one of these, you must be prepared to face exams on the more technical aspects of the hobby. For the Amateur Extra license, the top-of-the-line, there is an expert's code test at 20 words per minute.

To qualify for a higher-class license, you must pass the theory exam for each level up to that point. For example, suppose you want to go from Novice directly to Amateur Extra. You will have to take the Technician, General and Advanced theory exams before you take the Amateur Extra test. (You can just take the 20 WPM code test, however, without passing the 13 WPM General class code test.)

[You should turn to Chapter 12 now and study the questions with numbers that begin 2A-9, 2A-11 and 2A-13. Review the material in this section if you have difficulty with any of these questions.]

THE NOVICE LICENSE

A Novice license allows you to operate on portions of six Amateur Radio bands. We normally identify these bands by specifying the frequency range they cover or by listing the wavelength in meters. (You will learn more about the relationship between frequency and wavelength later in this book.) Table 2-1 lists the frequency range for each of the Novice bands. This table also serves as a comparison between Novice privileges and those given to higher class licensees.

Table 2-1
Amateur Operator Licenses*

Class	*Code Test*	*Written Examination*	*Privileges*
Novice	5 WPM (Element 1A)	Elementary theory and regulations. (Element 2)	Telegraphy in 3700-3750 kHz (5167.5 kHz Alaska only, emergency communications using single sideband), telegraphy in 7100-7150 kHz, 21,100-21,200 kHz, telegraphy and RTTY in 28,100-28,300 kHz, telegraphy and single-sideband voice on 28,300-28,500 kHz, all amateur privileges authorized on 222.1 to 223.91 kHz and 1270 to 1295 MHz.
Technician	5 WPM (Element 1A)	Elementary theory and regulations, Technician-level theory and regulations. (Elements 2 and 3A)	All amateur privileges above 50.0 MHz plus Novice HF privileges.
General	13 WPM (Element 1B)	Elementary theory and regulations, Technician and General theory and regulations. (Elements 2, 3A and 3B)	All amateur privileges except those reserved for Advanced and Amateur Extra class. 1500-watts PEP output maximum.
Advanced	13 WPM (Element 1B)	All lower exam elements, plus intermediate theory. (Elements 2, 3A, 3B and 4A)	All amateur privileges except those reserved for Amateur Extra class. 1500-watts PEP output maximum.
Amateur Extra	20 WPM (Element 1C)	All lower exam elements, plus special exam on advanced techniques. (Elements 2, 3A, 3B, 4A and 4B)	All amateur privileges. 1500-watts PEP output maximum.

*A licensed radio amateur will be required to pass only those elements that are not included in the examination for the amateur license currently held. In other words, if you hold a Novice license, you need not take another 5-WPM code test to qualify for a Technician class license.

Frequency Privileges

When operating, you must stay within your assigned frequency bands. Novice operators may transmit on portions of six bands in the radio spectrum. Amateurs usually refer to these frequency bands by their wavelength. Novices may operate in the 80, 40, 15 and 10-meter bands, and in the 1.25-meter (220 MHz) and 23-centimeter (1270 MHz) bands. These bands are further divided into subbands for the different classes of license. Novice operators have **frequency privileges** (or permission to operate) on these subbands:

3700-3750 kHz in the 80-meter band
(5167.5 kHz—Alaska only, emergency communications, single-sideband voice emissions only [R3E or J3E])
7100-7150 kHz in the 40-meter band
21,100-21,200 kHz in the 15-meter band
28,100-28,500 kHz in the 10-meter band
222.1-223.91 MHz in the 1.25-meter band
1270-1295 MHz in the 23-centimeter band

We use the metric system of measurement in electronics. Chapter 4 includes a full explanation of the metric system and terms like *kilo*, *mega* and *centi* that we have used here. For now it is important that you know that *kilohertz* and *megahertz* are measures of the frequency of a radio signal. The *hertz* (Hz) is the basic unit of frequency. Kilo means thousand and mega means million. We can list any of the Novice frequency bands in either kilohertz or megahertz. For example, the 80-meter band is 3.7 to 3.75 MHz, and the 40-meter band is 7.1 to 7.5 MHz. The 15-meter band is from 21.1 MHz to 21.2 MHz and the 10-meter band is from 28.1 to 28.5 MHz.

[You must memorize these Novice frequency privileges. Before you go on to the next section, turn to Chapter 12 and study the questions that begin with numbers 2A-10. Review these frequencies often as you study the remaining material in this book.]

Emission Privileges

Amateur Radio operators transmit a wide variety of signals. They transmit Morse code, radioteletype, several types of voice communications and even television pictures. An **emission** is any radio-frequency (RF) signal from a transmitter.

There is a system for describing the various types of signals (or emissions) found on the amateur bands. These symbols, called emissions designators, consist of a letter, a number and another letter. The designator for Morse code generated by keying a continuous wave (CW) transmitter is A1A. Each mode has its own special identifier.

It isn't important for you to know how to put these emissions designators together. You should know what several of them stand for, however. The FCC lists the **emission privileges** for each license class by giving the emissions designators each may use. An emission privilege is FCC permission to use a particular emission type, such as Morse code or single sideband voice.

Novice operators may only transmit Morse code (A1A) on 80, 40 and 15 meters. The transmitter produces this Morse code signal by keying (switching on and off) the signal from a continuous wave (CW) transmitter. On 10 meters, Novices may use A1A from 28.1 to 28.5 MHz. Novice operators may also transmit radioteletype (emission F1B) from 28.1 to 28.3 MHz, and single-sideband voice (emission J3E) from 28.3 to 28.5 MHz. On the frequencies 222.1-223.91 MHz and 1270-1295 MHz, Novices may use all the emissions authorized to higher-class licensees on these bands, including FM voice (emission F3E) and digital modes, such as packet radio. Table 2-2 summarizes the amateur band limits and operating modes for each license class.

The FCC uses a special term to describe radioteletype (RTTY) and packet radio: **digital communications.** These are signals intended to be received and printed or displayed on a computer screen automatically. Information transferred directly from one computer to another is an example of digital communications.

[You have to memorize these Novice emission privileges. You should turn to Chapter 12 now and study those questions with numbers that begin 2A-17 and 2A-20. Review the material in this section if you have difficulty with any of the questions.]

Novice Transmitter Power

The FCC has issued rules explaining how transmitter power should be measured at the output of the transmitter. Novice licensees may use a maximum of 5 W **peak envelope power (PEP)** output on the 1270-MHz band, 25 W PEP on the 220-MHz band, and 200 W PEP on the 80, 40, 15 and 10-meter bands. The 200 W limitation also applies to any other licensed radio amateur who operates in the 80, 40 and 15-meter Novice bands. Remember that Novices may use up to 200 W PEP on any Novice frequency below 30 MHz.

In addition to these Novice bands, higher-class licensees may use a maximum of only 200 W PEP on the 30-meter band. On all other bands, the rules limit the maximum transmitter output power in the Amateur Radio Service to 1500 W PEP output. There are special exceptions for old-style AM transmitters, however. Higher-class licensees may use 1500 W PEP in the Novice sections of 10 meters, 220 MHz and 1270 MHz.

So far, we have been talking about the *maximum* transmitter power that the rules allow. There is another rule to consider, though. According to FCC rules, an Amateur Radio station must use the *minimum* transmitter power necessary to maintain reliable communication. What this means is simple—if you don't *need* 200 W to contact your friend across town, don't use it!

[Before proceeding to the next section turn to Chapter 12 and study the questions with numbers that begin 2A-18 and 2A-19. Review this section if you have any difficulties.]

THE NOVICE EXAM

The FCC refers to the various exams for Amateur Radio licenses as exam *Elements*. For example, exam Elements 1A, 1B and 1C are the 5, 13 and 20 word-per-minute (WPM) code exams. Element 2 is the Novice written exam. Table 2-1 summarizes the exams and privileges that go with each license class.

The table shows that you must pass exam Element 1A and Element 2 for the Novice license. The purpose of Element 1A is to prove your ability to send and receive messages using the international Morse code at a speed of 5 words per minute. The Element 2 written exam covers basic FCC Rules and Novice operating procedures, along with basic electronics theory.

[Now turn to Chapter 12 and study questions 2A-12.1 through 2A-12.3. Review this section if you have difficulty with any of these questions.]

What Will My Novice Exam be Like?

Two amateurs with General class licenses or higher, who are 18 or older, can give the test for a Novice license. The examiners must not be relatives of anyone taking the exam. The exams can be given at the convenience of the candidates and the examiners, at any location they agree to.

There is probably an active Volunteer Examining Team somewhere in your area. These teams conduct tests for the higher classes of Amateur Radio licenses. Chances are that the Volunteer Examiners will give you the Novice exam at one of their regular exam sessions.

If you have any trouble locating examiners, write to:

Educational Activities Branch
ARRL Headquarters
225 Main Street
Newington, CT 06111

We will refer you to someone near you who can arrange your test. We will also supply a list of ARRL/VEC Volunteer Exam sessions in your area.

The code test is normally given first. The examiners will usually send five minutes of five-word-per-minute code, and then test your copy. The test usually takes one of two forms. The examiners may ask you 10 questions based on the contents of the transmission, and you must answer 7 of the 10 questions correctly. Or, the examiners may check your answer sheet for one minute of solid (perfect) copy out of the five-minute transmission.

After you pass the code test, you will be given the written test. The written test consists of 30 questions on general operating practices, rules and regulations, and basic radio theory. To pass, you must correctly answer 22 of the 30 questions on your exam.

Your exam must include 30 questions taken from the pool of 372 questions printed in Chapter 12 of this book. Your examiners should choose one question from each of the 30 question-pool subsections. Tests are also available from the ARRL Educational Activities Branch if they wish to use one of those.

The *questions* must be used exactly as they are printed in Chapter 12. Your examiners are free to decide the format for your answers, though. Your examiners may request essay answers, fill-in-the-blank, multiple choice, true-false, etc.

The ARRL Educational Activities Branch will supply your examiners with a multiple-choice test, if they want to request it. We appreciate receiving a self-addressed stamped envelope for sending out these exams. The examiners must send a photocopy of their licenses (or other proof) to show that they are eligible to give the exams. (The exams are not available for students to use as practice!) Chapter 12 has the complete pool of 372 questions with multiple-choice answers.

FILLING OUT YOUR FCC FORM 610

Congratulations! You have passed both the code and

Table 2-2

US AMATEUR BANDS

Revised November 15, 1989

160 METERS

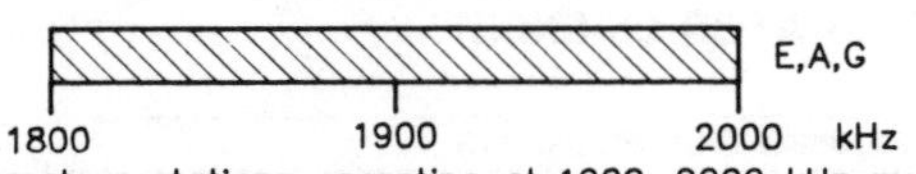

Amateur stations operating at 1900–2000 kHz must not cause harmful interference to the radiolocation service and are afforded no protection from radiolocation operations.

80 METERS

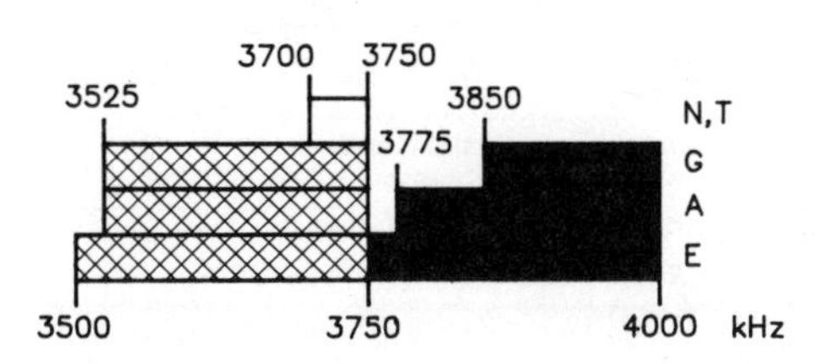

5167.5 kHz (SSB only): Alaska emergency use only.

40 METERS

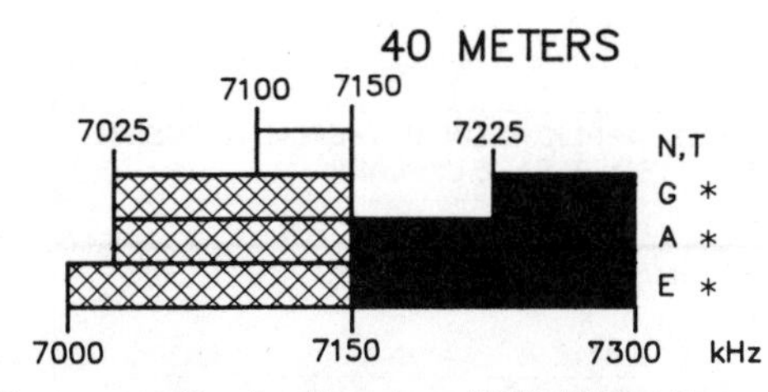

* Phone operation is allowed on 7075–7100 kHz in Puerto Rico, US Virgin islands and areas of the Caribbean south of 20 degrees north latitude; and in Hawaii and areas near ITU Region 3, including Alaska.

30 METERS

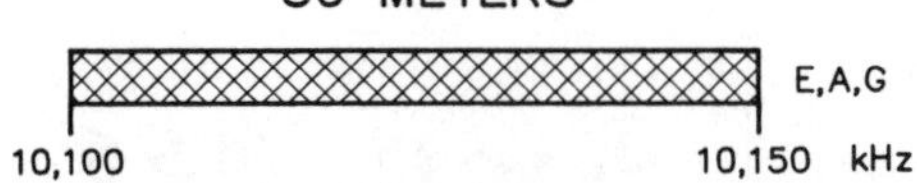

Maximum power on 30 meters is 200 watts PEP output. Amateurs must avoid interference to the fixed service outside the US.

20 METERS

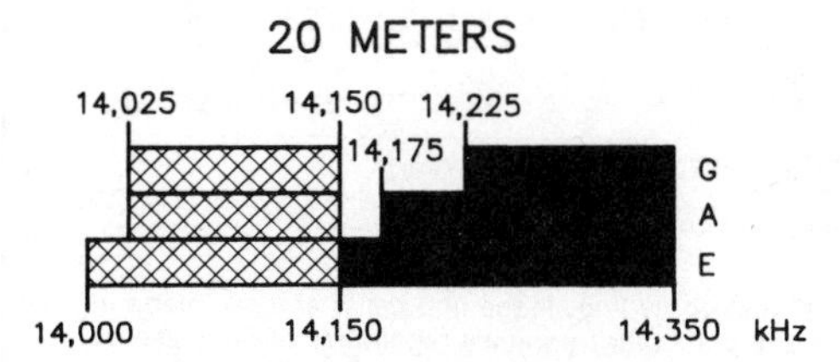

17 METERS

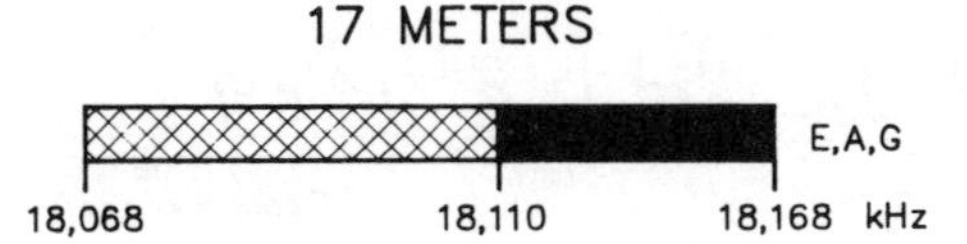

15 METERS

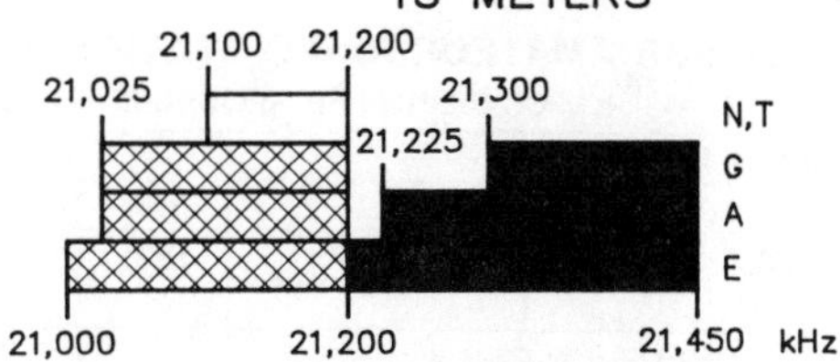

12 METERS

10 METERS

28,100 28,500

N,T

E,A,G

28,000 28,300 29,700 kHz

Novices and Technicians are limited to 200 watts PEP output on 10 meters.

6 METERS

50.1

E,A,G,T

50.0 54.0 MHz

2 METERS

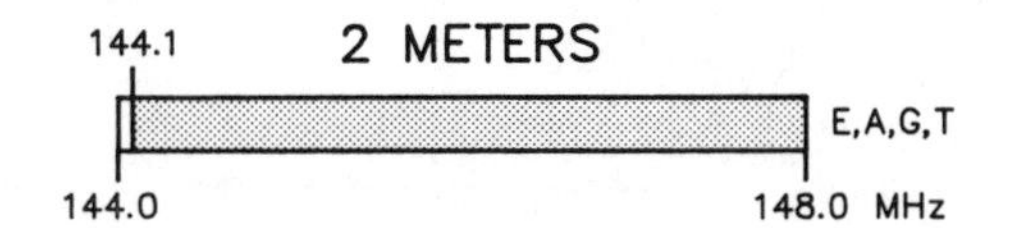

1.25 METERS

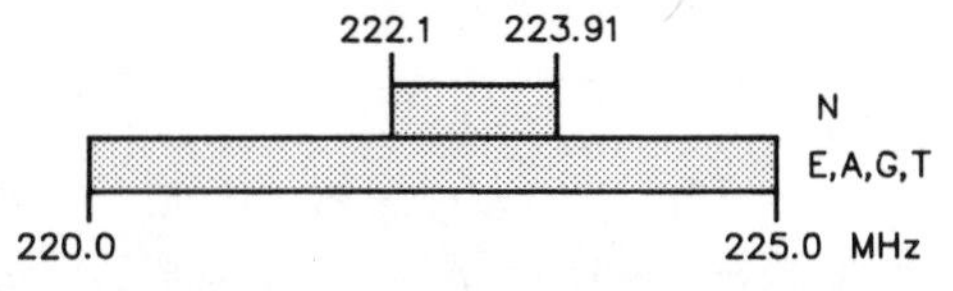

Novices are limited to 25 watts PEP output from 222.1 to 223.91 MHz.

70 CENTIMETERS

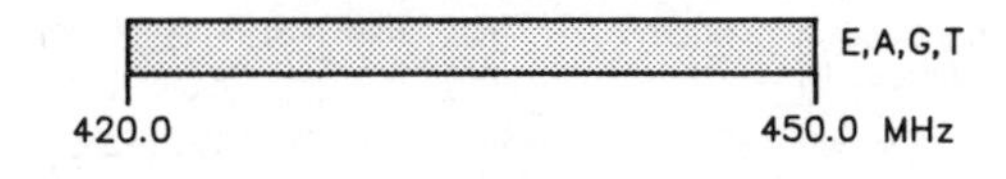

33 CENTIMETERS

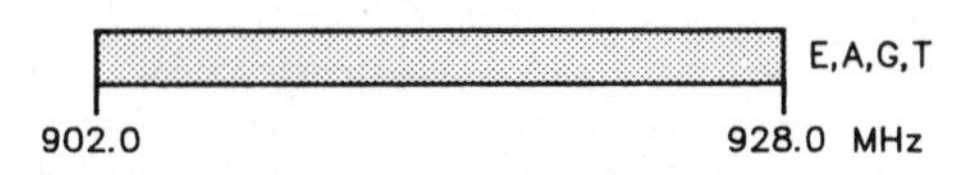

23 CENTIMETERS

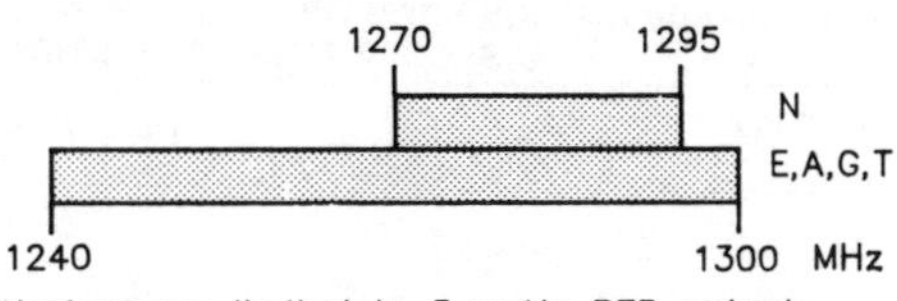

Novices are limited to 5 watts PEP output from 1270 to 1295 MHz.

US AMATEUR POWER LIMITS

At all times, transmitter power should be kept down to that necessary to carry out the desired communications. Power is rated in watts PEP output. Unless otherwise stated, the maximum power output is 1500 W. Power for all license classes is limited to 200 W in the 10,100–10,150 kHz band and in all Novice subbands below 28,100 kHz. Novices and Technicians are restricted to 200 W in the 28,100–28,500 kHz subbands In addition, Novices are restricted to 25 W in the 222.1–223.91 MHz subband and 5 W in the 1270–1295 MHz subband.

Operators with Technician class licenses and above may operate on all bands above 50 MHz. For more detailed information see The FCC Rule Book.

KEY

 = CW, RTTY and data

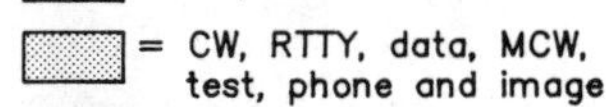 = CW, RTTY, data, MCW, test, phone and image

= CW, phone and image

= CW and SSB

= CW, RTTY, data, phone, and image

= CW only

E =AMATEUR EXTRA
A =ADVANCED
G =GENERAL
T =TECHNICIAN
N =NOVICE

FEDERAL COMMUNICATIONS COMMISSION
P.O. Box 1020
GETTYSBURG, PA 17326

Approved OMB
3060-0003
Expires 12/31/89

APPLICATION FOR AMATEUR RADIO STATION AND/OR OPERATOR LICENSE

NO FCC FILING FEE REQUIRED (see instruction H)

ADMINISTERING VEs' REPORT		EXAMINATION ELEMENTS							
Applicant is credited for:		1(A)	1(B)	1(C)	2	3(A)	3(B)	4(A)	4(B)
A. FCC Amateur license held (97.25(a)):	Class	(NT)	(GA)		(NTGA)	(TGA)	(GA)	(A)	
B. CERTIFICATE(S) OF SUCCESSFUL COMPLETION OF AN EXAMINATION HELD (97.25(b)):		Date Issued	Date Issued	Date Issued	Date Issued	Date Issued	Date Issued	Date Issued	Date Issued
C. FCC Commercial Radiotelegraph Operator License held (97.25(c)):	Number:			Exp Date					
D. Examination elements passed that were administered at this session:		X			X				

E. APPLICANT is qualified for operator license class: ☐ None:	H. Date of VEC coordinated examination session:
E1. ☒ NOVICE (Elements 1(A), 1(B), or 1(C) and 2)	
E2. ☐ TECHNICIAN (Elements 1(A), 1(B), or 1(C), 2 and 3(A)) ☐ GENERAL (Elements 1(B) or 1(C), 2, 3(A), and 3(B)) ☐ ADVANCED (Elements 1(B) or 1(C), 2, 3(A), 3(B) and 4(A)) ☐ AMATEUR EXTRA (Elements 1(C), 2, 3(A), 3(B), 4(A), and 4(B))	I. VEC Receipt Date:
F. Name of Volunteer-Examiner Coordinator: (VEC coordinated sessions only)	
G. Examination session location: (VEC coordinated sessions only)	

SECTION I

1. IF YOU HOLD A VALID LICENSE ATTACH THE ORIGINAL LICENSE OR PHOTOCOPY ON BACK OF APPLICATION. IF THE VALID LICENSE OR CERTIFICATE OF SUCCESSFUL COMPLETION OF AN EXAMINATION WAS LOST OR DESTROYED, PLEASE EXPLAIN.

2. CHECK ONE OR MORE ITEMS, NORMALLY ALL LICENSES ARE ISSUED FOR A 10 YEAR TERM.

2A. ☐ RENEW LICENSE—NO OTHER CHANGES → 2B. ☐ REINSTATE LICENSE EXPIRED LESS THAN 2 YEARS →	EXPIRATION DATE (Month, Day, Year)
2C. ☐ EXAMINATION FOR NEW LICENSE 2D. ☐ EXAMINATION TO UPGRADE OPERATOR CLASS 2E. ☐ CHANGE CALL SIGN (Be sure you are eligible—See Inst. 2E)	FORMER LAST NAME SUFFIX (Jr., Sr., etc.)
2F. ☐ CHANGE NAME (Give former name) → 2G. ☐ CHANGE MAILING ADDRESS 2H. ☐ CHANGE STATION LOCATION	FORMER FIRST NAME MIDDLE INITIAL
3. CALL SIGN (If you checked 2C above, skip items 3 and 4)	4. OPERATOR CLASS OF THE ATTACHED LICENSE:

5. CURRENT FIRST NAME	M.I.	LAST NAME	SUFFIX (Jr., Sr., etc.)	6. DATE OF BIRTH (Month, Day, Year)
William	B.	Svoboda		June 19, 1955

7. CURRENT MAILING ADDRESS (Number and Street)	CITY	STATE	ZIP CODE
458 A North Ninth Avenue	West Bend	WI	53095

8. CURRENT STATION LOCATION (Do not use a P.O. Box No., RFD No., or General Delivery. See Instruction 8)	CITY	STATE
458 A North Ninth Avenue	West Bend	WI

9. Would a Commission grant of your application be an action which may have a significant environmental effect as defined by Section 1.1307 of the Commission's Rules? See instruction 9. If you answer yes, submit the statement as required by Sections 1.1308 and 1.1311. ☐ YES ☒ NO

10. Do you have any other amateur radio application on file with the Commission that has not been acted upon? If yes, answer items 11 and 12. ☐ YES ☒ NO

11. PURPOSE OF OTHER APPLICATION	12. DATE SUBMITTED (Month, Day, Year)

CERTIFICATION

I CERTIFY THAT all statements herein and attachments herewith are true, complete, and correct to the best of my knowledge and belief and are made in good faith; that I am not a representative of a foreign government; that I waive any claim to the use of any particular frequency regardless of prior use by license or otherwise; and that the station to be licensed will be inaccessible to unauthorized persons.

WILLFUL FALSE STATEMENTS MADE ON THIS FORM OR ATTACHMENTS ARE PUNISHABLE BY FINE AND IMPRISONMENT U.S. CODE TITLE 18, SECTION 1001

13. SIGNATURE OF APPLICANT: (Must match item 5)	14. DATE SIGNED:
William B. Svoboda	July 9, 1989

FCC Form 610
September 1987

Figure 2-3—A completed FCC Form 610. All applications for new Amateur Radio licenses must be submitted using this form.

ATTACH THE ORIGINAL LICENSE OR PHOTOCOPY HERE

SECTION II—EXAMINATION INFORMATION

SECTION II-A FOR NOVICE OPERATOR EXAMINATION ONLY. To be completed by the Administering VEs after completing the Administering VE's Report on the other side of this form.

CERTIFICATION

I CERTIFY THAT I have complied with the Administering VE requirements stated in Part 97 of the Commission's Rules; THAT I have administered to the applicant and graded an amateur radio operator examination in accordance with Part 97 of the Commission's Rules; THAT I have indicated in the Administering VE's Report the examination element(s) the applicant passed; THAT I have examined documents held by the applicant and I have indicated in the Administering VE's Report the examination element for which the applicant is given examination credit in accordance with Part 97 of the Commission's Rules.

1A. VOLUNTEER EXAMINER'S NAME: (First, MI, Last, Suffix) *(Print or Type)*
Alice G. Brown

1B. VE'S MAILING ADDRESS: (Number, Street, City, State, ZIP Code)
P.O. Box 405 West Bend WI 53095

1C. VE'S OPERATOR CLASS: ☒ GENERAL ☐ ADVANCED ☐ AMATEUR EXTRA

1D. VE'S STATION CALL SIGN
N9XYZ

1E. LICENSE EXPIRATION DATE:
OCT 16, 1990

1F. IF YOU HAVE AN APPLICATION PENDING FOR YOUR LICENSE, GIVE FILING DATE:

1G. SIGNATURE: (Must match Item 1A)
Alice G. Brown

DATE SIGNED
July 9, 1989

2A. VOLUNTEER EXAMINER'S NAME: (First, MI, Last, Suffix) *(Print or Type)*
OLIVER W. ENTERHEIM

2B. VE'S MAILING ADDRESS: (Number, Street, City, State, ZIP Code)
35 CENTER STREET, KEWMSKUM, WI 53087

2C. VE'S OPERATOR CLASS: ☐ GENERAL ☐ ADVANCED ☒ AMATEUR EXTRA

2D. VE'S STATION CALL SIGN
N9OE

2E. LICENSE EXPIRATION DATE:
MAY 5, 1998

2F. IF YOU HAVE AN APPLICATION PENDING FOR YOUR LICENSE, GIVE FILING DATE:

2G. SIGNATURE: (Must match Item 2A)
Oliver W. Enterheim

DATE SIGNED
JULY 9, 1989

SECTION II-B FOR TECHNICIAN, GENERAL, ADVANCED, OR AMATEUR EXTRA OPERATOR EXAMINATION ONLY. To be completed by the Administering VEs after completing the Administering VE's Report on the other side of this form.

CERTIFICATION

I CERTIFY THAT I have complied with the Administering VE requirements stated in Part 97 of the Commission's Rules; THAT I have administered to the applicant and graded an amateur radio operator examination in accordance with Part 97 of the Commission's Rules; THAT I have indicated in the Administering VE's Report the examination element(s) the applicant passed; THAT I have examined documents held by the applicant and I have indicated in the Administering VE's Report the examination element(s) for which the applicant is given examination credit in accordance with Part 97 of the Commission's Rules.

1A. VOLUNTEER EXAMINER'S NAME: (First, MI, Last, Suffix) *(Print or Type)*

1B. VE'S STATION CALL SIGN:

1C. SIGNATURE: (Must match Item 1A)

DATE SIGNED:

2A. VOLUNTEER EXAMINER'S NAME: (First, MI, Last, Suffix) *(Print or Type)*

2B. VE'S STATION CALL SIGN:

2C. SIGNATURE: (Must match Item 2A)

DATE SIGNED:

3A. VOLUNTEER EXAMINER'S NAME: (First, MI, Last, Suffix) *(Print or Type)*

3B. VE'S STATION CALL SIGN:

3C. SIGNATURE: (Must match Item 3A)

DATE SIGNED:

FCC Form 610

U.S. GOVERNMENT PRINTING OFFICE: 1987 190-665 (m)

written tests. What's next? More paperwork! All applications for new amateur licenses, or modifications or renewals, are made on an FCC Form 610. Figure 2-3 shows a Form 610 completed for a successful Novice applicant.

Your examiners will fill out the Administering VE's report as shown at the top of the form in Figure 2-3. Your work begins in Section I, a third of the way down the form. Check box 2C, "EXAMINATION FOR NEW LICENSE." Skip items 3 and 4, and fill out items 5, 6, 7 and 8. Remember that you cannot use a post office box for your station location. You must show the actual location of your station in item 8.

Unless you plan to install an antenna over 300 feet high, or your station will be in a designated wilderness area, wildlife preserve or nationally recognized scenic and recreational area, check "NO" on item 9. Since you are applying for a new Novice license, check "NO" in box 10, and leave boxes 11 and 12 blank. Be sure to sign and date your application in boxes 13 and 14.

Your examiners should fill out Section II-A on the back of the form. This is where the examiners list their qualifications, and certify that you have passed the exam. Be sure to have *both* examiners fill in Section II-A. When all parts of the form are complete, the examiners will send it to: FCC, Box 1020, Gettysburg, PA 17326. After that comes the hard part—waiting for your new license!

The 610 Form serves as the application for your license. You will use the same form to renew your license before it expires. This form also serves another very important purpose. Line 7 provides the FCC with a mailing address where they can contact you. If you move, or your address changes for any reason, use an FCC Form 610 to notify them of your new address. It's important that you receive any mail sent to you by the FCC.

[Turn to Chapter 12 and study exam question 2A-14.1. If you have trouble answering this question, reread the last paragraph.]

NOW THAT YOU'RE A NOVICE

Well, the big day finally arrived, and your Novice "ticket" came in the mail from the FCC! You have permission to operate an Amateur Radio station. Now you are ready to put your very own Amateur Radio station on the air!

As the proud owner of a new Novice ticket, you'll soon be an "on-the-air" person instead of an "off-the-air" person. You probably can't wait to make your first contact! You'll be putting all the information you had to learn for the exam to good use.

First things first. When you get your new ticket, "hot off the press," make a few photocopies of it. Then put the original license in a safe place. You must have your license (or a photocopy) available whenever you operate, so put one of those copies in your wallet. That magic piece of paper is your **operator license**, which lets you operate a station within your Novice privileges. It is also your **station license**, and lists the call sign that identifies your station.

CALL SIGNS

The FCC issues call signs on a systematic basis. When they process your application, you get the next call sign to come out of the FCC computer. The FCC issues call signs from four "groups," depending on the class of license.

Novice class licensees are given "Group D" call signs. These have a "two-by-three" format—two letters followed by a number, followed by three letters. The letters before the number make up the call sign *prefix*, and the letters after the number are the *suffix*. An example of a Novice call sign is KA9OLS.

Other Amateur Radio license classes have different call-sign formats: Group C calls are given to Technician and General class amateurs. These are one-by-three calls, such as N1CZC. Advanced class amateurs get calls from Group B—the "two-by-two" group. KB9UE is an example of a call from this group. Group A calls are for Amateur Extra class

SOME EARLY AMATEUR INGENUITY

One young lad of seventeen, known to possess an especially efficient spark, CW and radio telephone station, was discovered to be the son of a laboring man in extremely reduced circumstances. The son had attended grammar school until he was able to work, and then he assisted in the support of his family. They were poor indeed. Yet despite this the young chap had a marvelously complete and effective station, installed in a miserable small closet in his mother's kitchen. How had he done it? The answer was that he had constructed every last detail of the station himself. Even such complicated and intricate structures as head telephones and vacuum tubes were homemade! Asked how he managed to make these products of specialists, he showed the most ingenious construction of headphones from bits of wood and wire. To build vacuum tubes he had found where a wholesale drug company dumped its broken test tubes, and where the electric light company dumped its burned-out bulbs, and had picked up enough glass to build his own tubes and enough bits of tungsten wire to make his own filaments. To exhaust the tubes he built his own mercury vacuum pump from scraps of glass. His greatest difficulty was in securing the mercury for his pump. He finally begged enough of this from another amateur. And the tubes were good ones—better than many commercially manufactured and sold. The greatest financial investment that this lad had made in building his amateur station was 25 cents for a pair of combination cutting pliers. His was the spirit that has made Amateur Radio.—Clinton B. Desoto, in *200 Meters and Down*

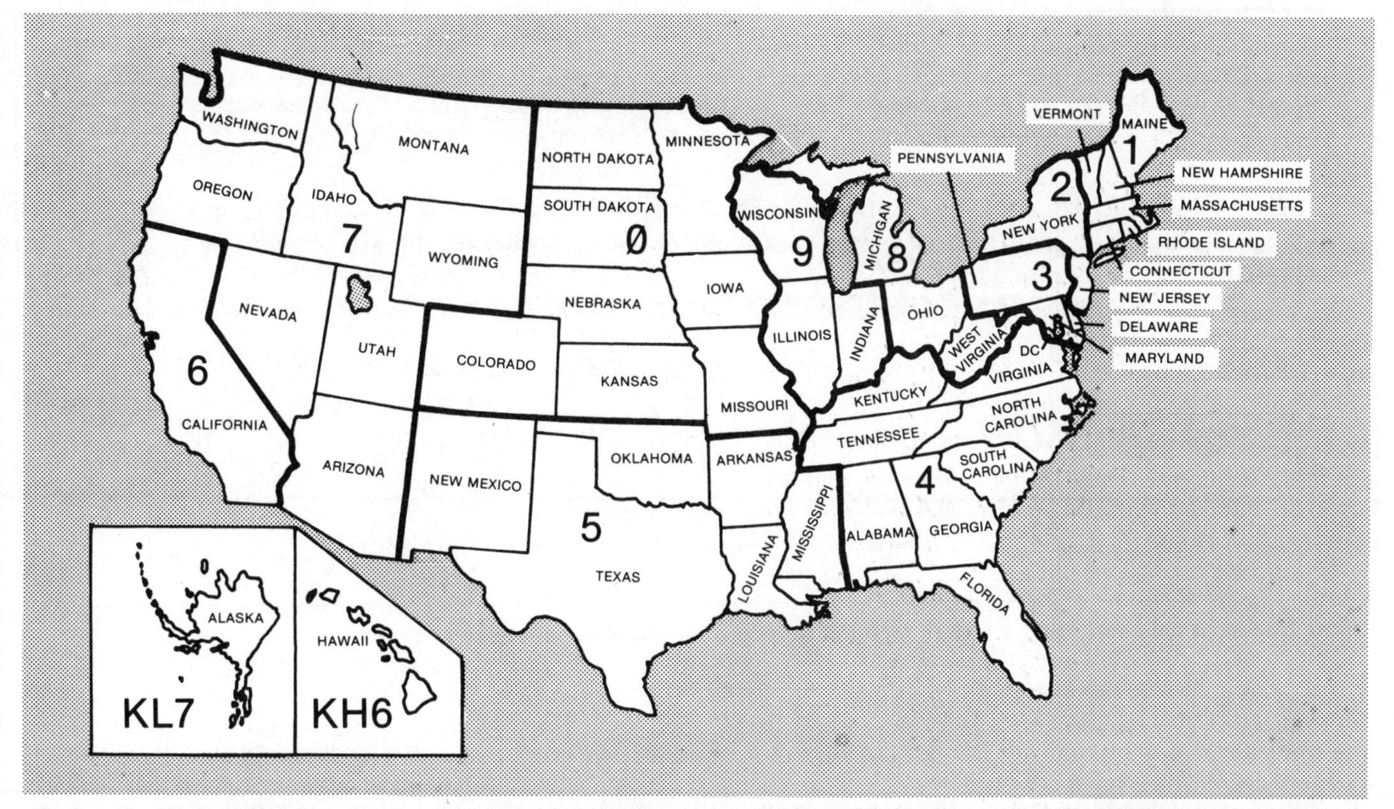

Figure 2-4—The 10 US call districts. An amateur holding the call sign KA9OLS lived in the ninth district when the FCC assigned him that call. Alaska is part of the seventh call district, but has its own set of prefixes: AL7, KL7, NL7 and WL7. Hawaii, part of the sixth district, has the AH6, KH6, NH6 and WH6 prefixes.

operators only. These calls are of the "one-by-two" or "Two-by-one" format. (K8CH is a one-by-two call; AA2Z is a two-by-one call.)

Once you have a call sign, you may keep it as long as you want to (unless your license expires or is revoked). In other words, there's no requirement to change your call sign after you upgrade to a higher license class. The FCC gives amateurs their choice in this matter.

The number in the call sign shows the district where the call was first issued. Figure 2-4 shows the 10 US call districts. Amateurs may keep their calls when they move from one district to another. This means the number is not always an indication of where an amateur is. It tells only where he or she was living when the license was first issued. For example, WB3IOS received her license in Pennsylvania, in the third call district, but she now lives in the first district.

The first letter of a US call will always be A, N, K or W. These letters are assigned to the United States as amateur call-sign prefixes. Other countries use other prefixes—LA2UA is a Norwegian call sign, and ON4SO is from Belgium.

[Now it's time to turn to Chapter 12 and study some questions. You should be able to answer all of the questions with numbers that begin 2A-15. If you have any difficulty, review this section.]

OPERATING GUIDELINES

Keep your operation "above board"—set an operating example that will make you proud. You should be familiar with the basic operating and technical rules. You should also know the standard operating practices used for the various modes you operate. Chapter 10 includes detailed information about operating your station. In this section we will study the FCC rules of operating and cover some very simple guidelines.

FCC rules do not require you to keep a station logbook, but many amateurs prefer to keep track of their station activities. Logbooks are useful for recording dates, calls, names and locations of those stations you contact. When

DATE	FREQ.	MODE	POWER	TIME	STATION WORKED	REPORT SENT	REPORT REC'D	TIME OFF	COMMENTS QTH NAME QSL VIA	QSL S	QSL R
16 NOV	3715	A1A	100	1800	KA1EBV	589	479	1835	MANOMET, MA - JOHN WILLIAMS	✓	
	"	"	"	1840	WB7TPY	599	599	1855	TUFTS UNIVERSITY - DAVE GOES TO DENTAL SCHOOL		
17 NOV	28.125	A1A	100	2000	VP2ML	579	589	2003	CARIBBEAN - CHOD - NICE WX! BIG PILE UP	✓	✓
	28.140	"	"	2010	KAØHJD	599	599	2015	DES MOINES, IA - KRISTEN AND DAD, WØSH		
					WØSH				THEY COLLECT OLD TELEGRAPH KEYS!	✓	✓
	3710	A1A	100	2033	KA1GQJ	579	589	2055	LINDA - NURSE IN A HOSPITAL	✓	
20 NOV	3715	A1A	100	1645	NAIL	559	559	1700	DALE IS A LAWYER; GOOD SIGNAL,	✓	✓
									KA1IXI, XYL - CHERYL		
	3722	A1A	100	1702	AA2Z	599	599	1705	EAST HAMPTON, CT - WOW!	✓	✓
									MARK WORKS AT ARRL HQ!!		
	3720	A1A	100	2315	WA1TBY			2325	NENN - QNI QTC 2		
	21.175	A1A	100	2332	N6TR	539	569	2345	OREGON - SAID HIS NAME		
									WAS "TREE"?!	✓	

Figure 2-5—Most amateurs keep a station logbook to record information about each contact.

you confirm contacts by sending QSL cards, a logbook is a convenient way to keep track of these exchanges. Your log will provide a useful and interesting history, if you choose to keep one. Figure 2-5 shows an example of the information many amateurs keep in a logbook.

TIME OUT FOR STATION IDENTIFICATION

The FCC regulations are very specific about station identification. You must identify your station every 10 minutes or less during a contact and at the end of the contact. You do not have to identify with every transmission, only at the required intervals. Identify your station by transmitting the station call sign listed on your license.

The rules prohibit **unidentified communications or signals** (where the transmitting station's call sign is not transmitted). Be sure you understand the proper station identification procedures, so you don't violate this rule.

You don't have to transmit your call sign at the beginning of a contact. In most cases it will help you establish the communications, though. Unless you give your call sign, the other stations won't know who you are! When you take part in a net or talk with a group of hams regularly, they may learn to recognize your voice. Then you may not have to send your call sign for them to know who you are.

You do not have to send both call signs when you are talking with another ham. You only have to transmit your own call sign. The exception is in cases of third-party communications with a station in a foreign country. (You'll learn more about third-party communications later in this chapter.) Then you do have to transmit both call signs at the end of the contact.

Let's look at an example of how to identify an Amateur Radio station properly. KA9OLS and KB1MW have been in communication for 45 minutes and are now signing off. Each operator has already transmitted his call a minimum of four times (once after each 10-minute interval). Each should now transmit his call one more time as they sign off, for a total of five times during the QSO. (QSO is ham talk for a two-way contact or communication with another ham.)

Suppose the QSO had lasted only 8 minutes. Each station would be required to transmit his call sign only once (at the end of the communication). You may identify more often than this to make it easier to communicate on a crowded band. But the rules specify only that you identify every 10 minutes, and at the end of a contact. In Morse code, identification takes the form DE KA9OLS ("DE" means "from" in French).

[You should turn to Chapter 12 now, and study those questions with numbers that begin 2A-27 and 2A-38. Review this section if you have any difficulty with those questions.]

POINTS OF COMMUNICATION

Who can you talk to with your new Novice license? The FCC defines "points of communication" to specify what kinds of radio stations they allow you to talk with. You may converse with all Amateur Radio stations at any time. This includes amateurs in foreign countries, unless either amateur's government prohibits the communications. (There are a few countries in the world that do not allow Amateur Radio. There are also times when a government will not allow its amateurs to talk with people in other countries.)

The FCC must authorize any communication with any stations not licensed in the Amateur Radio Service. An example of such authorization is when amateur stations communicate with the military communications stations on Armed Forces Day each year.

Another example of amateurs communicating with nonamateur stations is during Radio Amateur Civil Emergency Service (RACES) operation. During an emergency, a registered RACES station may conduct civil-defense communications with stations in the Disaster Communications Service. During such emergencies, RACES stations may also communicate with other US Government stations authorized to conduct civil-defense communications.

[Turn to Chapter 12 now and study questions 2A-28.1, 2A-28.2 and 2A-31.1. Review this section if you have difficulty with any of these questions.]

BROADCASTING

Amateur Radio is a two-way communications service. Amateur Radio stations may not broadcast information intended for reception by the general public. There are also restrictions regulating one-way transmissions of information of general interest to other amateurs. Amateur stations may transmit one-way signals while in beacon operation or radio-control operation. Novice control operators don't have these privileges, however.

[Now turn to Chapter 12 and study question 2A-33.1. If you are uncertain of the answer to this question, review this section.]

THIRD-PARTY TRAFFIC

Communication on behalf of anyone other than the operators of the two stations in contact is **third-party traffic**. For example, sending a message from your mother-in-law to her relatives in Scarsdale on Valentine's Day is passing third-party traffic. Third-party traffic of a personal nature is okay, but passing traffic involving business matters is not. The Amateur Service is not the place to conduct business.

You can pass third-party traffic to other stations in the United States. Outside the US, FCC rules strictly limit traffic to those countries that have third-party agreements with our country.

In many countries, the government operates the telephone system and other communications lines. You can imagine why these governments are reluctant to allow their Amateur Radio operators to pass messages. They would be in direct competition with the government-operated communications system.

The ARRL publishes a list of countries with which the US has a third-party traffic agreement in *QST* from time to time. The list does change, and sometimes there are temporary agreements to handle special events. Check such a list, or ask someone who knows, before you try to pass a message to another country for a nonham.

You may allow an unlicensed person to participate in Amateur Radio from your station. This is **third-party participation**, or **third-party communication**. It is another form of third-party traffic.

You (as control operator) must always be present to make sure the unlicensed person follows all the rules. You can allow your family members and friends to enjoy some of the excitement of Amateur Radio in this way. They can speak into the microphone or even send Morse code, as long as you are present to control the radio. They can also type messages on your computer keyboard to talk with someone using packet radio or radioteletype.

There is one important rule to keep in mind about third-party participation. If the unlicensed person was an Amateur Radio operator whose license was suspended or revoked by the FCC, that person may not participate in any Amateur Radio communication. You can't allow that person to talk into the microphone of your transmitter or operate your Morse code key or computer keyboard.

[Now turn to Chapter 12 and study the questions with numbers that begin 2A-34. Review this section as needed.]

BUSINESS COMMUNICATIONS

Amateur Radio communication is noncommercial radio communication between Amateur Radio stations, solely with a personal aim and without pecuniary or business interest. Pecuniary refers to payment of any type. This definition tells us that Amateur Radio stations should not conduct any type of business communications. This does not apply just to your business interests. It applies to anyone else's business as well.

Obviously, you would not use Amateur Radio to keep in touch with your company's delivery trucks. But what about using the repeater autopatch to call the local Pizza Palace and order dinner on your way home? Sure, it isn't *your* business, but it is a business call. Don't do it!

Some hams believe it is okay to send a message to ARRL Headquarters asking for some forms or other information. After all, the ARRL is a not-for-profit organization. It is still a business, however, and such requests *are* the business of the League. Messages like that are not allowed over Amateur Radio, because they are business messages.

As with most rules, there is a possible exception to the business communications rule. Your car has just broke down in the middle of a busy expressway. Can you use the autopatch to call a garage for help? Now you have an emergency. Your property (car), and possibly your life, are in immediate danger. In an emergency, you can use Amateur Radio to call a business for help.

There are many examples of times when personal property or lives are in immediate danger. You may use Amateur Radio to call for help in such a situation, even though you seem to be violating other rules. Just be sure you really have an emergency situation first.

No one can use an amateur station for monetary gain. You must not accept payment in any form for the use of your station at any time. This also means you may not accept payment for transmitting a message for anyone. Payment means more than just money here. It refers to any type of compensation, which would include materials or services of any type.

There is one exception to this rule. A club station intended primarily for transmitting Morse code practice and information bulletins of interest to all amateurs may employ a paid control operator. That person can be paid to serve as the control operator only when the station is actually transmitting code practice or bulletins. This exception allows the ARRL to pay a control operator for W1AW, the station at League Headquarters in Newington, Connecticut. The station is dedicated to Hiram Percy Maxim, the League's first president. The W1AW operator can't make general contacts with other hams after the code-practice sessions, however.

[Before you go on to the next section, turn to Chapter

BE ON THE FRONTIER OF TECHNOLOGICAL ADVANCES

For 90 years hams have carried on a tradition of learning by doing. From the earliest days of radio, hams have built their transmitters from scratch, wrapping strands of copper wire salvaged from Model T automobiles around oatmeal boxes. Through experimenting with building their own equipment, hams have pioneered advances in technology such as the techniques for single sideband. Hams were the first to bounce signals off the moon to extend signal range. Amateurs' practical experience has led to many technical refinements and cost reductions beneficial to the commercial radio industry. The photo at the right shows a complete transmitter constructed in a small package.

12 and study those questions with numbers that begin 2A-30 and 2A-32. If you have any difficulty with any of them, review this section.]

OTHER ASSORTED RULES

Under FCC rules, amateurs may not transmit music of any form. They may not use obscene, indecent or profane language. You can't use codes or ciphers to obscure the meaning of transmissions. This means you can't make up a "secret" code to send messages over the air to a friend.

Amateurs may not cause **malicious** (intentional) **interference** to other communications. You may not like the other operator's practices, or you may believe he or she is violating the rules. You have no right to interfere with their communications, however.

Amateurs may not transmit **false or deceptive signals**, such as a distress call when no emergency exists. You must not, for example, start calling MAYDAY (an international distress signal) unless you are in a life-threatening situation.

[Now it's time to turn to Chapter 12 again, and study a few questions. You should be able to answer those questions with numbers that begin 2A-35, 2A-36, 2A-37 and 2A-39. Review this section if you have difficulty with any of these questions.]

NOTICES OF VIOLATION

The FCC is the agency in charge of maintaining law and order in radio operation in the US. The International Telecommunication Union (ITU) sets up international rules that the government agencies of each country must follow. Both sets of rules provide the basic structure for Amateur Radio in the United States.

Suppose you receive an official notice from the FCC informing you that you have violated a regulation. Now what should you do? You must respond in writing within 10 days to the FCC office that issued the notice. If the notice relates to equipment problems, you must tell the FCC what steps you will take to prevent further violations. If the notice concerns improper operating procedures, you must tell them who the control operator was at the time of the violation.

Operating a station at your summer cottage is considered **portable operation**. The FCC defines portable operation as "radio communication conducted from a specific geographical location other than that shown on the station license." When you use a transceiver installed in your car, you're conducting a **mobile operation**. (FCC definition: "radio communication conducted while in motion or during halts at unspecified locations.") Amateurs do *not* have to notify the FCC before placing a station in portable or mobile operation.

Remember that you must still respond to an FCC Notice of Violation within 10 days, however. The FCC will mail any notices of violation to the permanent mailing address shown on your license. There may be times when you will operate your station away from home for a long period. Arrange to have your mail forwarded from your permanent mailing address to your temporary home in those cases.

There exists a philosophy in Amateur Radio that is deeply rooted in our history. This philosphy is as strong now as it was in the days of the radio pioneers. We are talking about the philosophy of self-policing our bands. Over the years, amateurs have become known for their ability to maintain high operating standards and technical skills. We do this without excessive regulation by the FCC. The Commission itself has praised the Amateur Service for its tradition of self-policing. Perhaps the underlying reason for this is the amateur's sense of pride, accomplishment, fellowship, loyalty and concern. Amateur Radio is far more than just a hobby to most amateurs.

As a new or prospective Amateur Radio operator, you will begin to discover the wide horizons of your new pastime. You will learn more about the rich heritage we all share. Take this sense of pride with you, and follow the Amateur's Code.

[This completes your study of Chapter 2 and the FCC rules and regulations regarding your Novice station operation. Turn to Chapter 12 and study the questions that begin 2A-29 and 2A-40. By now you should have no difficulty with the questions in subelement 2A on the question pool. Review the material in this chapter until you feel confident, and then move on to Chapter 3.]

The Amateur's Code

The Radio Amateur is:

CONSIDERATE. . . never knowingly operates in such a way as to lessen the pleasure of others.

LOYAL. . . offers loyalty, encouragement and support to other amateurs, local clubs, and the American Radio Relay League, through which Amateur Radio in the United States is represented nationally and internationally.

PROGRESSIVE. . . with knowledge abreast of science, a well-built and efficient station and operation above reproach.

FRIENDLY. . . slow and patient operating when requested; friendly advice and counsel to the beginner; kindly assistance, cooperation and consideration for the interests of others. These are the hallmarks of the amateur spirit.

BALANCED. . . radio is an avocation, never interfering with duties owed to family, job, school, or community.

PATRIOTIC. . . station and skill always ready for service to country and community.

—The original Amateur's Code was written by Paul M. Segal, W9EEA, in 1928.—

KEY WORDS

A1A emission—The FCC emission designator used to describe Morse code telegraphy (CW) by on/off keying of a radio-frequency signal.

Bandwidth—The range of frequencies in the radio spectrum that a radio transmission occupies.

Code key—A device used as a switch to generate Morse code.

Code-practice oscillator—A device that produces an audio tone, used for learning the code.

Continuous wave (CW)—A term used by amateurs as a synonym for Morse code communication. Hams usually produce Morse code signals by interrupting the continuous-wave signal from a transmitter to form the dots and dashes.

Dash—The long sound used in Morse code. Pronounce this as "dah" when verbally sounding Morse code characters.

Dot—The short sound used in Morse code. Pronounce this as "dit" when verbally sounding Morse code characters if the dot comes at the end of the character. If the dot comes at the beginning or in the middle of the character, pronounce it as "di."

Emission—RF signals transmitted from a radio station.

Emission privileges—Permissible types of transmitted signals.

Fist—The unique rhythm of an individual amateur's Morse code sending.

Phone—Voice communications.

Q signals—Three-letter symbols beginning with "Q," used in amateur CW work to save time and for better comprehension.

Traffic net—An on-the-air meeting of amateurs, for the purpose of relaying messages.

W1AW—The headquarters station of the American Radio Relay League. This station is a memorial to the League's cofounder, Hiram Percy Maxim. The station provides daily on-the-air code practice and bulletins of interest to hams.

Chapter 3

Learning Your New Language

When WA6INJ's jeep went over a cliff in a February snowstorm, he was able to call for help using his mobile rig. This worked well at first, but as the search for him continued, he became unable to speak. The nearly frozen man managed to tap out Morse code signals with his microphone push-to-talk button. That was all his rescuers had to work with to locate him. The fact that every ham knows the code saved this ham's life.

CODE IS FUN!

Using the code is an exciting way to communicate, even when it's not a life-or-death emergency. Many long-time hams beam with pride when they proclaim that they don't even own a microphone! You are fluent in another complete language when you know the code. You can chat with hams from all around the world using this common language. With the practice you will gain by making on-the-air contacts, your speed will increase quickly. With more advanced licenses, you'll be operating on more portions of the bands, increasing your realm of contacts.

All Amateur Radio operators must know the international Morse code if their license permits them to transmit on frequencies below 30 MHz. An international treaty sets this rule, and the United States must follow the treaty. Because of this, the Federal Communications Commission requires you to pass a code test to earn your Novice license.

There's a lot more to it than just satisfying the terms of an international treaty, however. Morse code goes back to the very beginning of radio, and is still one of the most effective radio-communication methods. We send Morse code by interrupting the **continuous-wave** signal generated by a transmitter, and so we call it **CW** for short.

For one thing, it takes far less power to establish reliable communications with CW than it does with voice (**phone**). On phone, we sometimes need high power and elaborate antennas to communicate with distant stations, or *DX* in

AS MUCH FUN AS "SESAME STREET"

Think Morse code is a difficult hurdle that takes weeks and weeks of hard work to master? Think again! Guy Mitchell, WDØDVX, of Buckingham, Iowa, picked up Morse code from tapes his parents were using for their Novice class. Soon, he began studying on his own, and mastered the alphabet and numbers with little difficulty.

What makes this story unique is that Guy Mitchell was four years old when he passed the five-word-per-minute code test! He had another birthday by the time he passed the written test, however. At five years of age, Guy was the youngest licensed ham radio operator in the US, and perhaps the world!

Guy Mitchell started kindergarten five months after he passed the Novice code test. That was in 1977. Now Guy is an A student in high school. He enjoys sports, music and other activities. He says, "Ham radio is something I enjoy doing with my family. I like it not just for the communications part, but for the science part as well."

the ham's lingo. On CW, less power and more modest stations will provide the same contacts. Finally, there is great satisfaction in being able to communicate using Morse code. This is similar to the satisfaction you might feel from using any acquired skill.

CONSERVE TIME AND SPECTRUM SPACE

Another advantage of CW over phone is its very narrow **bandwidth**. Using code is an efficient way to conserve spectrum space. The group of frequencies where the hams operate, the *ham bands*, are narrow portions of the whole spectrum. Many stations use the bands, and because they are so crowded, interference is sometimes a problem. A CW signal occupies about one-tenth the bandwidth of a phone signal and so takes up less space in the band. This means 10 CW signals can fit into the same part of a ham band that one phone signal would take up.

Over the years, radiotelegraph operators have developed a vocabulary of three-letter **Q signals**, which other radiotelegraphers throughout the world understand. For example, the Q signal *QRM* means "you are being interfered with." Just imagine how hard it would be to communicate that thought to someone who didn't understand one word of English. (Chapter 10 includes a list of common Q signals.)

There is another advantage to using Q signals: speed. It's much faster to send three letters than to spell out each word. That's why you'll be using these Q signals even when you're chatting by code with another English-speaking ham. Speed of transmission is also the reason radiotelegraphers use a code "shorthand." For example, to acknowledge that you heard what was transmitted to you, send the letter *R*. This is short for "roger," which means "I have received your transmission okay."

Standard Q signals and shorthand abbreviations reduce the total time necessary to send a message. When radio conditions are poor and signals are marginal, a short message is much more likely to get through.

Many hams prefer to use CW in **traffic nets**. (A net is a regular on-the-air meeting of a group of hams.) Using these nets, hams send messages across the country for just about anyone. When they send messages using CW, there is no confusion about the spelling of names such as Lee, Lea or Leigh.

CW: SOMETIMES THE ONLY CHOICE

For some types of transmissions, CW (also called type **A1A emission** by the FCC) is the only available choice. As

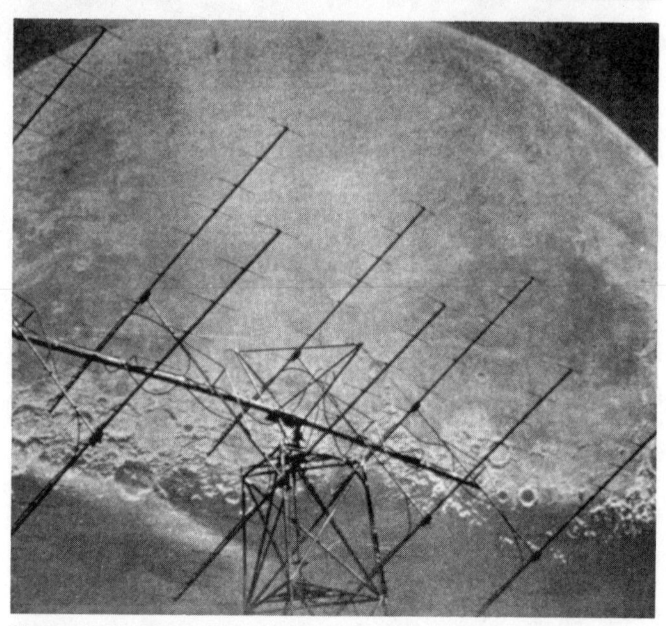

Figure 3-1—Even the moon isn't immune from ever-ambitious hams, and code is the most efficient type of signal to bounce off its surface and be reflected back to earth.

a Novice, CW is the only **emission privilege** you may use on the 80, 40 and 15-meter amateur bands. (One exception to this rule is if you are in Alaska, and involved in some emergency communications. In that case you may use single sideband [J3E or R3E] on 5167.5 kHz.) CW is the only mode allowed on *all* amateur frequencies.

Some hams like to bounce messages off the surface of the moon to another ham station on the earth. Because of its efficiency, hams use CW in most of this *moonbounce* work. They could use voice signals, but this increases the power and antenna gain requirements quite a bit. Few amateurs can afford this extra expense.

On some frequencies, amateurs communicate by bouncing their signals off an auroral curtain in the northern sky. (Stations in the Southern Hemisphere would use an auroral curtain in the southern sky.) Phone signals become so distorted in the process of reflecting off an auroral curtain that they are difficult or impossible to understand. CW is the most effective way to communicate using signals bounced off an aurora.

You will feel a special thrill and a warm satisfaction when you use the Morse code to communicate with someone. This feeling comes partly from sending messages to another part of the world without regard to language barriers. Sending and receiving Morse code is a skill that sets amateurs apart from other radio operators. It provides a common bond between all amateurs worldwide.

GETTING ACQUAINTED WITH THE CODE

The basic element of a Morse code character is a **dot**. The length of the dot determines how long a **dash** should be. The dot length also determines the length of the spaces between elements, characters and words. Figure 3-2 shows the proper timing of each piece. Notice that a dash is three times as long as a dot. The time between dots and dashes in a character is equal to the length of a dot. The time between letters in a word is equal to three dot lengths and the space between words is seven dot lengths.

The lengths of Morse code characters are not all the same, of course. Samuel Finley Breese Morse (1791-1872) developed the system of dots and dashes in 1838. He assigned the shortest combinations to the most-used letters in plain-language text. The letter E has the shortest sound, because it is the most-used letter. T and I are the next most-used characters, and they are also short. Character lengths get longer for letters used less often.

An analysis of English plain-language text shows that the average word (including the space after the word) is 50 units long. By a unit we mean the time of a single dot or space between the parts of a character. The word PARIS is 50 units long, so we use it as a standard word to check code speed accurately. For example, to transmit at 5 words per minute (WPM), adjust your code-speed timing to send PARIS five times in one minute. To transmit at 10 WPM, adjust your timing to send PARIS 10 times in one minute.

As you can see, the correct dot length (and the length of dashes and spaces) changes for each code speed. As a result, the characters sound different when the speed changes. This leads to problems for a person learning the code. Also, at slower speeds, the characters seem long and drawn out. The slow pace encourages students to count dots and dashes, and to learn the code through this counting method.

Unfortunately, learning the code by counting dots and dashes introduces an extra translation in your brain. (Learning the code by memorizing Morse-code-character dot/dash patterns from a printed copy introduces a similar extra trans-

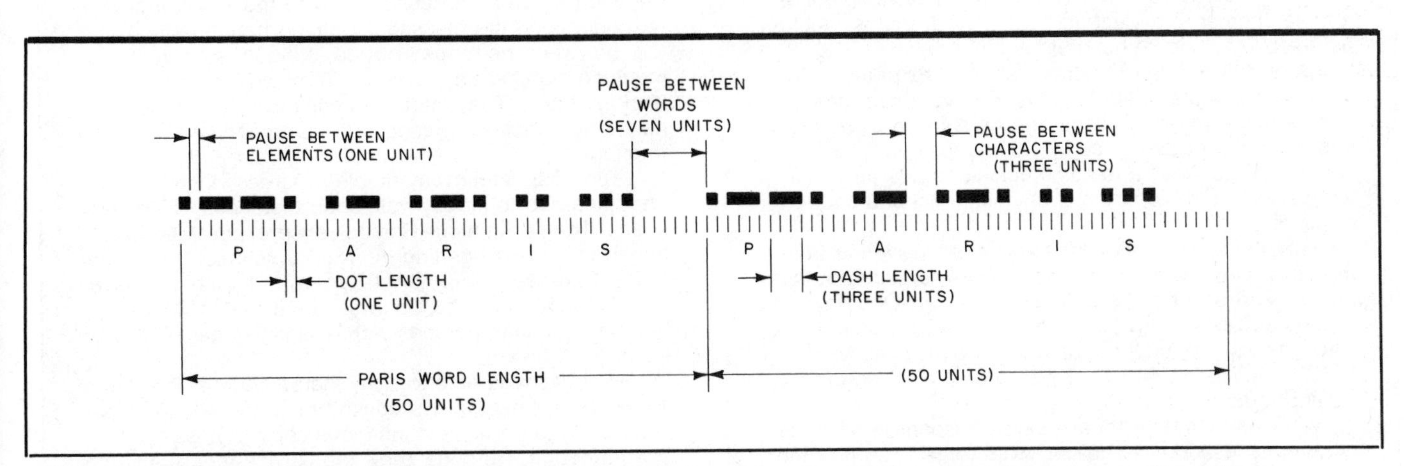

Figure 3-2—Whether you are a beginner or an expert, good sending depends on maintaining proper time ratios or "weight" among the dots, dashes and spaces, as shown.

lation.) That extra translation may not seem so bad at first. As you try to increase your speed, however, you will soon find out what a problem it is. You won't be able to count the dots and dashes and then make the translation to a character fast enough!

People learning the Morse code with either of these methods often reach a learning plateau at about 10 words per minute. Unfortunately, this is just below the 13 WPM speed required to upgrade to the General class license. That frustration (I just can't copy faster than 10 WPM!) is overcome by other methods, however.

Many studies have been done, and various techniques tried, to overcome this learning plateau. The method that has met with the most success is the Farnsworth method. That is the one used on the ARRL code-teaching tapes. With this technique, we send each character at a faster speed (we use 18 WPM). At speeds in this range, the characters—and even some short words—begin to take on a distinctive rhythmic pattern.

With this faster character speed, we use longer spaces between characters and words to slow the overall code speed. The ARRL code-teaching tapes send code at an overall speed of 5 WPM. (You can still measure this timing by using the word PARIS, as described earlier.) Once you learn the character sounds, and can copy at 5 WPM, it will be easy to increase your speed. Just decrease the spaces between letters and words, and your code speed increases without changing the rhythmic pattern of the characters.

Learn to recognize that rhythmic pattern, and associate it directly with the character. That way, you'll learn the code in the shortest possible time. Increasing your code speed to upgrade to the General class license will be easy. Decreasing the space between characters and words provides a natural progression to increase your code speed. The ARRL offers code-practice tapes from 5 to 10 WPM, 10 to 15 WPM and 15 to 22 WPM using this technique.

Morse Code: the Essential Language, by L. Peter Carron, Jr, W3DKV (published by ARRL) contains other suggestions for learning Morse code. This book also describes the history of Morse code. There are several stories about lives saved because of emergency messages transmitted over the radio. It is an interesting and informative book.

USING THE CODE TAPES

If you purchased the *Tune in the World with Ham Radio* kit, it came with two 90-minute cassette tapes. With instruction from the first cassette tape in the set, *Your Introduction to Morse Code*, you'll learn the code by listening. Using this method, you hear the sound "didah" and associate that sound with the letter A. With practice, you'll learn all the sounds and associate them with the correct letters, numbers and punctuation.

Learn the code this way, instead of by mentally picturing the long dashes and short dots. You'll be able to increase your speed sooner. When you hear the sound, you'll immediately associate the correct character and write it down. Translating the sound you hear into visual dots and dashes in your mind only slows the process.

Listen to the difference between the sounds you make saying the word "dit" and saying "dah." If you can tell the difference between those sounds, you have all the ability you need to learn the code. Being able to receive Morse code is really nothing more than being able to recognize a sound. Try it yourself. Say "dididah." Now say "didah." Hear the difference? Congratulations! You're on your way to learning the Morse code.

The 26 letters of the alphabet and the numbers 0 through 9 each have a different sound. There are also different sounds for the period, comma, question mark, double dash (=), fraction bar (/) and some procedural signals that hams use. The + sign, which hams call $\overline{AR}$, means "end of message." $\overline{SK}$ means "end of work," or

AMATEUR RADIO—MORE THAN A HOBBY FOR SOME

Tony Padula was a ham radio operator, a computer trouble shooter and electronic whiz. Was, that is, until an automobile accident changed his life.

After his accident, Padula was on a respirator and showed few signs of life for several days. Even after Tony's brain waves changed and he showed signs of improvement, he remained in a coma for nearly two years. He was unable to communicate, made no facial expressions and didn't move anything except a couple of his fingers.

Tony's doctors said his chances of recovering from the coma were slim. His mother, Madeline, stood by him and worked with him persistently, however. One day while she was visiting Tony, she met a radio amateur who was visiting another patient at the hospital. Mrs Padula mentioned that Tony was a ham also, so the visiting radio amateur stopped in to see Tony.

The visitor didn't get any verbal responses or signs of communication while talking with Tony. Then he tried tapping Morse code on Tony's forehead. He tapped out the word HI, and then asked Tony to move his fingers if he understood the message. Nothing happened.

But when he tapped out CQ, a signal sent by hams meaning "calling any station," Tony was able to tap back HI. Later, Tony started to communicate by holding hands and squeezing code with his thumb and forefinger.

Tony has made remarkable progress since then. He has relearned to sit up, walk short distances, write and speak. Tony's Amateur Radio license expired, so he began working on getting his Novice license. Instructors at the Courage HANDI-HAM System in Minnesota have been helping him. The Courage HANDI-HAM System has been helping persons with disabilities earn amateur licenses for years.

Now, Tony is able to copy Morse code at the rate of 17 words per minute. He doesn't have the communications skills to organize continuous copy to pass the exam, though. He does know the Novice theory, though, and says he hopes in time to be back on the air.—*courtesy Courage HANDI-HAM System*.

"end of contact." You will learn each of these sounds by listening to the first tape, *Your Introduction to Morse Code*.

Side one of tape two, *Morse Code Practice*, contains 45 minutes of practice. This practice includes words, numbers and common terms and abbreviations that radio amateurs use. The practice sessions gradually increase speed from 5 WPM to 6.5 WPM during the 45 minutes of practice. This ensures a margin for error when you take your code test.

Table 3-2 is a transcript of the code on tape one and on tape two, side one. (This table appears at the end of the chapter.)

If a particular section of a tape gives you a problem, you can compare your copy with the printed version. Don't look at the transcript before trying to copy the code on the tapes, though. You shouldn't use the transcript to check your copy every time you listen to the tapes, either. You are more likely to memorize the text that way, and the tapes won't be as effective. Refer to the text printed in Table 3-2 only as a last resort.

The second side of the *Morse Code Practice* tape contains 45 minutes of practice copying Amateur Radio contacts. (We call them QSOs.) This practice is very similar to the code exams given by the ARRL/VEC. Table 3-3 has a 10-question quiz to go with each of the six practice sessions on that tape. This table also appears at the end of the chapter. These quizzes are very similar to the code tests used by the ARRL/VEC. Many Novice examiners use tests like these.

Save the practice sessions on the second side of tape 2 until you are fairly sure you can copy the code accurately. Then copy a practice QSO and try to answer the quiz questions. When you're confident that you've answered all the questions correctly, you are ready for the Novice code exam.

We didn't include a copy of the text from the second side of *Morse Code Practice* in Table 3-2. We didn't print the answers to the quiz questions in Table 3-3, either. That way, you won't be tempted to look up the answers before writing them down.

You may have bought the *Tune in the World with Ham Radio* book, without the Morse code teaching cassettes. You can still buy the tapes, if you've decided you need them now. The cost is $10, plus a shipping and handling charge. Contact the ARRL, 225 Main St., Newington, CT 06111, and order the *Tune in the World* Morse-code teaching tapes.

You may also have received this book with the AEA Morse University program for the Commodore 64™ or C128™ computer. In that case, you have one of the best Morse-code-teaching systems at your fingertips. You probably won't need the code cassettes if you follow the instructions for learning Morse code included with the Morse University program. Whether you use the *Tune in the World* code tapes or the AEA Morse University computer program, you'll soon be ready to pass the 5-WPM code test for your Novice license!

There are many computer programs around for teaching Morse code, and some of them may be very helpful. The ARRL sells *Morse Tutor* for the IBM® PC and compatibles. That is an excellent teaching program. Beware of programs written in BASIC. Because of the way BASIC operates, the timing is often not quite right. Also beware of programs written "to teach myself Morse code." While such writers may be excellent programmers, they are often unaware of the finer points of code training.

There are also some visual code-teaching methods. If you seem unable to learn the code using the *Tune in the World* cassette tapes, one of these may be what you need. Please give the tapes a thorough try first, however. Practice faithfully with the tapes every day for at least three weeks. Then, if you haven't learned many of the characters, you may want to try a visual method.

Learning to Write

To learn the code, you train your hand to write a certain letter, number or punctuation mark whenever you hear a specific sound. It's the same as acquiring a habit. After all, acquiring a habit is nothing more than doing something the same way time after time. Eventually, whenever you want to do that thing, you automatically do it the same way; it has become a habit.

You will need lots of practice writing the specific characters each time you hear a sound. Eventually you'll copy the code without thinking about it. In other words, you'll respond automatically to the sound by writing the corresponding character.

As you try copying code sent at faster speeds, you may find that your ability to write the letters limits you. Practice writing the characters as quickly as possible. If you normally print, look for ways to avoid retracing lines. Don't allow yourself to be sloppy, though, because you may not be able to read your writting later.

Many people find that script, or cursive writing is faster than printing. Experiment with different writing methods, and find one that works for you. Then practice writing with that method so you don't have to think about forming the characters when you hear the sounds.

Figure 3-3 shows a systematic method of printing numbers and the letters of the alphabet. This system requires a minimum of pencil movement or retraced lines. You may find this technique helpful to increase your writing speed. Whatever method you choose, practice so it becomes second nature.

Figure 3-3—A method of hand-printing letters and numbers with minimum effort and maximum speed.

ASSEMBLING A CODE-PRACTICE OSCILLATOR

It is not difficult to construct a code-practice oscillator. A complete oscillator that mounts on a small piece of wood is shown in Figure A. Figure B shows all the parts and tools you will need for this project. The circuit board alone or the entire parts kit can be ordered from Circuit Board Specialists, PO Box 951, Pueblo, CO 81002, tel. 719-542-4525.

Please read all instructions carefully before mounting any parts. Check the parts placement diagram for the location of each part.

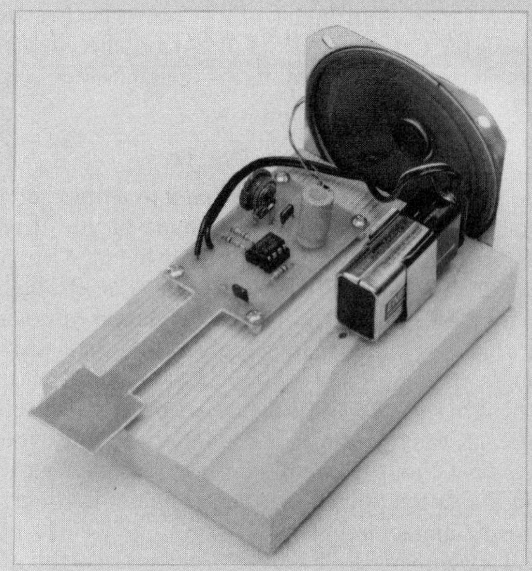

Figure A

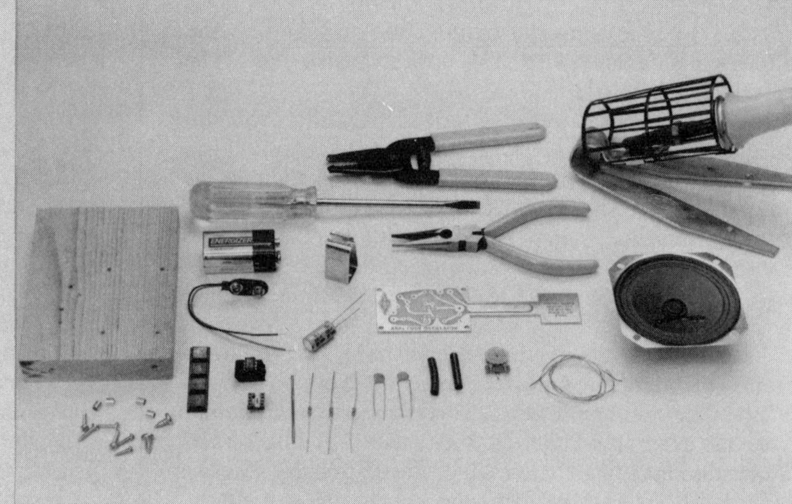

Figure B

Parts List

[ ] Check box as each part is installed and soldered

	Quantity	Description	Radio Shack Part Number	Component Number	Used in Step Number
Capacitors					
[ ]	1	0.01-μF	272-131	C1	3
[ ]	1	0.01-μF	272-131	C2	6
[ ]	1	220-μF, 35-V electrolytic	272-1029	C3	5
Resistors					
[ ]	1	10-kilohm, ¼ W (brown-black-orange stripes)	271-1335	R2	4
[ ]	1	68-kilohm, ¼ W (blue-gray-orange stripes)	271-1345	R3	8
[ ]	1	10-kilohm, ¼ W (brown-black-orange stripes)	271-1335	R1	9
Miscellaneous					
[ ]	1	100-kilohm potentiometer	271-220	R4	7
[ ]	1	7555 CMOS IC Timer	276-1718	U1	2
[ ]	1	Loudspeaker—3-inch, 8-ohm	40-245	LS1	11
[ ]	1	9-V Battery connector	23-553	BT1	10
[ ]	1	9-V Battery			
[ ]	1	Brass rod 2 inches long, approximately 18 gauge (about the diameter of a wire coat hanger). (Available at hobby shops.)			
[ ]	4	¼-inch spacers	64-3024		
[ ]	1	U-shaped battery holder	270-326		
[ ]	1	2 × 4 × ½-inch piece of wood for base			
[ ]	6	No. 6 wood screws			

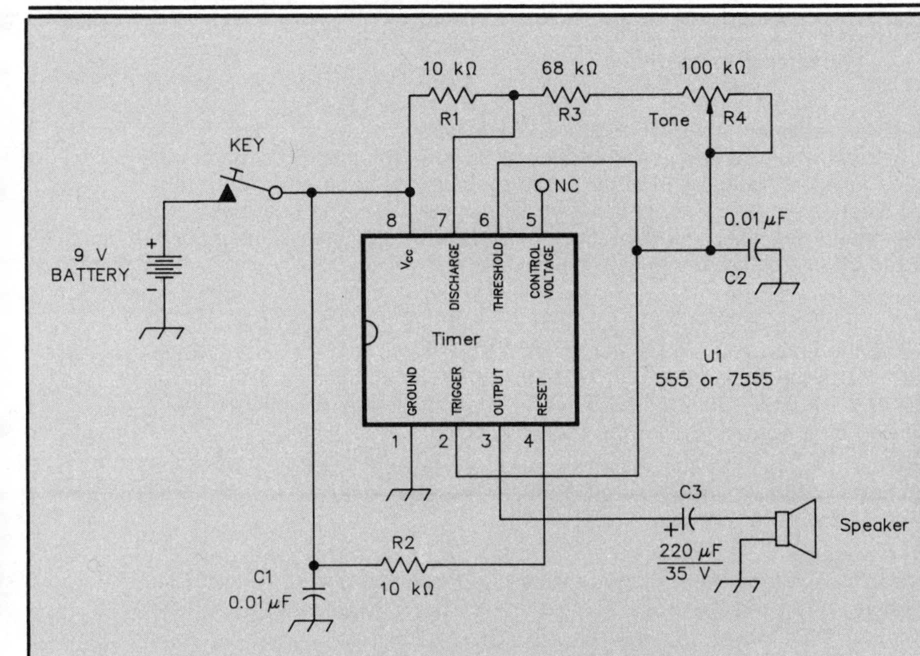

Figure C—Schematic diagram of a code-practice oscillator.

What You Need

Besides the circuit board and parts, you will need a low- power (25-30 watt) soldering iron, some thin rosin-core solder, a straight-blade screwdriver and 000-gauge steel wool.

The circuit board has one blank side for components and one side with metal foil tracings. The foil tracings are the shiny metal paths on the board. Before you solder any part to the board, clean the part leads (long thin wires) with the steel wool. Rub lightly.

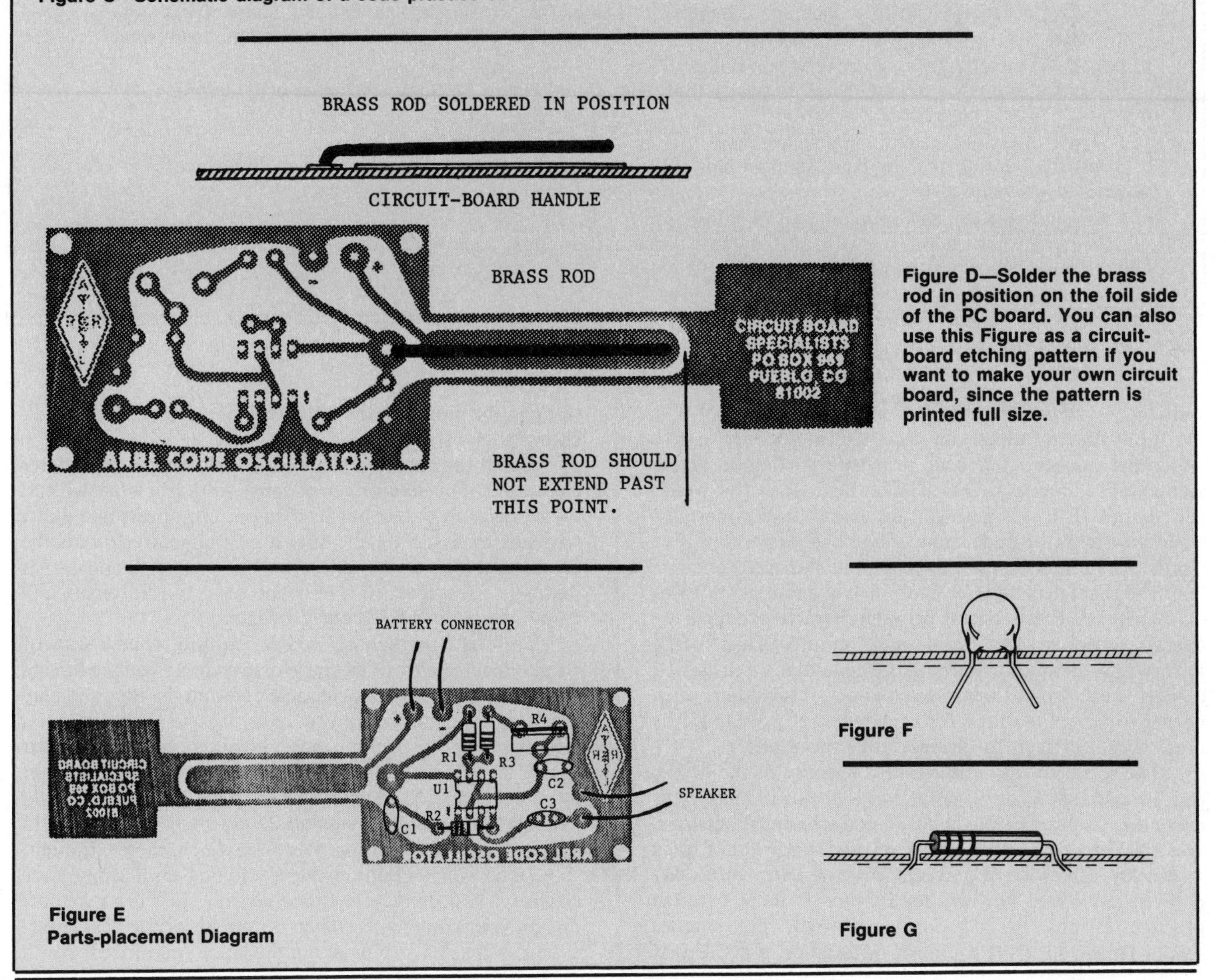

Figure D—Solder the brass rod in position on the foil side of the PC board. You can also use this Figure as a circuit-board etching pattern if you want to make your own circuit board, since the pattern is printed full size.

Figure E
Parts-placement Diagram

Figure F

Figure G

Assembly Instructions

(Check box as each step is completed)

[ ] Step 1: Attach the brass rod. Check parts-placement diagram (Figure E) for location.

Clean the brass rod with sandpaper or steel wool. Bend one end of the rod slightly less than 90 degrees. Lay the circuit board on table with foil side up. Place the hooked end of the brass rod over the large hole near the handle (see Figure D). Make sure the rod extends out over the handle area. Solder the rod to the board on the foil side. The end of the brass rod should not extend past the marked oval on the handle. This is your contact point. If it does extend beyond this point, cut the rod off just before the end of the oval.

[ ] Step 2: Solder the IC socket to the board.

The socket for the IC is placed on the component side of the board first. Do not plug the IC into the socket now. After all the other parts are soldered to the board you will be instructed to plug the IC into the socket (Step 13). Identify the notched end of the socket. Insert the socket into the circuit board on the component side. Make sure the notched end is facing the handle of the board. Turn the board over and gently spread the pins on the socket so they make contact with the foil side of the board. Solder the socket in place.

[ ] Step 3: Place C1 (0.01-μF capacitor)

Thread the wire leads on C1 through holes on board. (See Figure F.) Solder the wires onto the foil side of the foil side of the board. Cut the extra wire off above the solder joint.

[ ] Step 4: Place R2 (10-kilohm resistor) on board.

Prepare resistors for mounting by bending each lead (wire) of the resistor to approximately a 90° angle. (See Figure G). Insert the leads into the board holes and bend them over to hold the resistor in place. Solder the leads to the foil and trim them close to the foil.

[ ] Step 5: Place C3 (220-μF, 35-volt electrolytic capacitor) on board.

This capacitor has a plus (+) side and a negative (−) side. The (−) side is placed on the board facing away from the handle. Insert the capacitor leads into the circuit board holes, solder in place and trim off the extra wire.

[ ] Step 6: Place C2 (0.01-μF capacitor) on board.

Thread C2 wire leads through holes on board. (See Step 3, and Figure F.) Solder the wires onto the board. Cut the extra wire off above the solder joint.

[ ] Step 7: Place R4 (100-kilohm potentiometer)

This component has three pins. All three pins must be plugged into the holes on the board. (It fits only one way.) Solder them in place.

[ ] Step 8: Place R3 (68-kilohm resistor) on board.

Bend the wires on the resistor to plug it into the board. (See Step 4 and Figure G.) Plug the resistor into the board, spread the wires and solder in place.

Suggestions For Study

Your cassette tapes play code sent at the rate of five words per minute. (The code practice on side one of the second tape increases to a speed a bit faster than five words per minute. This will give you an extra "safety margin" when you take the code exam.) You'll need to copy five words per minute (WPM) to pass your Novice exam.

The characters themselves are sent at a rate of 18 WPM on these tapes. Extra spacing between characters reduces the overall speed to five words per minute. This is the Farnsworth method of learning the code that we discussed earlier. If each sound were stretched out, it would be much harder to learn the sound for each letter. It would also be much more difficult to increase your speed later.

The secret to easy and painless mastery of the Morse code is regular practice. Set aside two 15 to 30 minute periods every day to practice the code. If you try longer sessions, you are likely to become over tired, and you will not learn as quickly. Likewise, if you only practice every other day or even less often, you will tend to forget more between practice sessions. It's a good idea to work your practice sessions into your daily routine. For instance, practice first thing in the morning and before dinner. Daily practice will show quick results.

Learn the sound of each letter. Morse code character elements sound like dits and dahs, so that's what we call them. Each character has its own pattern of dits and dahs, so learn to associate the sound of that pattern with the character. Don't try to remember how many dots and dashes make up each character. Practice until you automatically recognize each Morse code character.

Feel free to review. You're learning a new way of communicating, by using the Morse code. If you are having trouble with a particular character, rewind the tape and play that section again. After you've listened to the practice on one character two or three times, however, go on to the next one. You will get more practice on the problem character later in the tape. You can even review the problem character again in a later practice session. Don't spend your entire 15 minutes of practice going over a single character, though.

Don't worry about mistakes. If you hear a Morse code character you don't immediately know, just draw a short line on your copy paper. Then get ready for the next letter. If you sit there worrying about the letter you missed, you'll

[ ] Step 9: Place R1 (10-kilohm resistor) on board.

Bend the wires on the resistor to plug it into the board. (See Step 4 and Figure G.) Plug the resistor into the board, spread the wires and solder in place.

[ ] Step 10: Hook up the battery connector leads.

The battery connector consists of two wires, one red and one black, attached to a snap-on cap. Remove ¼ inch of plastic insulation from the end of both wires. The black wire is negative and the red wire is positive. The positive and negative battery connections are marked on the solder side of the board. Be sure the red wire goes in the hole marked "+", and the black wire goes in the hole marked "–". Solder the wires in place.

[ ] Step 11: Hook up the speaker.

Cut the speaker wire into two equal lengths. Remove ¼ inch of plastic insulation from each end of both wires. Solder one piece of wire to one of the speaker terminals and the other wire to the other speaker terminal. Solder the other end of each wire to the board at the points marked on the parts placement diagram.

[ ] Step 12: Attach the circuit board to the wood base.

Place the completed circuit board on the wood. Trace through the four holes with a pencil. Take the board off the wood and lay it aside. Place the spacers on the wood, standing upright. Carefully put the circuit board on top of the spacers. Put the screws through the holes in the circuit board and through the spacers and screw them into the board until snug. Be sure not to overtighten the screws, which could crack the circuit board. Attach the speaker to the front end of the board opposite the handle with the two screws left over. Attach the U-shaped metal battery holder to the wooden base. You can either use glue or a screw to attach it to the base.

[ ] Step 13: Plug the integrated circuit (IC) into the socket.

CAUTION—The static electricity from your body could destroy the IC. Before touching the IC, be sure you have discharged any static that may be built up on your body. While sitting at your table or workbench, touch a metal pipe or other large metal object for a few seconds. Carefully remove the IC from its foam padding. Hold it by the black body and avoid touching the wires. Plug it into the socket, being sure that the notched end of the IC is facing towards the handle. The notch on the IC should line up with the notch on the socket.

Attach the battery to the snap-on battery connector, and place it in the U-shaped metal battery holder. This unit only uses electricity when the telegraph key handle is pushed down. No ON/OFF switch is necessary and you may leave the battery connected at all times.

You're done! The oscillator should produce a tone when you press the key. If your oscillator does not work, check all your connections carefully. Make sure the IC is positioned correctly in the socket, and that you have a fresh battery.

Once you have the oscillator working, you're ready to use it to practice Morse code. If you are studying with a friend, you can use the oscillator to send code to each other. If you are studying alone, tape record your sending and play it back later. Can you copy what you sent? How would it sound on the air? Good luck!

miss a lot more! If you ignore your mistakes now, you'll make fewer of them. Don't worry about perfect copy while you are learning the code. That will come with more practice.

Don't anticipate what the next letter will be. When you think you've recognized a word after copying a few letters, concentrate all the more on the actual code sent. Don't anticipate what comes next, because your guess may turn out to be wrong. When that happens, you'll probably get confused and miss the next few letters as well. Write each letter just after it is sent. With more practice, you'll learn to "copy behind," writing whole words at one time.

Practice sending code. To communicate with Morse code you must be able to send it as well as receive it! You will need a **code-practice oscillator** and a **code key**. The sidebar, "Assembling A Code-Practice Oscillator" describes a simple oscillator that you can build. It even includes a simple key that you can use to practice sending.

One trick that some people use while learning the code is to whistle or hum the code while walking or driving. Send the words on street signs, billboards and store windows. This extra practice may be just the help you need to master the code!

Morse code is a language. Eventually you'll begin to recognize common syllables and words. With practice you will know many complete words, and won't even listen to the individual letters. When you become this familiar with the code, it will really start to be fun!

Don't be discouraged if you don't seem to be breaking any speed records. Some people have an ear for code and can learn the entire code in a week or less. Others require a month or more to learn it. Be patient, continue to practice and you will reach your goal.

To pass the Novice class code examination, you must be able to understand a plain-language message sent at 5 WPM. As you know, in the English language some words are just one or two letters long, others 10 or more letters long. To standardize the code test, the FCC defines a word as a group of five letters. Numbers and punctuation marks normally count as two characters for this purpose.

ABOUT THE TAPES

The cassette tapes teach you the Morse code and give you plenty of practice to help you pass the 5 WPM exam. To use the tapes, you will need a cassette player. If you don't own one, you may find the purchase of a small portable tape recorder worthwhile. You will use a tape recorder in many other aspects of radio. You'll want to record class notes, **W1AW** code practice, your sending, on-the-air contacts and more. Even an inexpensive model is adequate for listening to code practice; high fidelity is not essential.

You may want to consider using a stereo cassette player to listen to your tapes. Then you can take advantage of a special feature of the tapes that come with the *Tune in the World* kit. The tapes have code on both stereo channels, but voice on only one channel. You can use the tape for extra practice without having the announcer tell you what all the letters are. Just turn down the voice-track volume.

Insert the tape in your cassette player, gather pencil and paper and find a quiet, comfortable place where you can listen. You must be able to concentrate, but also relax. When you're ready, turn on the tape player, listen to the introduction, and follow the instructions.

Figure 3-4—Whether you're 5 or 55, the only way to learn international Morse code is through listening. These cassette tapes use the modern, efficient, listening method of teaching the code characters. One 90-minute cassette is dedicated to teaching you all the Morse code characters you need to know to pass any Amateur Radio exam. A second 90-minute cassette provides plenty of practice at the 5 WPM speed you'll need to pass the Novice test.

Care of Your Cassettes

Any precision-engineered item requires some care to work properly and code cassettes are no exception. If you store and handle your cassettes properly, they might outlast the tape recorder!

Heat, dust or magnetic fields are poison to recording tape. Heat warps the cassette itself (causing jamming), softens the tape (allowing stretching) and increases random noise. One of the hottest places to store tapes is on the dashboard or in the glove compartment of a car. Take them inside to school or work with you. You wouldn't leave your pet in the car all day; don't leave your cassettes there, either.

Dust is another enemy of cassettes. Manufacturers carefully engineer the tape in cassettes to very close tolerances. Even a small amount of dust within the cassette can cause binding and jamming, possibly stretching or even breaking the tape. Store your cassettes in plastic storage boxes, or in another dust-free place. Pockets are not appropriate.

You should also keep your tape machine clean. Some of the magnetic oxide coating comes off a tape each time you play it. This coating sticks to the rollers and tape heads in the machine. Clean the tape player whenever you notice a brown residue on the rollers or tape heads. There are many supplies sold to clean tape players. It doesn't matter which you use, as long as you keep the machine clean. If you allow too much oxide to build up on these surfaces, the tape may stick to a roller and jam in the machine. This will probably destroy the tape and may damage the cassette player.

Cassettes store information magnetically, and external magnetic fields can erase the tape. Keep your cassettes away from electric motors, transformers, bulk tape erasers and other sources of magnetic fields.

The first time you use the cassettes, adjust the tension of the tape to your own personal recorder. Play each tape through completely at normal playing speed, without rewinding. You can turn down the volume and just let the player run, if you like. This process ensures that the tape and your recorder are compatible. Play the tape all the way through occasionally to reduce problems caused by frequent rewinding.

Be sure that the tape is completely wound on the spools before putting the cassette into the player. A little slack in the cassette is a prime cause of jamming and stretching. Turn one hub from the inside with a pencil or a finger to take up all the slack before playing it. Hub locks (a plastic bar with ends that fit into the cassette holes) help prevent this slackening when you aren't using the tapes. Plastic storage boxes also have built-in hub locks to prevent this slackening.

Print-Through

If you listen to the cassettes on a high-quality sound system, you may hear an "echo" of the code between letters. This is particularly evident on slower-speed tapes, with the longer pause between code characters. This soft code echo, called print-through, is a characteristic of even the highest-quality tapes.

The code character forms a highly magnetized layer on the tape. This tends to magnetize the adjacent layer (turn of tape on the spool) slightly, which produces the echo. This print-through is far below the level of the code and is hidden by the random noise in most cassette players. High-fidelity machines may pick it up, however. If it is annoying, turn down the volume and it should disappear into the background noise.

Treat these quality cassettes properly, and they will give you many hours of trouble-free service. Clean the record and playback heads and the capstan roller of your tape player regularly. A dirty tape player can quickly destroy your tapes.

CODE IS THEIR COMMON BOND

When Anne Harlan asked, on the air, how she could join a CW net Scott Prather responded. (A net is a group of hams who meet on a certain frequency on a regular schedule to pass messages.) Scott said, "Since I was the resident expert of sorts on CW nets, I called Anne and gave her the information. I then asked how I could contact her in the future, and she gave me a list of her operating frequencies."

From there, it wasn't long before Scott and Anne were working side by side passing messages during the annual Simulated Emergency Test. Their common interest in Morse code had led to a mutual interest in each other. Anne (KA9EHV) and Scott (KB9Y) Prather now compete for operating time at the same Amateur Radio station as husband and wife.

Scott runs an electronics-repair business part time. He recently arranged radio communications for a church-sponsored bike trip. Anne's operating interests are in CW ragchewing and weather nets. She says, "No matter what else we do, it seems that CW and public service are eventually a part of it."

COMFORTABLE SENDING

There's more to the code than just learning to receive it; you'll also have to learn to send it. To accomplish this, you'll need a telegraph key and a code-practice oscillator. You can get these items at most electronics-parts stores, or you can build your own simple oscillator. The "Assembling A Code-Practice Oscillator" sidebar gives you step-by-step instructions for building one simple oscillator that even includes a key for you to practice with.

Figure 3-5 shows two different standard straight key models (one in the foreground and one to the left). There is a semiautomatic "bug" in the background and a popular Bencher paddle used with modern electronic keyers on the right. You'll want to obtain some type of code key to practice with before you're ready for your Novice code test. You'll need the key for some of your on-the-air operating after the license arrives from the FCC as well! Most new hams start with an inexpensive straight key.

Just as with receiving, it is important that you be comfortable when sending. It helps to rest your arm on the table, letting your wrist and hand do all the work. Grasp the key lightly with your fingertips. Don't grip it tightly. If you do, you'll soon discover a few muscles you didn't know you had, and each one will ache. With a light grasp, you'll be able to send for long periods without fatigue. See Figure 3-6.

Another important part of sending code is the proper adjustment of your telegraph key. There are only two adjustments to make on a straight key, but you'll find them to be very important. Figure 3-7 illustrates these adjustments.

The first adjustment is the spacing between the contacts, which determines the distance the key knob must move to send a letter. Adjust the contacts so that the knob moves about the thickness of a dime (1/16 inch). Try it. If you're not satisfied with this setting, try widening the space. If that doesn't do it, try narrowing the space. Eventually, you'll find a spacing that works best for you. Don't be surprised if your feelings change from time to time, however, especially when your sending speed increases.

The second adjustment you must make is the spring tension that keeps the contacts apart. Just as there is no "correct" contact spacing, there is no correct tension adjustment for everyone. You will find, however, that adjusting the spacing may also require a tension adjustment. Adjust the tension to provide what you think is the best "feel" when sending.

Some straight keys have ball bearing pivot points on either side of the crossarms. These normally need no adjustment, but you should be sure the crossarms move freely in these pivots. If the bearings are too tight the key will bind or stick. If they are too loose there will be excessive play in the bearings. The side screws also adjust the crossarms from side to side so the contact points line up.

All of these adjustment screws have lock nuts, so be sure you loosen them before you make any adjustments. Tighten the lock nuts securely after making the adjustment.

Some better-quality keys have a shorting bar like the one shown in Figure 3-7. This shorting bar can be used to close the key contacts for transmitter tuning or adjustments.

You'll probably want to fasten the key to a piece of

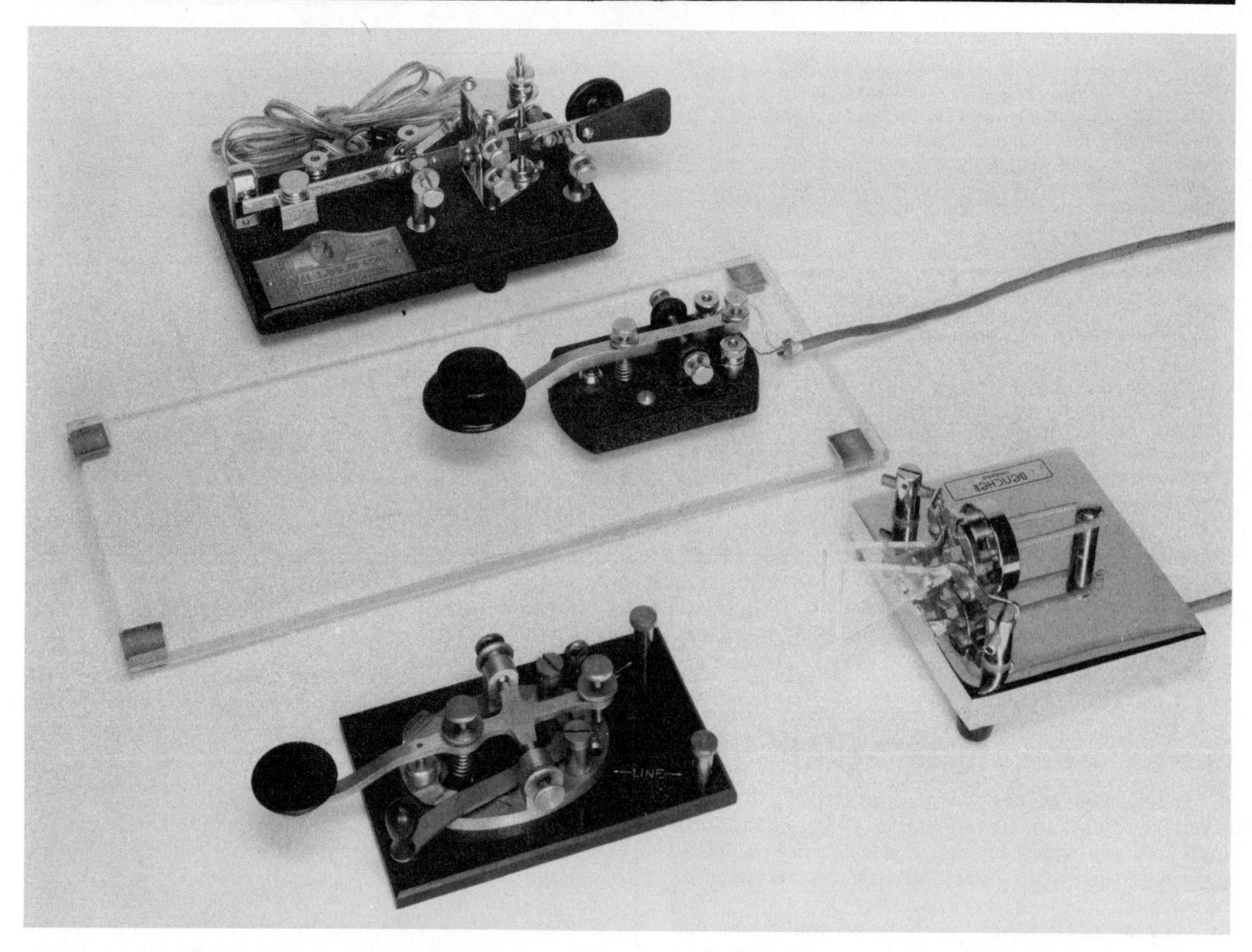

Figure 3-5—Older than radio itself, code still reigns as the most efficient and effective communications mode; many hams use it almost exclusively. A modern straight key, the device most beginners use, is in the center.

wood or other heavy weight to prevent it from sliding around as you send. You might even want to fasten the key directly to the table. It's best to experiment with different positions before permanently attaching the key to any surface, however.

Some operators use a board that extends under their forearm and allows their arm to hold the key in position. This way the key can be moved aside for storage, or to clean off the table for other activities. A piece of ¼-inch Plexiglas® about 12 inches long works well for this type of key base. This technique also allows you to reposition the key to find the most comfortable sending position.

Learning to send good code, like learning to receive, requires practice. A good way to start is to send along with your code tapes. Try to duplicate the sounds as much as possible. Send along with the first tape, *Your Introduction to Morse Code*, to get the feel for the rhythm of each character. Later you can send along with the second tape, *Morse Code Practice*, for some additional practice.

Always remember that you're trying to send a complete sound, not a series of dots and dashes. With that in mind, take a moment to think about the sound you're trying to send. It consists of dits, dahs and pauses (or spaces). The key to good sending, then, is timing. If you're not convinced, listen to some code on the Novice bands. What transmissions would you prefer to receive? Why? That's right, because they have good timing.

Each person develops a unique rhythm in his or her sending. It's almost impossible to describe what makes one person's rhythm different from another person's. No two people send code exactly the same way, though. An experienced CW operator learns to identify the person

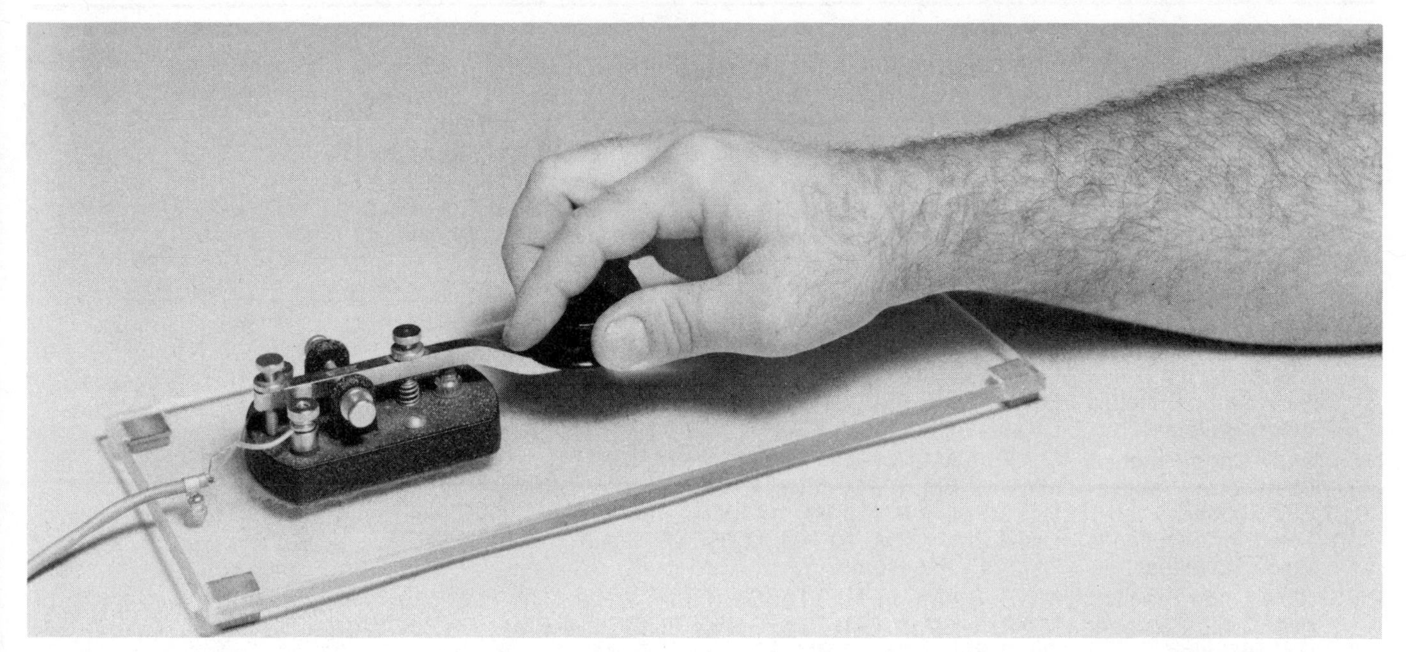

Figure 3-6—Proper forearm support, with the wrist off the table, a gentle grip and a smooth up-and-down motion make for clean, effortless sending.

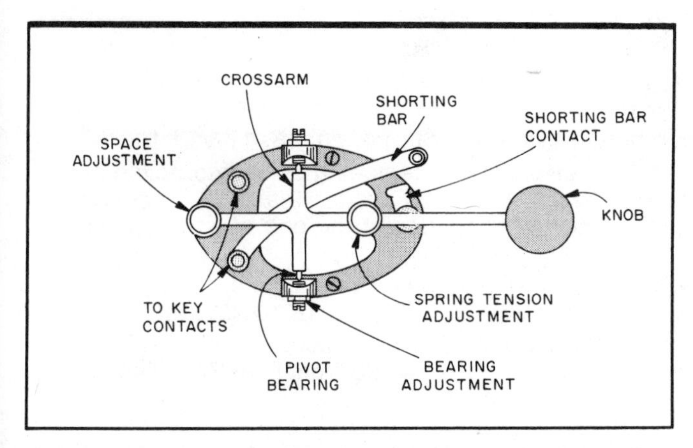

Figure 3-7—A few simple adjustments to suit your style allow you to send for hours without fatigue. The contact spacing and spring tension should be set to provide the most comfortable feel.

sending code by this unique rhythm, or **fist**.

Some people send code that is very easy and enjoyable to copy. Others are not so good, and require a lot of concentration to understand. Since your fist is your on-the-air signature, try to make it as easy to read as possible. Learn to send code so that it sounds like the code on your practice tapes. Then you'll have a near-perfect fist, and you can be proud of that. It can make a big difference in your success as an amateur operator.

One of the best ways to learn to send and receive code is to work with another person. That person may be another member of your own family. If you're attending an organized class, you may be able to get together with another student several times a week.

If you must work alone, make good use of your tape recorder. Try recording your sending. After waiting a day or two, try receiving what you sent. The wait between sending and receiving will help prevent you from writing down the message from memory. Not only will this procedure provide practice in receiving, but it will also let you hear exactly how your sending sounds. If your timing is off, you'll hear it. If you're having trouble with a specific letter or number or punctuation mark, you'll soon know about it. If you can't understand your own sending, neither will anyone else!

Regardless of whether you're sending or receiving, the key to your success with Morse code is regular practice. Make use of the tapes provided with this book. After you've learned all the characters you may even want to obtain some more practice tapes. They can provide a wider variety of practice material.

If you can, listen to actual contacts between hams. Make use of the code-practice material transmitted daily by **W1AW**, the ARRL station heard nationwide from Newington, Connecticut. Table 3-1 lists the code-practice schedule for W1AW. When we change from Standard Time to Daylight Saving Time, the code-practice is sent at the same local times.

Table 3-1
W1AW Schedule

MTWThFSSn = Days of Week   Dy = Daily

W1AW code practice and bulletin transmissions are sent on the following schedule:

EST	Slow Code Practice	MWF: 9 AM, 7 PM; TThSSn: 4 PM, 10 PM
	Fast Code Practice	MWF: 4 PM, 10 PM; TTh: 9 AM; TThSSn: 7 PM
	CW Bulletins	Dy: 5 PM, 8 PM, 11 PM; MTWThF: 10 AM
	Teleprinter Bulletins	Dy: 6 PM, 9 PM, 12 PM; MTWThF: 11 AM
	Voice Bulletins	Dy: 9:45 PM, 12:45 AM
CST	Slow Code Practice	MWF: 8 AM, 6 PM; TThSSn: 3 PM, 9 PM
	Fast Code Practice	MWF: 3 PM, 9 PM; TTh: 8 AM; TThSSn: 6 PM
	CW Bulletins	Dy: 4 PM, 7 PM, 10 PM; MTWThF: 9 AM
	Teleprinter Bulletins	Dy: 5 PM, 8 PM, 11 PM; MTWThF: 10 AM
	Voice Bulletins	Dy: 8:45 PM, 11:45 PM
MST	Slow Code Practice	MWF: 7 AM, 5 PM; TThSSn: 2 PM, 8 PM
	Fast Code Practice	MWF: 2 PM, 8 PM; TTh: 7 AM; TThSSn: 5 PM
	CW Bulletins	Dy: 3 PM, 6 PM, 9 PM; MTWThF: 8 AM
	Teleprinter Bulletins	Dy: 4 PM, 7 PM, 10 PM; MTWThF: 9 AM
	Voice Bulletins	Dy: 7:45 PM, 10:45 PM
PST	Slow Code Practice	MWF: 6 AM, 4 PM; TThSSn: 1 PM; 7 PM
	Fast Code Practice	MWF: 1 PM, 7 PM; TTh: 6 AM; TThSSn: 4 PM
	CW Bulletins	Dy: 2 PM, 5 PM, 8 PM; MTWThF: 7 AM
	Teleprinter Bulletins	Dy: 3 PM, 6 PM, 9 PM; MTWThF: 8 AM
	Voice Bulletins	Dy: 6:45 PM, 9:45 PM

Code practice, Qualifying Run and CW bulletin frequencies: 1.818, 3.58, 7.0475, 14.0475, 21.08, 28.08, 50.08, 147.555 MHz.

Teleprinter bulletin frequencies: 3.625, 7.095, 14.095, 21.095, 28.095, 147.555 MHz.

Voice bulletin frequencies: 1.89, 3.99, 7.29, 14.29, 21.39, 28.59, 50.19, 147.555 MHz.

On Monday, Wednesday and Friday, 9 AM through 5 PM EST, transmissions are beamed to Europe on 14, 21 and 28 MHz; on Wednesday at 6 PM EST they are beamed south.

Slow code practice is at 5, 7½, 10, 13 and 15 WPM. Fast code practice is at 35, 30, 25, 20, 15, 13 and 10 WPM.

Code practice texts are from *QST*, and the source of each practice is given at the beginning of each practice and at the beginning of alternate speeds. For example, "Text is from August 1989 *QST*, pages 9 and 76" indicates that the main text is from the article on page 9 and the mixed number/letter groups at the end of each speed are from the contest scores on page 76.

Some of the slow practice sessions are sent with each line of text from *QST* reversed. For example, "Last October, the ARRL Board of Directors" would be sent as DIRECTORS OF BOARD ARRL THE, OCTOBER LAST.

Teleprinter bulletins are 45.45-baud Baudot, 110-baud ASCII and 100-baud AMTOR, FEC mode. Baudot, ASCII and AMTOR (in that order) are sent during all 11 AM EST transmissions, and 6 PM EST on TThFSSn. During other transmission times, AMTOR is sent only as time permits.

CW bulletins are sent at 18 WPM.

W1AW is open for visitors Monday through Friday from 8 AM to 1 AM EST and on Saturday and Sunday from 3:30 PM to 1 AM EST. If you desire to operate W1AW, be sure to bring a copy of your license with you. W1AW is available for operation by visitors between 1 and 4 PM Monday through Friday.

In a communications emergency, monitor W1AW for special bulletins as follows: voice on the hour, teleprinter at 15 minutes past the hour, and CW on the half hour.

THE CODE TEST

To pass the Novice code test, you must demonstrate your ability to send and receive Morse code at 5 words per minute. Your Novice test examiners will give you the code test. The FCC specifies that the exam message should be at least 5 minutes long. The message must contain all letters of the alphabet, all numbers, period, comma, question mark and fraction bar (/). The exam message must also include some common procedural signals. These are the + sign ($\overline{\mathrm{AR}}$) the = sign ($\overline{\mathrm{BT}}$) and $\overline{\mathrm{SK}}$.

The two hams giving you the test have several options

TWO LICENSES PAY OFF

It was hardly a routine distress call on 27.065 MHz, CB channel 9. But was it genuine?

Dave McCollum wasn't sure. Dave was an officer of Santa Barbara (California) REACT as well as a new ham, call sign WA6RGJ. He was trained in emergency procedures. All he could hear over the emergency channel was a click, but that was enough to get his attention. "Is any station calling Santa Barbara REACT?" he asked. Again, the signal-strength meter on his CB transceiver moved for a few seconds, then dropped to zero.

"This is Santa Barbara REACT KSX7459," he called. "If you have an emergency press your mike button twice for yes and once for no." Hearing two clicks, Dave envisioned a stranded motorist. "Can you flag a passing car or get to a telephone?" The response was one click—negative. Dave had an idea. "Are you a marine station?" Two clicks—yes. "Do you know Morse code?" Again, two clicks. Tuning his amateur receiver to 27.065, Dave began a unique rescue operation—an Amateur Radio/CB operator using Morse code on the citizens band.

When informed of the situation, the Coast Guard was skeptical. A Guard lieutenant asked Dave how he could talk to the marine station if he couldn't hear the station. As Dave described the CW conversation, the lieutenant commented, "This is certainly different from most of the CB calls we get!"

A little later the story unfolded, when the grateful skipper, Army Major Dick Tarr, called Dave on the telephone. While on a cruise to Santa Cruz Island, a hole had developed in the craft's discharge hose. Taking on water rapidly, the boat would have sunk if a pump had failed.

"Okay," Dave said, "but where did you learn Morse code?" The skipper's reply startled him: "As a Scout, when I was 17 years old, and I've never used it since!"

The potential tragedy—and Morse code's role in preventing it—was not lost on Dave's family. Daughter Elizabeth holds the Novice call sign WA6RGK, and wife Kay has also joined them on the amateur bands. She quickly upgraded to Technician (WD6BJG) and has been very active in emergency work and public service events.

The incident showed graphically how cooperation between three services—amateur, citizens band and the Coast Guard, can avert potential tragedies. But if Dick Tarr hadn't learned Morse code, that cooperation would never have materialized.

Dave has recently become interested in Packet Radio. He has a complete, portable packet station that he put to good use in the July 1985 California forest fire. Public service communications are a big part of his Amateur Radio interest.

Dave says CW is still the purest form of our hobby. He finds it very relaxing and enjoyable to get into a CW ragchew. He often gets on the Novice bands for a CW ragchew. Dave says, "I find a certain satisfaction in talking to new hams and sharing some of my experience . . ."

in deciding if you pass the exam. If you have one minute of solid (perfect) copy out of a five-minute test, you pass. The examiners may also give you a 10-question multiple-choice or fill-in-the-blank exam about the exam content. A passing grade on such a written exam is 74%. Your examiners may devise other fair and reasonable standards for taking the code test. Again, the examiners are simply certifying that you have proven your ability to send and receive code at 5 words per minute.

EXAMINATION NERVES

Probably more people fail the code examination because of the jitters than because of inadequate command of the code. If you have prepared properly, it should be easy—a pleasant experience, not an agonizing one. The key to a successful exam is to relax.

The exam isn't *that* important. You can always take the test again soon. It won't kill you even if you don't pass. But you are going to pass! You can copy 5 WPM at home with only a couple of errors every few minutes. The code test will be even easier.

Think positively: "I will pass!" Relax. Before you know it, you'll be grinning from ear to ear as you hear the words, "You've passed!"

Table 3-2
Transcript of Code Tapes

Side 1 of *Your Introduction to Morse Code*

A A AA AAAAA AAAAA AAAAA E EEE EEEEE EEEEE EEEEE R R RR RRRRR RRRRR RRRRR AE R ARE ARE ARE EAR EAR EAR RARE RARE RARE N N NN NNNNN NNNNN NNNNN NEAR NEAR NEAR EARN EARN EARN + + + + + + + + + + + + + + T TTT TTTTT TTTTT TTTTT TEN TEN TEN RATE RATE RATE TENT TENT TENT I I II IIIII IIIII IIIII IN IN IN TRAIN + TRAIN + TRAIN + O OOO OOOOO OOOOO OOOOO OAR OAR OAR TON TON TON NONE NONE NONE + + S SSS SSSSS SSSSS SSSSS SEA SEA SEA SENSE SENSE SENSE D DDD DDDDD DDDDD DDDDD DIODE DIODE DIODE NODE NODE NODE H H HH HHHHH HHHHH HHHHH HAT HAT HAT DASH DASH DASH C CCC CCCCC CCCCC CCCCC CAN CAN CAN CHAIR CHAIR CHAIR CHEER CHEER CHEER U UUU UUUUU UUUUU UUUUU USE USE USE TRUE TRUE TRUE OCCUR OCCUR OCCUR Y YYY YYYYY YYYYY YYYYY NAY NAY NAY TRY TRY TRY YEARN YEARN YEARN .... SEND CODE SOON. NO SEASON IS SET. L LLL LLLLL LLLLL LLLLL LUNCH LUNCH LUNCH ILL ILL ILL M MMM MMMMM MMMMM MMMMM MUSIC MUSIC MUSIC CHARMS CHARMS CHARMS MOM MOM MOM.. P P PP PPPPP PPPPP PPPPP PEPPER PLANTS PRODUCE PEPPERS. G GGG GGGGG GGGGG GGGGG GARDENERS GATHER GARLIC. + + + F F FF FFFFF FFFFF FFFFF FISH FISH FISH FLUFF FLUFF FLUFF FROGS FLIP. , ,,, ,,,,, ,,,,, CODE IS HEARD, NOT SEEN. CODE IS HEARD, NOT SEEN. + +

Side 2 of *Your Introduction to Morse Code*

W WWW WWWWW WWWWW WWWWW WHY WHY WHY NOW NOW NOW WOLF WOLF WOLF B B BB BBBBB BBBBB BBBBB BIG BEEF BEGS BUTCHERING. = = = = = = = = = = = = = = J J JJ JJJJJ JJJJJ JJJJJ JET JET JET JOWL JOWL JOWL = = JUMBLE JUMBLE JUMBLE / / / ///// ///// ///// ///// KKKK KKKKK KKKKK KKKKK KING KING KING JACK JACK JACK KHAKI KHAKI KHAKI Q QQQ QQQQQ QQQQQ QQQQQ QUAIL QUAIL QUAIL AQUA AQUA AQUA = = X X XX XXXXX XXXXX XXXXX XRAY XRAY XRAY AXIOM AXIOM AXIOM OXIDE OXIDE OXIDE / / V V VV VVVVV VVVVV VVVVV VERY VERY VERY VIVID VIVID VIVID Z Z ZZ ZZZZZ ZZZZZ ZZZZZ ZERO ZEBRAS ZING. = ZIG ZAG ZESTFULLY. + ? ? ? ? ????? ????? TNX BRI ?? BRUCE SO HOW COPY NOW? = $\overline{SK}$ $\overline{SK}$ $\overline{SK}$ $\overline{SK}$ $\overline{SK}$ $\overline{SK}$ $\overline{SK}$ $\overline{SK}$ $\overline{SK}$ $\overline{SK}$ $\overline{SK}$ $\overline{SK}$ = 1 11 11111 11111 2 22 22222 22222 3 33 33333 33333 4 44 44444 44444 5 55 55555 55555 215 324 234 125 415 521 431 2151 4531 5324 1543 3154 6 66 66666 66666 7 77 77777 77777 8 88 88888 88888 9 99 99999 99999 0 00 00000 00000 760 879 789 670 960 076 986 7606 9086 0879 6098 8609 196 245 837 604 932 758 103 821 073 3874 1928 5603 7495 1620 4835 2071 3778 1952 4394 1762

Side 1 of *Morse Code Practice*

KAYAK KAYAK SWEET SWEET JUICE JUICE OXYGEN OXYGEN FLAG FLAG QUICK QUICK MAGAZINE MAGAZINE WAVE WAVE CQ CQ FREQUENCY FREQUENCY ABT ABT QSL QSL SKIP SKIP QTH QTH 73 73 TNX TNX FIELD FIELD CANOE LABEL LADY PIANO VIEW HUNGRY YAWN HOUSE FROG AXLE MAIL JEWEL KAZOO CRAZY BUBBLE QUIET MAGNET MIXER VIVID QUARTZ QRM 88 FILTER YL RCVR LISSAJOUS 349 OM CW PSE COAXIAL HW 256 QSO XMTR B4 MAGNETIC CARBON FB 579 WX TU ANODE K8CH/1 DE WA3VIL R FB CHUCK = NAME HR IS LARRY = UR RST 57S ? 579 = QTH GORDON, PA. RIG IS HW 5400 = ANT IS 120 FT WIRE UP 60 FT = WX IS COLD 3 DEG = JUST HAD LUNCH ES MUST GET BK TO WRK SO 73 DE WA3VIL + $\overline{SK}$ HORIZON GND 73 QRN INDUCTOR CUL RFI QRS R NR 469 XYL ANT QRZ DE 357 QSB WUD RIG IS TS440 THE EXPLORER WAS QUICKLY FROZEN IN HIS BIG KAYAK JUST AFTER MAKING A DISCOVERY. SIX JAVELINS THROWN BY THE QUIET SAVAGES WHIZZED FORTY PACES BEYOND THE MARK. CQ CQ CQ DE KA1DYZ KA1DYZ K KA1DYZ DE W1YL/4 K W1YL DE KA1DYZ TNX FER CALL = UR RST 479 = QTH NEW BRITAIN, CT. = NAME MAUREEN = SO HW CPY? + FB MAUREEN = UR RST 589 = QTH HOMESTEAD, FL = NAME ELLEN = WX HR VY GUD = TEMP 80 DEG ES SUNNY = SO HW CPY MAUREEN ? BK FB ELLEN = SOLID = WX HR IS COLD ES SNOWY 30 DEG = MAYBE ILL COME VISIT HI = RIG HR IS TS520 ANT IS DIPOLE UP 60 FT = SRI PHONE MUST GO W1YL DE KA1DYZ $\overline{SK}$

HEADING ON UP

Once you obtain your Novice ticket, you'll soon set your sights on a Technician or General class license. The ARRL has other code-practice tapes available to help increase your code speed. There are tapes to go from 5 to 10 WPM, from 10 to 15 WPM and from 15 to 22 WPM. Each set of tapes includes two 90-minute cassettes, which provide practice with words, sentences, random code groups, related text and QSO practice. The space between characters decreases gradually to increase the overall code speed. You'll be copying faster code before you know it! W1AW also provides practice at a variety of speeds. Contact the ARRL, 225 Main Street, Newington, CT 06111 for the tapes you'll need to reach your desired code speed.

Table 3-3

Code Exam Questions for use with QSO Practice on Side 2 of *Morse Code Practice*

QSO 1 Test Questions

1. What is the receiving station's call sign?
2. What is the transmitting station's call sign?
3. What is the signal report? (all three digits)
4. The transmitting station is in what city?
5. What state is the transmitting station located in?
6. What is the transmitting operator's name?
7. How long has the transmitting operator been a ham?
8. What rig is the transmitting operator using?
9. How much power is the transmitting operator using?
10. How high is the antenna at the transmitting station?

QSO 2 Test Questions

1. What is the receiving operator's name?
2. What is the signal report? (all three digits)
3. What is the transmitting operator's name?
4. What rig is the transmitting operator using?
5. How much power is the transmitting operator using?
6. What kind of antenna is the transmitting operator using?
7. How high is the antenna at the transmitting station?
8. What is the receiving station's call sign?
9. What is the transmitting station's call sign?
10. Will this QSO continue after this transmission?

QSO 3 Test Questions

1. What is the receiving station's call sign?
2. What is the transmitting station's call sign?
3. What is the signal report? (all three digits)
4. What city is the transmitting station located in?
5. What state is the transmitting station located in?
6. What is the transmitting operator's name?
7. How long has the transmitting operator been a ham?
8. How old is the transmitting operator?
9. What is the temperature at the transmitting station?
10. Will this QSO continue after this transmission?

QSO 4 Test Questions

1. What is the receiving operator's name?
2. How old is the transmitting operator?
3. How long has the transmitting operator been a ham?
4. How long has the transmitting operator had an Amateur Extra Class license?
5. What is the signal report? (all three digits)
6. What problem has the transmitting operator indicated with the other operator's signals?
7. What is the weather at the transmitting station?
8. What is the temperature at the transmitting station?
9. What is the receiving station's call sign?
10. What is the transmitting station's call sign?

QSO 5 Test Questions

1. What is the receiving station's call sign?
2. What is the transmitting station's call sign?
3. What is the signal report? (all three digits)
4. What state is the transmitting station located in?
5. What is the transmitting operator's name?
6. What rig is the transmitting operator using?
7. How much power is the transmitting operator using?
8. What type of antenna is the transmitting operator using?
9. How long is the antenna at the transmitting station?
10. What is the receiving operator's name?

QSO 6 Test Questions

1. What is the receiving station's call sign?
2. What is the transmitting station's call sign?
3. What is the signal report? (all three digits)
4. What is the name of the transmitting operator?
5. Where is the transmitting station located? (city and state)
6. What rig is the transmitting operator using?
7. What type of antenna is the transmitting operator using?
8. How high is the antenna at the transmitting station?
9. How long has the transmitting operator been a ham?
10. What type of operating does the transmitting operator say he does most often?

KEY WORDS

Alternating current (ac)—Electrical current that flows first in one direction in a wire and then in the other. The applied voltage is also changing polarity. This direction reversal continues at a rate that depends on the frequency of the ac.

Alternator—A machine used to generate alternating-current electricity.

Ampere (A)—The basic unit of electrical current, equal to 6.24×10^{18} electrons moving past a point in one second.[1] We abbreviate amperes as *amps*.

Atom—A basic building block of all matter. Inside an atom, there is a positively charged, dense central core, surrounded by a "cloud" of negatively charged electrons. There are the same number of negative charges as there are positive charges, so the atom is electrically neutral.

Audio frequency (AF)—The range of frequencies that the human ear can detect. Audio frequencies are usually listed as 20 Hz to 20,000 Hz.

Battery—A device that stores electrical energy. It provides excess electrons to produce a current and the voltage or EMF to push those electrons through a circuit.

Breakdown voltage—The voltage that will cause a current in an insulator. Different insulating materials have different breakdown voltages. Breakdown voltage is also related to the thickness of the insulating material.

Centi—The metric prefix for 10^{-2}, or divide by 100.

Conductor—A material that has a loose grip on its electrons, so that an electrical current can pass through it.

Current—A flow of electrons in an electrical circuit.

Direct current (dc)—Electrical current that flows in one direction only.

Electromotive force (EMF)—The force or pressure that pushes a current through a circuit.

Electron—A tiny, negatively charged particle, normally found in an area surrounding the nucleus of an atom. Moving electrons make up an electrical current.

Energy—The ability to do work; the ability to exert a force to move some object.

Frequency—The number of complete cycles of an alternating current that occur per second.

Giga—The metric prefix for 10^9, or times 1,000,000,000.

Hertz (Hz)—An alternating-current frequency of one cycle per second. The basic unit of frequency.

Insulator—A material that maintains a tight grip on its electrons, so that an electrical current cannot pass through it.

Ion—An electrically charged particle. An electron is an ion. Another example of an ion is the nucleus of an atom that is surrounded by too few or too many electrons. An atom like this has a net positive or negative charge.

Kilo—The metric prefix for 10^3, or times 1000.

Mega—The metric prefix for 10^6, or times 1,000,000.

Metric prefixes—A series of terms used in the metric system of measurement. We use metric prefixes to describe a quantity as compared to a basic unit. The metric prefixes indicate multiples of 10.

Metric system—A system of measurement developed by scientists and used in most countries of the world. This system uses a set of prefixes that are multiples of 10 to indicate quantities larger or smaller than the basic unit.

Micro—The metric prefix for 10^{-6}, or divide by 1,000,000.

[1]Numbers written as a multiple of some power are expressed in *exponential notation.* This notation is explained in detail on page 4-2.

Milli—The metric prefix for 10^{-3}, or divide by 1000.

Negative Charge—One of two types of electrical charge. The electrical charge of a single electron.

Neutral—Having no electrical charge, or having an equal number of positive and negative charges.

Nucleus—The dense central portion of an atom. The nucleus contains positively charged particles.

Ohm—The basic unit of electrical resistance, used to describe the amount of opposition to current.

Ohm's Law—A basic law of electronics. Ohm's Law gives a relationship between voltage, resistance and current ($E = IR$).

Open circuit—An electrical circuit that does not have a complete path, so current can't flow through the circuit.

Parallel circuit—An electrical circuit where the electrons follow more than one path.

Pico—The metric prefix for 10^{-12}, or divide by 1,000,000,000,000.

Positive charge—One of two types of electrical charge. A positive charge is the opposite of a negative charge. Electrons have a negative charge. The nucleus of an atom has a positive charge.

Power—The rate of energy consumption. We calculate power in an electrical circuit by multiplying the voltage applied to the circuit times the current through the circuit.

Power supply—That part of an electrical circuit that provides excess electrons to flow into a circuit. The power supply also supplies the voltage or EMF to push the electrons along. Power supplies convert a power source (such as the ac mains) to a useful form.

Radio frequency (RF)—The range of frequencies that can be radiated through space in the form of electromagnetic radiation. We usually consider RF to be those frequencies higher than the audio frequencies, or above 20 kilohertz.

Resistance—The ability to oppose an electric current.

Resistor—Any material that opposes a current in an electrical circuit. An electronic component especially designed to oppose current.

Series circuit—An electrical circuit where the electrons must all flow through every part of the circuit. There is only one path for the current to follow.

Short circuit—An electrical circuit where the current does not take the desired path, but finds a shortcut instead. Often the current goes directly from the negative power-supply terminal to the positive one, bypassing the rest of the circuit.

Sine wave—A smooth curve, usually drawn to represent the variation in voltage or current over time for an ac signal.

Subatomic particles—The building blocks of atoms. Electrons, protons and neutrons are the most common subatomic particles.

Transformer—A device that changes ac voltage levels.

Volt (V)—The basic unit of electrical pressure or EMF.

Voltage—The EMF or pressure that causes electrons to move through an electrical circuit.

Voltage source—Any source of excess electrons. A voltage source produces a current and the force to push the electrons through an electrical circuit.

Watt (W)—The unit of power in the metric system. The watt describes how fast a circuit uses electrical energy.

Wavelength—The distance an ac signal will travel during the time it takes the signal to go through one complete cycle.

Chapter 4

Understanding Basic Theory

The purpose of this chapter is to introduce you to basic electronics and radio theory. We present the theory in this chapter that you'll need to pass your Novice exam. This basic theory will be a foundation for you to build on. You'll find the basics useful when you start studying for your Technician or higher class license. In addition, the theory you learn here will help you to assemble and operate your amateur station.

There is much for you to learn, and most of the material presented here will be new to you. To make your study as easy as possible, Chapter 4 is divided into three main sections. We start with basic electrical principles. Once you've learned the basics, we cover resistance, Ohm's Law and power. Finally, you'll learn about direct and alternating currents. To get the most from this chapter, you should take it one section at a time. Study the material in each section and really know it before you go on to the next section. The sections build on each other, so you may find yourself referring to sections you've already studied from time to time.

There are many technical terms used in electronics. We have provided definitions that are as simple and to-the-point as possible. You will want to refer often to the **Key Words** at the beginning of this chapter. Don't be afraid to turn back to any section you have already studied. This review is helpful if you come across a term that you are not sure about. You will probably not remember every bit of this theory just by reading the chapter once.

There are many drawings and illustrations presented in this chapter to help you learn the material. Pay attention to these graphics, and you'll find it easier to understand the text. We'll direct you to Chapter 12 at appropriate points in the text. Use these directions to help you study the Novice Question Pool. When you think you've learned the section, you are ready to move on.

If you have trouble understanding parts of this chapter, ask your instructor or another experienced ham for help. There are many other books that can help, too. The booklet "First Steps in Radio" is a collection of Doug DeMaw's *QST* series. "First Steps" covers a wide range of technical topics written especially for a beginner. To study more advanced theory, you may want to purchase a copy of *The ARRL Handbook*. Both publications are available from your local ham dealer or from ARRL Headquarters.

Remember! Take it slowly, section by section. Before you know it, you'll have learned what you need to know to pass your test and get on the air. Good luck!

BASIC ELECTRICAL PRINCIPLES

In this section, you will learn what electricity is and how it works. We'll introduce you to the atom and the electron, the basic elements of electricity. There are no questions about atoms and electrons on the Novice exam, but understanding them will help you understand the rest of this chapter.

THE METRIC SYSTEM

We'll be talking about the units used to describe electrical pressure and other electrical units later in this chapter. Before we do that, let's take a few minutes now to become familiar with the **metric system**. This simple system is a standard system of measurement used all over the world. All the units used to describe electrical quantities are part of the metric system.

In the US, we use a measuring system known as the US Customary system. In this system there is no logical progression between the various units of length, weight, volume or other quantities. For example, we have 12 inches in 1 foot, 3 feet in 1 yard and 1760 yards in 1 mile. For measuring the volume of liquids we have 2 cups in 1 pint, 2 pints in 1 quart and 4 quarts in 1 gallon. To make things even more difficult, we use some of these same names for different volumes when we measure dry materials! As you can see, this system of measurements can be very confusing. Even those who are very familiar with the system do not know all the units used for different types of measurements.

It is exactly this confusion that led scientists to develop the orderly system we know as the metric system today. This system uses a basic unit for each different type of measurement. For example, the basic unit of length is the meter (sometimes spelled metre). The basic unit of volume is the liter (or litre). The unit for mass (or quantity of matter) is the gram. The newton is the metric unit of force, or weight, but we often use the gram to indicate how "heavy" something is. We can express larger or smaller quantities by multiplying or dividing the basic unit by factors of 10 (10, 100, 1000, 10,000 and so on). These multiples result in a standard set of prefixes, which can be used with all the basic units. Table 4-1 summarizes the most-used **metric prefixes**. These same prefixes can be applied to any basic unit in the metric system. Even if you come across some terms you are unfamiliar with, you will be able to recognize the prefixes.

We can write these prefixes as powers of 10, as shown in the table. The power of 10 (called the *exponent*) shows how many times you must multiply (or divide) the basic unit by 10. For example, we can see from the table that **kilo** means 10^3. Let's use the meter as an example. If you multiply a meter by 10 three times, you will have a *kilo*meter. (1 meter $\times 10^3 = 1 \text{ m} \times 10 \times 10 \times 10 = 1000$ meters, or 1 kilometer.) If you multiply 1 meter by 10 six times, you have a **mega**meter. (1 meter $\times 10^6 = 1\text{m} \times 10 \times 10 \times 10 \times 10 \times 10 \times 10 = 1{,}000{,}000$ meters or 1 megameter.)

Notice that the exponent for some of the prefixes is a negative number. This indicates that you must *divide* the basic unit by 10 that number of times. If you divide a meter by 10, you will have a **deci**meter. (1 meter $\times 10^{-1} = 1 \text{ m} \div 10 = 0.1$ meter, or 1 decimeter.) When we write 10^{-6}, it means you must divide by 10 six times. (1 meter $\times 10^{-6} = 1 \text{ m} \div 10 \div 10 \div 10 \div 10 \div 10 \div 10 = 0.000001$ meter, or 1 **micro**meter.)

Table 4-1

International System of Units (SI)—Metric Units

Prefix	*Symbol*		*Multiplication Factor*
tera	T	10^{12} =	1,000,000,000,000
giga	G	10^{9} =	1,000,000,000
mega	M	10^{6} =	1,000,000
kilo	k	10^{3} =	1,000
hecto	h	10^{2} =	100
deca	da	10^{1} =	10
(unit)		10^{0} =	1
deci	d	10^{-1} =	0.1
centi	c	10^{-2} =	0.01
milli	m	10^{-3} =	0.001
micro	μ	10^{-6} =	0.000001
nano	n	10^{-9} =	0.000000001
pico	p	10^{-12} =	0.000000000001

We can easily write very large or very small numbers with this system. We can use the metric prefixes with the basic units, or we can use powers of 10. Many of the quantities used in basic electronics are either very large or very small numbers, so we use these prefixes quite a bit. You should be sure you are familiar at least with the following prefixes and their associated powers of 10: **giga** (10^9), **mega** (10^6), **kilo** (10^3), **centi** (10^{-2}), **milli** (10^{-3}), **micro** (10^{-6}) and **pico** (10^{-12}).

Let's try an example. We have a receiver dial calibrated in megahertz (MHz), and it shows a signal at a frequency of 3.725 MHz. Where would a dial calibrated in kilohertz show the signal? From Table 4-1 we see that kilo means times 1000, and mega means times 1,000,000. That means that our signal is at 3.725 MHz $\times$ 1,000,000 = 3,725,000 hertz. There are 1000 hertz in a kilohertz, so 3,725,000 divided by 1000 gives us 3725 kHz.

How about another one? If we have a current of 3000 milliamps, how many amps is this? From Table 4-1 we see that milli means multiply by 0.001 or divide by 1000. Dividing 3000 milliamps by 1000 gives us 3 amps. The metric prefixes make it easy to use numbers that are a convenient size simply by changing the units. 3.725 MHz is certainly easier to work with than 3,725,000 hertz!

[Before you go on to the next section, turn to Chapter 12. Be sure you can answer all of the questions in the question pool that begin with numbers 2E-1. Review this section if you have any difficulty.]

ELECTRICITY

The word is a spine-tingling mystery. It's the force behind our space-age civilization. It's one of nature's greatest powers. We love it; we fear it. We use it in our work and play. But what is it?

It's a mystery only in our minds. Actually, electricity is the marvelous stuff which, when untamed, we call lightning. One lightning bolt produces enough electricity to supply your needs for a lifetime. In another form, electricity is the power in a battery that cranks the engine to start your car. Electricity also ignites the gasoline in the engine. Yet, with all its power, electricity is the careful messenger carrying information from your brain to your muscles, enabling you to move your arms and legs. You can buy a small container of electricity no bigger than a dime (a battery). Electric utilities generate and transmit huge amounts of electricity every day. From lightning bolts to brain waves, it's all the same stuff: **electrons**.

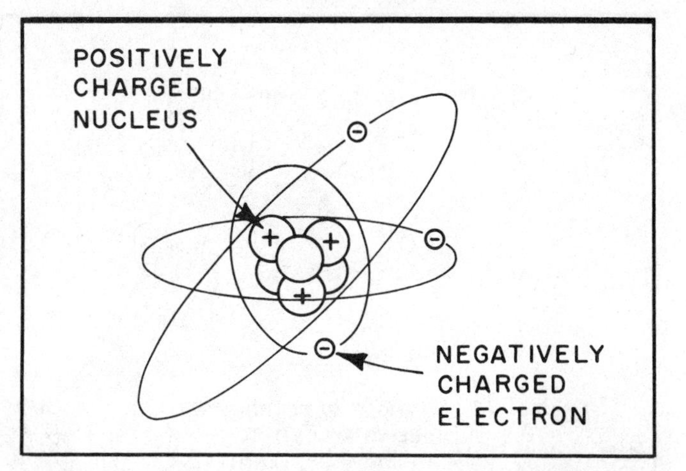

Figure 4-2—Each atom is a microscopic particle composed of a central, dense, positively charged nucleus, surrounded by tiny negatively charged electrons. There are the same number of positive particles in the nucleus and negative electrons outside the nucleus.

Figure 4-1—A phenomenon that has fascinated mankind for ages, lightning is simply a natural source of electricity.

INSIDE ATOMS

Everything you can see and touch is made up of **atoms**. Atoms are the building blocks of nature. Atoms are too small to see, but the **subatomic particles** inside atoms are even smaller.

Each atom has a **nucleus** in its center. Other particles orbit around this central core. Think of the familiar maps of our solar system: planets orbit the sun. In an atom, charged particles orbit the central core (the nucleus). Other charged particles make up the nucleus. Figure 4-2 is a simplified illustration of the structure of an atom.

Some particles have **negative charges** while others have **positive charges**. The core of an atom contains positively charged particles. Negative particles, called electrons, orbit around the nucleus. Scientists have identified more than 100 different kinds of atoms. The number of positively and negatively charged particles in an atom determines what type of element that atom is. Different kinds of atoms combine to form various materials. For example, a hydrogen atom has one positively charged particle in its nucleus and one electron around the outside. An oxygen atom has eight positive particles in the nucleus and eight electrons around the outside. When two hydrogen atoms combine with one oxygen atom, we have water.

Have you ever tried to push the north poles of two magnets together? Remember that soft, but firm, pressure holding them apart? Similar poles in magnets repel each other, opposite poles attract each other. You can see this if you experiment with a pair of small magnets.

Charged particles behave in a way similar to the two magnets. A positively charged particle and a negatively charged particle attract each other. Two positive or two negative particles repel each other. Like repels like; opposites attract.

Electrons stay near the central core, or nucleus, of the atom. The positive charge on the nucleus attracts the negative electrons. Meanwhile, since the electrons are all negatively charged, they repel each other. This makes the electrons move apart and fill the space around the nucleus. (Scientists sometimes refer to this area as an electron "cloud.") An atom has an equal balance of negative and positive charges and shows no electrical effect to the outside world. We say the atom is **neutral**.

ELECTRON FLOW

In many materials, especially metals, it's easy to dislodge an electron from an atom. When the atom loses an electron, it upsets the stability and electrical balance of the atom. With one particle of negative charge gone, the atom has an excess of positive charges. The free electron has a negative charge. We call the atom that lost the electron a positive **ion** because of its net positive charge. (An ion is a charged particle.) If there are billions of similar ions in

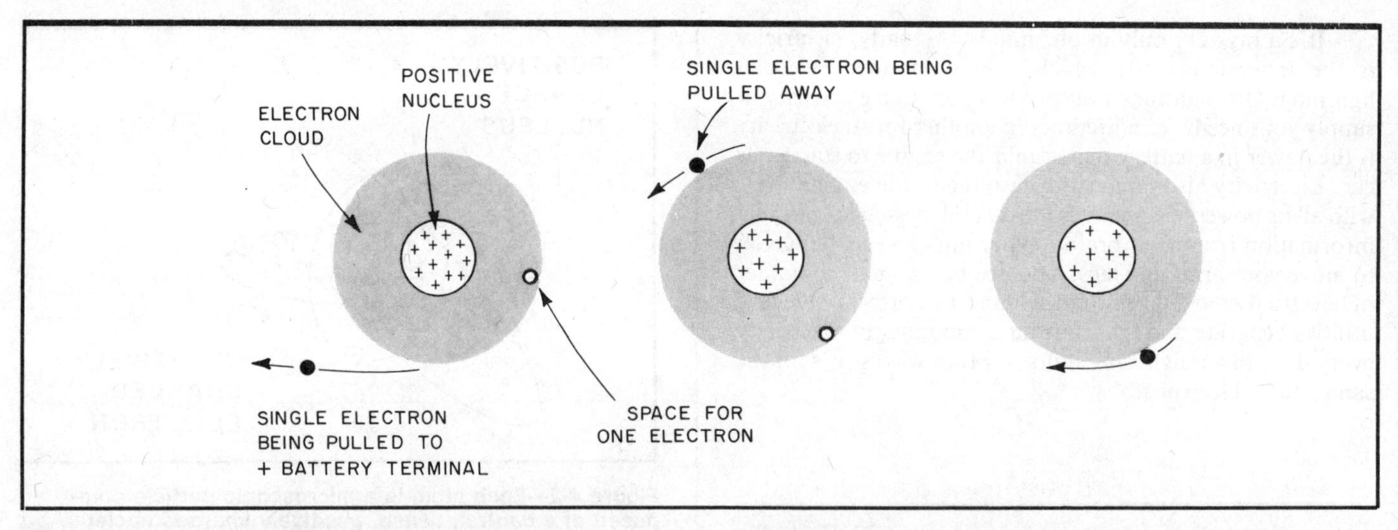

Figure 4-3—An ion is an electrically charged particle. When an electron (a negative ion) moves from one atom to another, the atom losing the electron becomes a positive ion. This electron movement represents an electric current. The electrons are moving from right to left in this drawing.

one place, the quantity of charge becomes large enough to cause a noticeable effect.

Positively charged ions can pull negative particles (electrons) from neutral atoms. These electrons can move across the space between the atom and the ion and orbit the positively charged ion. Now the positively charged ion has become a neutral atom again, and another atom has become a positively charged ion! If this process seems confusing, take a look at Figure 4-3. Here we show a series of atoms and positive ions, with electrons moving from one atom to the next. The electrons are moving from right to left in this diagram. We call this flow of electrons *electricity*. Electricity is nothing more than the flow of electrons.

How difficult is it for a positive ion to rip an electron away from an atom? That depends on the individual atoms making up a particular material. Some atoms hold firmly to their electrons and won't let them flow away easily. Other atoms keep only a loose grip on electrons and let them slip away easily. This means that some materials carry electricity better than others. **Conductors** are materials that keep only

DEMONSTRATING ELECTRON FLOW

Here's an easy way to see the power of electrical charges. On a dry day hang some metal foil from dry thread. Rub one end of a plastic comb on wool (or run it through your hair). Then bring the end of the comb near the metal foil.

Rubbing the comb on wool detaches electrons from the wool fibers, depositing them on the surface of the comb. When you bring the charged end of the comb near the foil, free electrons in the foil will be repelled by the negative charge on the comb. The free electrons will move as far from the comb as they can—on the edge farthest from the comb. This leaves the near edge of the foil with a shortage of electrons (in other words, with a positive charge). The near edge of the foil is then attracted to the negatively charged comb.

As soon as the foil touches the comb, some of the excess electrons on the comb flow onto the foil (electricity!). The foil will then have a net negative charge, as the comb does, and the foil will be repelled by the comb. This simple experiment shows the attraction, repulsion and flow of electrons—the heart of electricity.

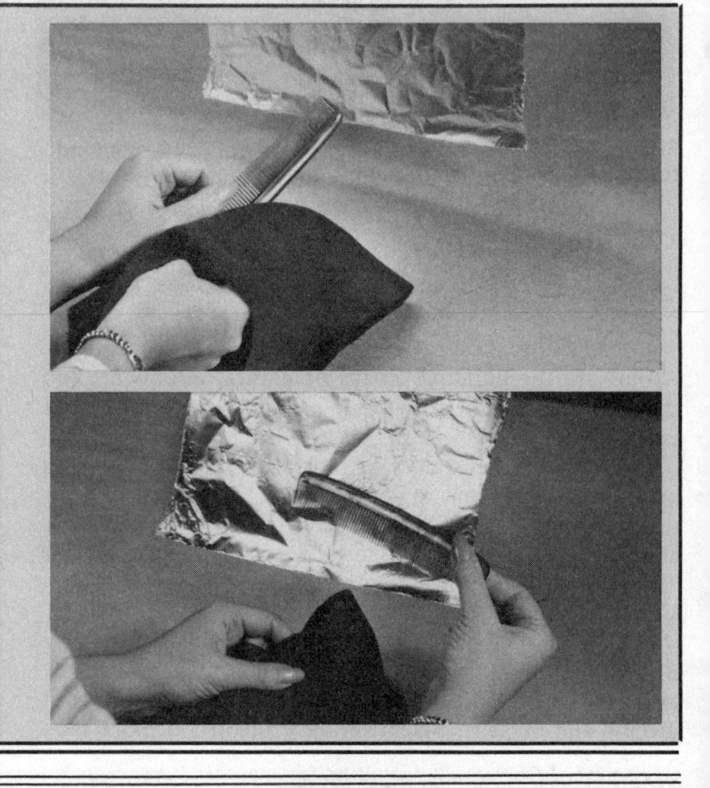

a loose grip on their electrons. We call those materials that try to hold on to their electrons **insulators**.

Why Do Electrons Flow?

There are many similarities between electricity flowing in a wire and water flowing through a pipe. Most people are familiar with what happens when you open a faucet and water comes out. We can use this to make a useful comparison between water flow and electron flow (electricity). Throughout this chapter we use examples of water flowing in a pipe to help explain electronics.

Do you know how your town's water system works? Chances are, there is a large supply of water stored somewhere. Some towns use a lake or river. Other towns get their water from wells and store it in a tank or reservoir. The system then uses gravity to pull the water down from the tank or reservoir. The water travels through a system of pipes to your house. Because the force of gravity is pulling down on the water in the tank, it exerts a pressure on the water in the pipes. This makes the water flow out of the faucets in your house with some force. If you have a well, you probably have a storage tank. The storage tank uses air pressure to push the water up to the top floor of your house.

In these water systems, a pump takes water from the large supply and puts it into a storage tank. Then the system uses air pressure or the force of gravity to push the water through pipes to the faucets in your home.

We can compare electrons flowing through wire to water flowing through a pipe. We need some force to make water flow through a pipe; what force exerts pressure to make electrons flow through a wire?

VOLTAGE

The amount of pressure that it takes to push water to your house depends on how far you live from the reservoir. If the water has to travel over hills along the way, even more pressure will be required. The pressure required to make electrons flow in an electrical circuit also depends on the opposition that the electrons must overcome. The pressure that forces the electrons through the circuit is known as **electromotive force** or, simply, **EMF**.

EMF is similar to water pressure. More pressure moves more water. Similarly, more EMF moves more electrons. We measure EMF in a unit called the **volt (V)**, so we sometimes refer to the EMF as a **voltage**. If more voltage is applied to a circuit, more electrons will flow. We measure voltage with a device called a *voltmeter*.

When you write about volts, use "12 V" if you mean twelve volts or "120 V" if you are saying one-hundred-twenty volts. Using our system of metric prefixes, we can express a thousand volts as "1 kV," or 1 kilovolt.

Another way to think about this electrical voltage pushing electrons through a circuit is to remember that like-charged objects repel. If we have a large group of electrons, the negative charge of these electrons will act to repel, or push, other electrons through the circuit. In a similar way, a large group of positively charged ions attract, or pull, electrons through the circuit.

Because there are two types of electric charge (positive and negative), there are also two polarities associated with a voltage. A **voltage source** always has two terminals, or poles; the positive terminal and the negative terminal. The negative terminal repels electrons (negatively charged particles) and the positive terminal attracts electrons. If we connect a piece of wire between the two terminals of a voltage source, electrons will flow through the wire. We call this flow an electrical **current**.

The basic unit of electromotive force (EMF) is the volt. The volt was named in honor of Allesandr Giuseppe Antonio Anastasio Volta (1745-1827). This Italian physicist invented the electric battery.

Batteries

In our water system, a pump supplies pressure to pull water from the source and force it into the pipes. Similarly, electrical circuits require an electron source and a "pump" to move the electrons along.

A **battery** is one example of a **power supply**. We use a battery as both the source of electrons and the pump that moves them along. The battery is like the storage tank in our water system. A battery provides pressure to keep the electrons moving.

There is an excess of electrons at the negative terminal of a power supply. At the positive terminal there is an excess of positive ions. With a conducting wire connected between the two terminals, the electrical pressure (voltage) generated by the power supply will cause electrons to move through the conductor.

Batteries come in all shapes and sizes. Some batteries are tiny, like those used in hearing aids and cameras. Other batteries are larger than the one in your car. A battery is one kind of voltage source.

[Now turn to Chapter 12 and study the questions with numbers that begin 2E-3. Review this section if you have any difficulty.]

CURRENT

You have probably heard the term "current" used to describe the flow of water in a stream or river. Similarly, we call the flow of electrons an electric current. Each electron is extremely small. It takes quintillions and quintillions of electrons to make your toaster heat bread or your TV draw pictures. (A quintillion is a one with 18 zeros after it—1,000,000,000,000,000,000. Using powers of 10, as described earlier, we could also write this as 1×10^{18}.)

When water flows from your home faucet, you don't try to count every drop. The numbers would be very large and unmanageable, and the drops are coming out much too fast to count! To measure water flow, you count larger quantities such as gallons and describe the flow in terms of gallons per minute. Similarly, we can't deal easily with large numbers of individual electrons, nor can we count them conveniently. We need a shorthand way to measure the number of electrons. So, as with gallons per minute of water, we have amperes of electric current. We measure current with a device called an *ammeter*.

The action of electric current on a magnet was first applied to telegraphy by Andre Marie Ampere (1775-1836) in 1820. An ampere is the basic unit of electrical current.

One **ampere** of current flows in a circuit when 6,240,000,000,000,000,000 or 6.24×10^{18}, electrons move past a point in one second. (Don't worry! You won't have to remember this number.) Rather than write such huge numbers, it's much easier to express the number of electrons flowing in a circuit in amperes. Write "2 A" for two amperes or "100 mA" (milliamps) for 0.1 ampere (sometimes also abbreviated amp or amps). See Table 4-1 to review the list of metric prefixes.

[Now study the questions in Chapter 12 that begin with numbers 2E-2. Review this section if you have trouble answering these questions.]

CONDUCTORS

As we pointed out earlier, some atoms have a firm grasp on their electrons and other atoms don't. More current can flow in materials made up of atoms that have only a weak hold on their electrons. Some materials, then, conduct current better than others.

Silver is an excellent conductor. The loosely attached electrons in silver atoms require very little voltage (pressure) to produce an electric current. Copper is much less expensive than silver and conducts almost as well. We can use copper to make wire needed in houses, and in radios and other electronic devices. Steel also conducts, but not as well as copper. In fact, most metals are fairly good conductors, so aluminum, mercury, zinc, tin and gold are all conductors.

INSULATORS

Other materials keep a very firm grip on their electrons. These materials do not conduct electricity very well. Materials such as glass, rubber, plastic, ceramic, mica, wood and even air are poor conductors. Pure distilled water is a fairly good insulator. Most tap water is a good conductor, however, because it has minerals and other impurities dissolved in it. Figure 4-4 lists some common insulators and conductors.

The electric power company supplies 120 V on the wires into your home. That voltage is available for your use at electrical outlets or sockets. Why don't the electrons spill out of the sockets? The insulation between the two sides of the outlet prevents the electrons from flowing from one side to the other. The air around the socket acts as an insulator to stop them from flowing into the room.

Because insulators are *poor* conductors rather than *non*conductors, every insulator has a **breakdown voltage**. Beyond the breakdown voltage, the insulator will start to conduct electricity. Depending on the material, it may be damaged if you exceed the breakdown voltage. Better insulators have higher breakdown voltages. Breakdown

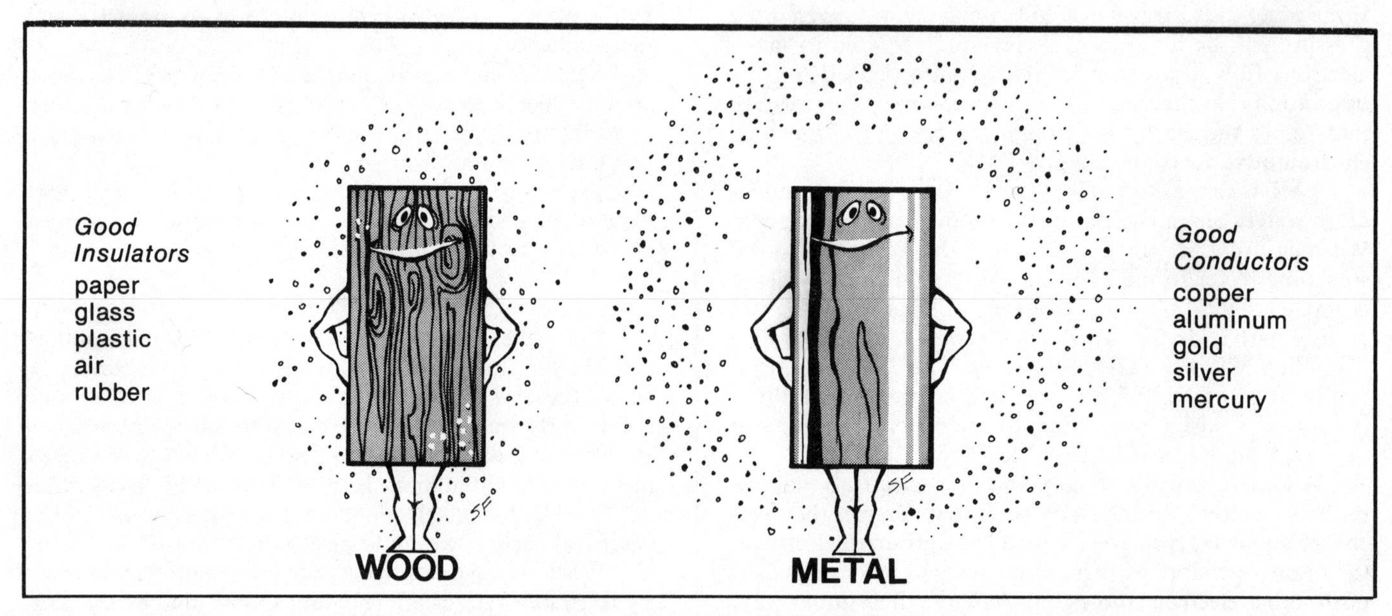

Figure 4-4—Here's one way to show how an insulator differs from a conductor. Wood, on the left, holds onto its electrons pretty tightly, keeping them from flowing between its atoms. Metals, on the other hand, are more generous with their electrons. Electrons are more easily pulled away from the metal atoms, and the metal atoms are then left with a positive charge. If the metal atoms attract extra electrons from neighboring atoms, they become negatively charged.

voltage also depends on the thickness of the insulating material. A thin layer of one insulating material (like Teflon® or mica) may be just as good as a much thicker layer of another material (like paper or air.)

A good example of an insulator that will conduct at very high voltage is air. Air is a fine insulator at the voltages normally found in homes and industry. When a force of millions of volts builds up, however, there's enough pressure to send a bolt of electrons through the air—lightning.

When you are insulating wires or components, always be sure to use the right insulating material. Make sure you use enough insulation for the voltages you're likely to encounter. Heat-shrinkable tubing or other insulating tubing is often convenient for covering a bare wire or a solder connection. You can also wrap the wire with electrical tape. Several layers of tape, wrapped so it overlaps itself, will provide enough insulation for up to a few hundred volts.

[Before you go on to the next section, turn to Chapter 12 and study questions 2E-4.1 and 2E-5.1. If you have any difficulty, review this section.]

ELECTRONICS FUNDAMENTALS

In this section, you'll learn about resistance. You'll see how resistance fits into one of the most fundamental laws of electronics, Ohm's Law. We will also briefly discuss series and parallel circuits. While there are no questions about these circuits on the Novice exam, understanding them will make it much easier for you to understand some of the other concepts you *will* be tested on!

This section shows you how to do some basic circuit calculations. We have tried to keep the arithmetic simple, and to explain all of the steps in the solution. Associating numbers with a concept often makes it easier to understand. Read these examples and then try the calculations yourself. Be sure you can work the problems in the question pool when you study those questions.

RESISTANCE AND OHM'S LAW

What if you partially blocked a water pipe with a sponge? Eventually, the water would get through the sponge, but it would have less pressure than before. It takes pressure to overcome the resistance of the sponge.

Similarly, materials that conduct electrical current also present some opposition, or **resistance**, to that flow. **Resistors** are devices that are especially designed to make use of this opposition. Figure 4-5 shows some common resistors.

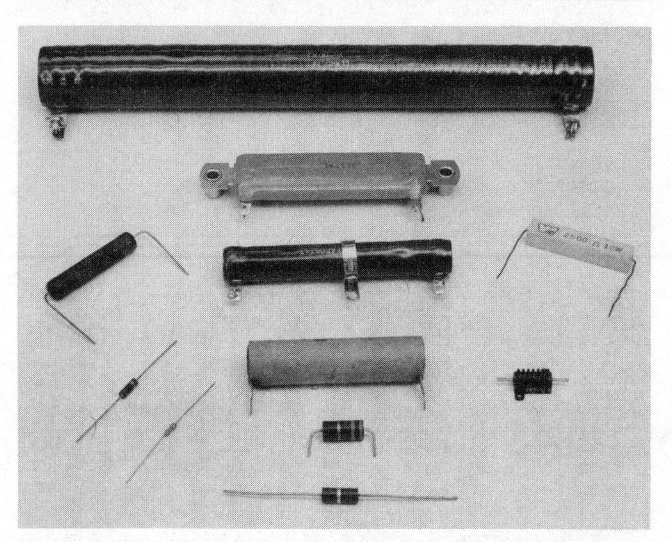

Figure 4-5—This photograph shows some of the many types of resistors. Large power resistors are at the top of the photo. The small resistors are used in low-power transistor circuits.

In a water pipe, increasing the pressure forces more water through the sponge (the resistance). In an electrical circuit, increasing the voltage forces more current through the resistor. The relationship between voltage, current and resistance is predictable. We call this relationship **Ohm's Law** and it is a basic electronics principle.

The amount of water flowing through a pipe increases as we increase the pressure and decreases as we increase the resistance. If we replace "pressure" with "voltage," we can write a mathematical relationship for an electric circuit:

$$\text{current} = \frac{\text{voltage}}{\text{resistance}} \qquad \text{(Eq 4-1)}$$

If the voltage stays constant but more current flows in the circuit, we know there must be less resistance. The relationship between current and voltage is a measure of the resistance:

$$\text{resistance} = \frac{\text{voltage}}{\text{current}} \qquad \text{(Eq 4-2)}$$

Finally, we can determine the voltage if we know how much current is flowing and the resistance in the circuit:

$$\text{voltage} = \text{current} \times \text{resistance} \qquad \text{(Eq 4-3)}$$

Scientists are always looking for shorthand ways of writing these relationships. They use symbols to replace the words: E represents voltage (remember EMF?), current is I (from the French word *intensité*) and resistance is R. We measure resistance in units called **ohms**. The abbreviation for ohms is Ω, the Greek capital letter omega. We can now express the above relationship in a couple of letters:

$$E = IR \text{ (volts} = \text{amperes} \times \text{ohms)} \qquad \text{(Eq 4-4)}$$

This is the most common way to express Ohm's Law, but we can also write it as:

THERE REALLY WAS AN OHM (1789-1854)

Although we take Ohm's Law for granted, it wasn't always so widely accepted. In 1827, Georg Simon Ohm finished his renowned work, *The Galvanic Circuit Mathematically Treated*. Scientists at first resisted Ohm's techniques. The majority of his colleagues were still holding to a non-mathematical approach. Finally, the younger physicists in Germany accepted Ohm's information in the early 1830s. The turning point for this basic electrical law came in 1841 when the Royal Society of London awarded Ohm the Copley Medal.

As the oldest son of a master locksmith, Georg received a solid education in philosophy and the sciences from his father. At the age of 16 he entered the University of Erlangen (Bavaria) and studied for three semesters. His father then forced him to withdraw because of alleged over-indulgences in dancing, billiards and ice skating. After 4½ years, his brilliance undaunted, Ohm returned to Erlangen to earn his PhD in mathematics.

Enthusiasm ran high at that time for scientific solutions to all problems. The first book Ohm wrote reflected his highly intellectualized views about the role of mathematics in education. These opinions changed, however, during a series of teaching positions. Oersted's discovery of electromagnetism in 1820 spurred Ohm to avid experiments, since he taught Physics and had a well-equipped lab. Ohm based his work on direct scientific observation and analyses, rather than on abstract theories.

One of Ohm's goals was to be appointed to a major university. In 1825 he started doing research with the thought of publishing his results. The following year he took a leave of absence from teaching and went to Berlin.

In Berlin, Ohm made his now-famous experiments. He ran wires between a zinc-copper battery and mercury-filled cups. While a Coulomb torsion balance (voltmeter) was across one leg of the series circuit, a "variable conductor" completed the loop. By measuring the loss in electromagnetic force for various lengths and sizes of wire, he had the basis for his formulas.

Because *The Galvanic Circuit* produced near-hostility, Ohm withdrew from the academic world for nearly six years before accepting a post in Nuremberg. English and French physicists do not seem to have been aware of the profound implications of Ohm's work until the late 1830s and early 1840s.

Following his belated recognition, Ohm became a corresponding member of the Berlin Academy. Late in 1849 he went to the University of Munich and in 1852, only two years before his death, he achieved his lifelong dream of a full professorship at a major university.

$$I = \frac{E}{R} \text{ (amperes = volts divided by ohms)} \qquad \text{(Eq 4-5)}$$

and

$$R = \frac{E}{I} \text{ (ohms = volts divided by amperes)} \qquad \text{(Eq 4-6)}$$

E is EMF in volts. I is the number of amperes of current, and R is the number of ohms, the unit we use to measure resistance. If you know two of the numbers, you can calculate the third. Figure 4-6 shows a diagram to help you solve Ohm's Law problems. Simply cover the symbol of the quantity that you do not know. If the remaining two are side-by-side, you must multiply them. If one symbol is above the other, then you must divide the quantity on top by the one on the bottom.

The basic unit of resistance is the ohm, named in honor of Georg Simon Ohm (1787-1854).

If you know current and resistance in a circuit, Ohm's Law will give you the voltage (Eq 4-4). For example, what is the voltage applied to the circuit if 5 amperes of current flows through 20 ohms of resistance? From Eq 4-4 or Figure 4-6, we see that we must multiply 5 amperes times 20 ohms to get the answer, 100 volts. The EMF in this circuit is 100 volts.

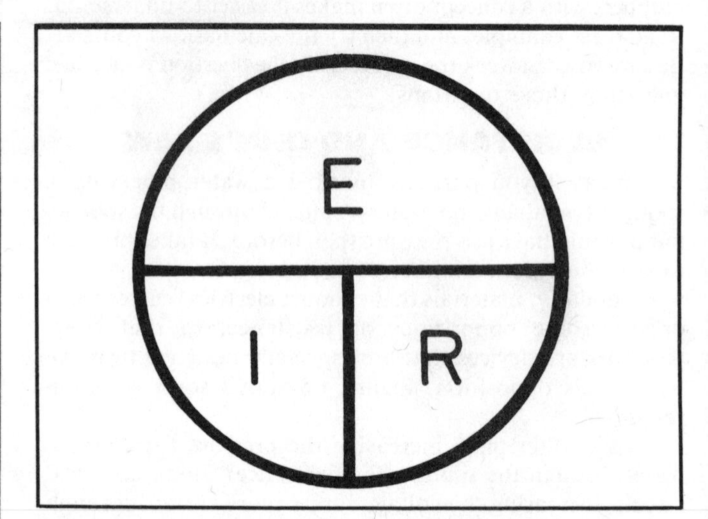

Figure 4-6—A simple diagram to help you remember the Ohm's Law relationships. To find any quantity if you know the other two, simply cover the unknown quantity with your hand or a piece of paper. The positions of the remaining two symbols show if you have to multiply (if they are side by side) or divide (if they appear one over the other as a fraction).

$$E = IR \qquad \text{(Eq 4-4)}$$
$$E = 5 \text{ amperes} \times 20 \text{ ohms}$$
$$E = 100 \text{ volts}$$

Suppose you know voltage and resistance—100 volts

in the circuit to push electrons against 20 ohms of resistance? Eq 4-5 gives the correct equation, or you can use the diagram in Figure 4-6. You must divide 100 volts by 20 ohms to find that 5 amperes of current is flowing.

$$I = \frac{E}{R} \qquad \text{(Eq 4-5)}$$

$$I = \frac{100 \text{ volts}}{20 \text{ ohms}}$$

$$I = 5 \text{ amperes}$$

If you know voltage and current in a circuit, you can calculate resistance. Using our example, 100 volts is pushing 5 amperes through the circuit. This time Eq 4-6 is the one to use, and you can also find this from Figure 4-6. 100 divided by 5 equals 20, so the resistance is 20 ohms.

$$R = \frac{E}{I} \qquad \text{(Eq 4-6)}$$

$$R = \frac{100 \text{ volts}}{5 \text{ amperes}}$$

$$R = 20 \text{ ohms}$$

If you know E and I, you can find R. If you know I and R, you can calculate E. If E and R are known, you can find I.

Put another way, if you know volts and amperes, you can calculate ohms. If amperes and ohms are known, volts can be found. Or if volts and ohms are known, amperes can be calculated. Figure 4-7 illustrates some simple circuits and how Ohm's Law can be used to find an unknown quantity in the circuit. Make up a few problems of your own and test how well you understand this basic law of electricity. You'll soon find that this predictable relationship, symbolized by the repeatable equation of Ohm's Law, makes calculating values of components in electrical circuits easy!

[Before you read further, turn to Chapter 12 and study the questions with numbers that begin 2E-6 and 2E-7. Review the material in this section if you have difficulty with any of these questions.]

SERIES AND PARALLEL CIRCUITS

There are two basic ways that you can connect the parts in an electric circuit. If we hook several resistors together in a string, we call it a **series circuit**. If we connect several resistors side-by-side to the same voltage source, we call it a **parallel circuit**.

In our water-pipe example, what would happen to the current through the pipe if we placed another sponge in it? You're right: The second sponge would further reduce the flow. We could do the same thing with a single, larger sponge. The total resistance in a series circuit is the sum of all the resistances in the circuit.

In a series circuit, the same current, I, flows through each resistor, since it has no other path to follow. When a voltage source (like a battery) is hooked to our string of resistors, we can calculate the voltage across each resistor. How? Using Ohm's Law, of course.

Remember E = IR? The voltage across any resistor in the circuit will be its value in ohms multiplied by the current in amperes. We call the voltage across the resistor the *voltage drop*. If we calculate the voltage drop across each resistor in the circuit and add them together, we will get the battery voltage. Figure 4-8 shows that with more resistors in a series circuit, there will be less current.

Now let's look at the case of a parallel circuit. This is similar to having two water pipes running side by side. More water can flow through two parallel pipes of the same size than through a single one. With two pipes, the current is greater for a given pressure.

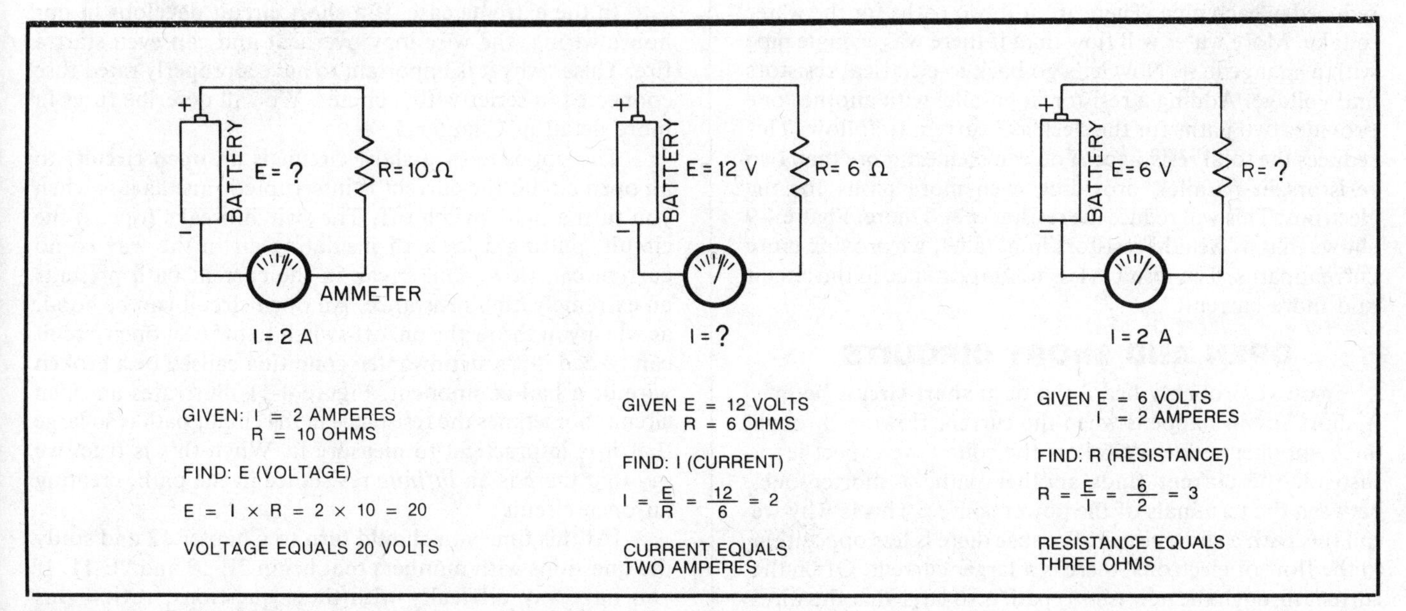

Figure 4-7—Some Ohm's Law problems and solutions.

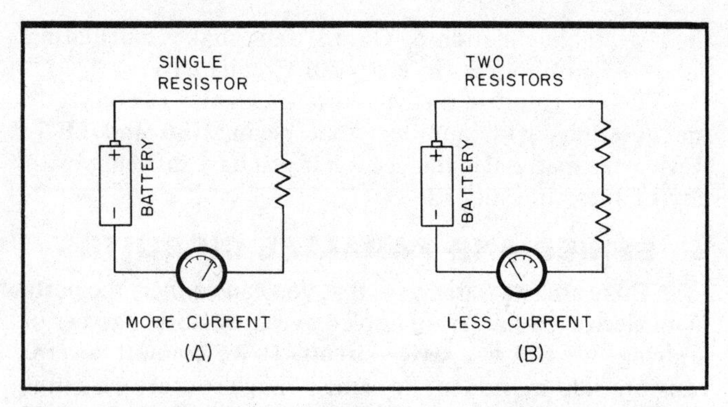

Figure 4-8—Resistance limits the amount of current that can flow in a circuit. Adding a second resistance reduces the current because the total resistance is larger.

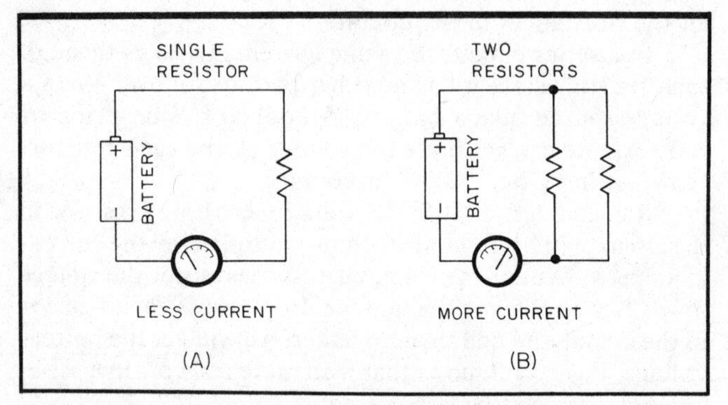

Figure 4-9—When we connect two equal-value resistors in parallel there is a second path for the electrons. This reduces the effective circuit resistance. This means more current will flow than if only one of the resistors were in the circuit.

If these pipes had sponges in them, the flow would be reduced in each pipe. There are still two paths for the water to take. More water will flow than if there was a single pipe with a sponge in it. Now let's go back to electrical resistors and voltage. Adding a resistor in parallel with another one provides two paths for the electrical current to follow. This reduces the total resistance. You can connect more than two resistors in parallel, providing even more paths for the electrons. This will reduce the resistance still more. Figure 4-9 shows that as we add resistors in parallel, we provide more current paths. The result is less total resistance in the circuit and more current.

OPEN AND SHORT CIRCUITS

You've probably heard the term **short circuit** before. A short circuit happens when the current flowing through the components doesn't follow the course we expect it to. Instead, the current finds another path, a shorter one, between the terminals of the power source. This is why we call this path a short circuit. Because there is less opposition to the flow of electrons, there is a larger current. Often the current through the new (short) path is so large that the wires or components can't handle it. When this happens, the wires

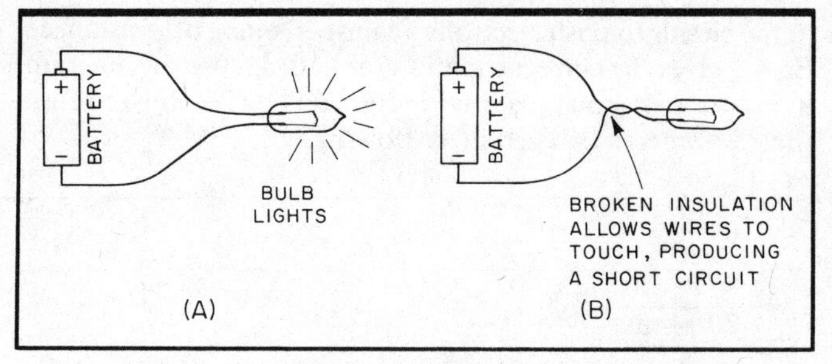

Figure 4-10—Part A shows a light bulb in a working circuit. In B, the insulation covering the wires has broken, and the two wires are touching, so we have a short circuit.

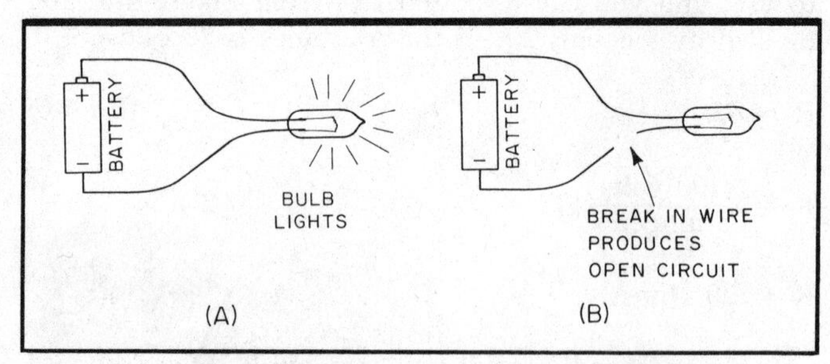

Figure 4-11—Part A shows a light bulb in a working circuit. In B, one wire has broken, preventing the current from flowing through the bulb. This is an example of an open circuit.

and components can be damaged. Figure 4-10 shows a bare wire causing a short circuit.

Some people think that a short circuit occurs when there is no resistance in the connection between the positive and negative terminals of the power supply. Actually there will always be *some* resistance but it may be such a small amount that it can be ignored.

In the extreme case, if a short circuit develops in our house wiring, the wire may overheat and can even start a fire. This is why it is important to have a properly rated fuse connected in series with a circuit. We will describe fuses in more detail in Chapter 5.

The opposite of a short circuit is an **open circuit**. In an open circuit the current is interrupted, just as it is when you turn a light switch off. The switch breaks (opens) the circuit, putting a layer of insulating air in the way so no current can flow. This break in the current path presents an extremely high resistance. An open circuit can be good, as when you throw the on/off switch to off. An open circuit can be bad if it's an unwanted condition caused by a broken wire or a bad component. Figure 4-11 illustrates an open circuit. Sometimes the resistance in the circuit path is so large that it is impractical to measure it. When this is true, we say that there is an *infinite* resistance in the path, creating an open circuit.

[At this time you should turn to Chapter 12 and study the questions with numbers that begin 2E-10 and 2E-11. If you have any difficulty with these questions, review this section about open and short circuits.]

ENERGY AND POWER

Energy can be defined as the ability to do work. An object can have energy because of its position (like a rock ready to fall off the edge of a cliff). An object in motion also has energy (like the same rock as it falls to the bottom of the cliff). In electronics a power supply or battery is the source of electrical energy. We can make use of that energy by connecting the supply to a light bulb, a radio or other circuit.

We discussed voltage drops in the last section. As electrons move through a circuit and go through resistances, there is a voltage drop. These voltage drops occur because energy is "used up" or "consumed." Actually, we can't lose the electrical energy; it is just changed to some other form. The resistor heats up because of the current through it; the resistance converts electrical energy to heat energy. More current produces still more heat, and the resistor becomes warmer. If too much current flows, the resistor might even catch fire!

As electrons flow through a light bulb, the resistance of the bulb converts some electrical energy to heat. The filament in the bulb gets so hot that it converts some of the electrical energy to light energy. Again, more current produces more light and heat.

You should get the idea that we can "use up" a certain amount of energy by having a small current go through a resistance for a long time or by having a larger current go through it for a shorter time. When you buy electricity from a power company, you pay for the electrical energy that you use. You might use all of the power in one day, or use a small amount every day. It doesn't matter to the power company. The electric meter on your house just measures how much energy you use and the power company sends someone around to read the meter each month.

Sometimes it is important to know how fast a circuit can use energy. You might want to compare how bright two different light bulbs will be. If you're buying a new freezer, you might want to know how much electricity it will use in a month. You will have to know how fast the freezer or the light bulbs use electrical energy. We use the term **power** to define the rate of energy consumption. The basic unit for measuring power in the metric system is the **watt**. You have probably seen this term used to rate electrical appliances. You know that a light bulb rated at 75 watts will be brighter than one rated at 40 watts. (Sometimes we abbreviate watts with a capital W.)

The basic unit of power is the watt. This unit is named after James Watt (1736-1819), the inventor of the steam engine.

In electronics, there is a very simple way to calculate power. The equation is:

$$P = IE \qquad \text{(Eq 4-7)}$$

where

P is power, measured in watts
I is the current in amperes
E is the EMF in volts

To calculate the power in a circuit, multiply amperes times volts. For example, suppose a 12-V battery is pushing 3 A of current through a light bulb that is operating normally. Using Eq 4-7, we can find the power rating for this light bulb.

$$P = IE = 3\text{ A} \times 12\text{ V} = 36\text{ W}$$

If you know the power in a system and the voltage applied to the circuit, you can compute the current. Or, if you know power and current, you can find voltage. In other words, if you know any two parts you can find the third. Figure 4-12 shows a diagram to help you with calculations like these, or we can write the equations:

$$I = \frac{P}{E} \qquad \text{(Eq 4-8)}$$

$$E = \frac{P}{I} \qquad \text{(Eq 4-9)}$$

If you apply a 10-V EMF to a 20-W circuit, divide 20 by 10 to find that there is a 2-A current.

$$I = \frac{P}{E} = \frac{20\text{ W}}{10\text{ V}} = 2\text{ A}$$

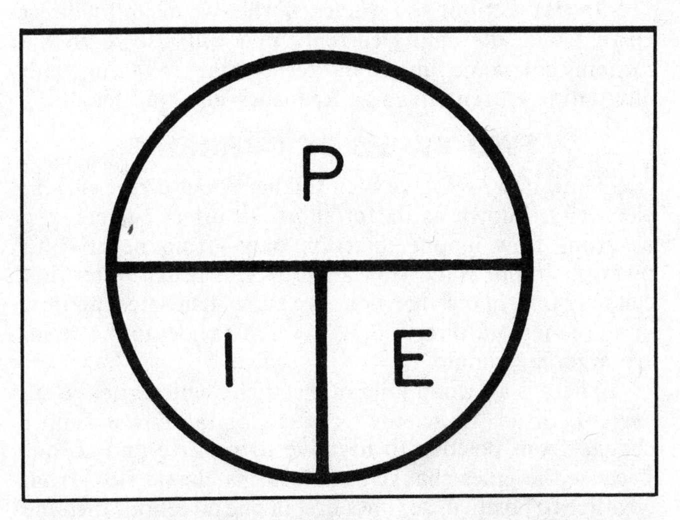

Figure 4-12—A simple diagram to help you remember the power equation relationships. To find any quantity given the other two, simply cover the unknown quantity with your hand or a piece of paper. The remaining two symbols will indicate if you have to multiply (if they are side by side) or divide (if they appear one over the other as a fraction).

If a 5-A current flows in a 60-W circuit, the EMF is 12 V. To find voltage you simply divide power by current.

$$E = \frac{P}{I} = \frac{60\ W}{5\ A} = 12\ V$$

Suppose you turn on a single light. One-half ampere of current flows through the bulb with 120 volts applied. What's the power rating of the light bulb? Eq 4-7 shows us how to find this power. Multiply 120 volts times ½ (0.5) ampere to calculate that the light is a 60-watt bulb.

$$P = IE = 0.5\ A \times 120\ V = 60\ W$$

By the way, the E and I in the power equation are the same E and I in Ohm's Law. So if we know any two of the four quantities voltage, current, resistance and power, we can calculate the other two.

Remember that a light bulb converts electrical energy into heat and light. The bulb has resistance; it opposes the flow of electrons. Is it possible to find the resistance of the wire inside the light bulb?

Yes. We know the voltage applied to the wire and the amount of current through it. This means that we can use Ohm's Law to calculate the resistance of the wire in the light bulb. Eq 4-6 gives the form of Ohm's Law that we need to find this answer. To calculate the resistance inside the bulb, divide 120 V by 0.5 A.

$$R = \frac{E}{I} = \frac{120\ V}{0.5\ A} = 240\ \Omega$$

The bulb is the equivalent, then, of a 240-ohm resistor in the circuit.

[Before you go on to the last section in this Chapter, turn to Chapter 12 and study those questions with numbers that begin 2E-8 and 2E-9. Review this section if you have any problems.]

DIRECT AND ALTERNATING CURRENT

In this section, you will learn what we mean by direct current and alternating current. You will also learn the meaning of some important terms that go along with alternating current, such as frequency and wavelength.

TWO TYPES OF CURRENT

Until now, we have been talking about **direct current** electricity, known as **dc** for short. In direct current, the electrons flow in one direction only—from negative to positive. In our water-flow analogy, this is like water that can only flow in one direction. We know that water can flow in more than one direction, however. The tides in the ocean are a good example.

There is a second kind of electricity called **alternating current**, or **ac**. In ac, the terminals of the power supply change from positive to negative to positive and so on. Because the poles change and electrons always flow from negative to positive, ac flows first in one direction, then the other. The current alternates in direction.

We call one complete round trip a *cycle*. The **frequency** of the ac is the number of complete cycles, or alternations, that occur in one second. We measure frequency in hertz (abbreviated Hz). One cycle per second is 1 Hz. 150 cycles per second is 150 Hz. One thousand cycles per second is one kilohertz (1 kHz). One million cycles per second is one megahertz (1 MHz).

The basic unit of frequency is the hertz. This unit is named in honor of Heinrich Rudolf Hertz (1857-1894). This German physicist was the first person to demonstrate the generation and reception of radio waves.

[Now study the questions in Chapter 12 that begin with 2E-12-1, 2E-12-2 and 2E-12-3. Review this section if you have difficulty answering any of these questions.]

MORE AC TERMINOLOGY

Batteries provide direct current. To make an alternating current from this direct-current source, you would have to switch the polarity of the voltage source rapidly. Imagine trying to turn the battery around so the plus and minus terminals changed position very rapidly. This would not be a very practical way to produce ac! You must have a power supply where the polarity is constantly changing. The terminals must be positive and negative one moment, and

WHY USE ALTERNATING CURRENT IN OUR HOMES?

Why do power companies use alternating current in the power lines that run to your home? The main reason is so they can use transformers to change the voltage. This allows the company to use an appropriate voltage for each part of their distribution system. In this way the power company can minimize the power losses in the transmission lines. The generator at the power station produces ac by moving a wire (actually many turns of wire) through a magnetic field in an alternator. The resulting output has a relatively low voltage. Why don't the power-companies send this directly through the power lines to your house? At first, this seems like a good idea. It would eliminate the many transformers and power stations that often clutter our landscape.

The answer can be found in Ohm's Law. Even a very good conductor, such as the copper used in the power company's high-voltage lines, has a certain amount of resistance. This factor becomes very important when we consider the very long distances the generated electricity must travel.

Remember that the voltage drop across a resistance is given by the formula $E = IR$, where I is the value of current and R is the value of resistance. If we can reduce either the resistance of the wire or the value of the current through the wire, we can reduce the voltage drop. The resistance of the wire is relatively constant, although we can reduce it somewhat by using a very large diameter wire. If we increase the voltage, a smaller current will be required for the same power transfer from the generating station to your home.

Using a very high voltage also provides more "overhead." If the power company starts with 750,000 volts, and the voltage has dropped to 740,000 volts by the time it reaches the first substation, they just use a transformer rated for 740,000-V input to give the desired output. If they send 50,000 volts on to the next substation, there is still plenty of overhead. By the time it gets to the power lines outside your house, the voltage has dropped to around 3000 volts. A pole transformer then steps it down to 240 volts to supply power to your house. This voltage is normally split in half to provide two 120-V circuits to your house.

then negative and positive, constantly switching back and forth.

The power company has a more practical way to create ac: they use a large machine called an **alternator**. The ac supplied to your home goes through 60 complete cycles each second. Thus, the electricity from the power company has a frequency of 60 Hz.

In ac electrical circuits, current builds slowly to a peak flow in one direction, then reverses to build to a peak flow in the opposite direction. If you plot these changes in current on a graph, you get a gentle up-and-down curve. We call this curve a **sine wave**. The ac voltage applied to a simple circuit also varies in this same manner. The voltage gradually builds to a maximum voltage in one direction, then decreases to zero and gradually increases to a peak in the opposite direction (or with the opposite polarity). Figure 4-13 shows several cycles of a sine-wave ac signal.

Alternating current can do things direct current can't. For instance, a 120-V ac source can be increased to a 1000-V source with a **transformer**. Transformers can change the value of an ac voltage, but not a dc voltage. The power company supplies 120-V ac to your house.

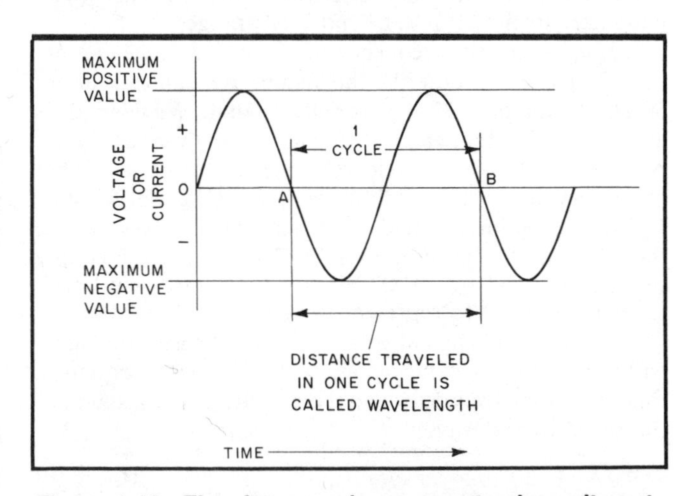

Figure 4-13—The sine wave is one way to show alternating current. Let's follow one cycle starting on line "0" at point A, indicated near the center of the graph. The wave goes in a negative direction to its most negative point, then heads back up to zero. After the wave goes through zero, it becomes more and more positive, reaches the positive peak, then goes back to zero again. This is one full cycle of alternating current.

Frequency and Wavelength

We discussed the frequency of an ac signal earlier. From this discussion, you must realize that alternating currents and voltages can change direction at almost any rate imaginable. Some signals have low frequencies, like the 60-Hz-ac electricity the power company supplies to your house. Other signals have higher frequencies, for example, radio signals can alternate at more than several million hertz.

When we talk about such a wide range of frequencies, it is common to break the wide range into several smaller ranges. One common breaking point is the difference between **audio-** and **radio-frequency** signals. If you connected an ac signal having a frequency anywhere between 20 Hz and 20,000 Hz (20 kHz) to a loudspeaker, you would hear a sound. Because these signals can produce sounds, they are called audio-frequency (AF) signals. The higher the frequency of the signal, the higher the pitch of the sound you would hear. (Caution—DO NOT connect the 60-Hz power from a household receptacle to a speaker even though 60 Hz is in the audio-frequency range! You may be seriously injured or KILLED by the voltage of this signal!)

Not all people can hear this full range of signals from 20 Hz to 20 kHz. Some people hear the low frequencies better than the high frequencies, and others hear the high frequencies better. This is the general range of frequencies that humans can expect to hear, however. (Dogs can hear signals at a much higher frequency, which is why dog whistles, used for training, don't produce a sound you can hear.)

If a signal has a frequency above the audio-frequency range (20 kHz), we call it a radio-frequency (RF) signal. Signals in the RF range can also be broken into smaller groups, such as very-low frequency (VLF), high frequency (HF), very-high frequency (VHF), ultra-high frequency (UHF) and so on. Don't worry about the names of these ranges for your exam. You will probably hear the terms as you listen in on other hams' discussions, though. The Novice bands are in the HF, VHF and UHF ranges.

If we know the frequency of an ac signal, we can use that frequency to describe the signal. We can talk about 60-Hz power or a 3745-kHz radio signal. **Wavelength** is another quality that can be associated with every ac signal. As its name implies, wavelength refers to the distance that the wave will travel through space in a single cycle. All such signals (sometimes called electromagnetic waves) travel though space at the speed of light, 300,000,000 meters per second (3.00×10^8 m/s). We use the lower-case Greek letter lambda (λ) to represent wavelength.

The faster a signal alternates, the less distance the signal will be able to travel during one cycle. There is an equation that relates the frequency and the wavelength of a signal to the speed of the wave:

$$c = f\lambda \qquad \text{(Eq 4-10)}$$

where

c is the speed of light, 3.00×10^8 meters per second
f is the frequency of the wave in hertz
λ is the wavelength of the wave in meters

We can solve this equation for either frequency or wavelength, depending on which quantity we want to find.

Table 4-2
Novice-Band Frequencies and Wavelengths

Frequency Range (megahertz)	*Approximate Wavelength (meters)*
3.7— 3.75	80
7.1— 7.15	40
21.1— 21.2	15
28.1— 28.5	10
222.1— 223.91	1.25
1270 —1295	0.23 (23 centimeters)

$$f = \frac{c}{\lambda} \qquad \text{(Eq 4-11)}$$

and

$$\lambda = \frac{c}{f} \qquad \text{(Eq 4-12)}$$

From these equations you may realize that as the frequency increases the wavelength gets shorter. As the frequency decreases the wavelength gets longer. Suppose you are transmitting a radio signal on 7.125 MHz. What is the wavelength of this signal? We can use Eq 4-12 to find the answer. First we must change the frequency to hertz: 7.125 MHz = 7,125,000 Hz.

$$\lambda = \frac{c}{f} = \frac{3.00 \times 10^8 \text{ m/s}}{7.125 \times 10^6 \text{ Hz}}$$

$$\lambda = \frac{300{,}000{,}000 \text{ m/s}}{7{,}125{,}000 \text{ Hz}} = 42 \text{ meters}$$

Of course, you already knew that this frequency was in the 40-meter Novice band, so this answer should not surprise you.

As another example, what is the wavelength of a signal that has a frequency of 3.725 MHz? (3.725 MHz = 3,725,000 Hz.)

$$\lambda = \frac{c}{f} = \frac{3.00 \times 10^8 \text{ m/s}}{3.725 \times 10^6 \text{ Hz}}$$

$$\lambda = \frac{300{,}000{,}000 \text{ m/s}}{3{,}725{,}000 \text{ Hz}} = 80.5 \text{ meters}$$

Even if you have trouble with this arithmetic, you should be able to learn the frequency and wavelength relationships for the six Novice bands, as shown in Table 4-2.

[Congratulations! You have made it all the way through the most difficult chapter in this book. You are now well on your way to knowing all the electronics you will need to pass your Novice exam. Before you move on to Chapter 5, turn to Chapter 12 and study the questions that begin 2E-12-4, 2E-12-5 and 2E-13. Don't hesitate to come back to this chapter to review any sections that you are still a little uncertain about.]

KEY WORDS

Battery—A device that converts chemical energy into electrical energy.

Chassis ground—The common connection for all parts of a circuit that connect to the negative side of the power supply.

Double-pole, double-throw (DPDT) switch—A switch that has six contacts. The DPDT switch has two center contacts. The two center contacts can each be connected to one of two other contacts.

Earth ground—A circuit connection to a cold-water pipe or to a ground rod driven into the earth.

Fuse—A thin strip of metal mounted in a holder. When too much current passes through the fuse, the metal strip melts and opens the circuit.

Grid—The control element (or elements) in a vacuum tube.

Potentiometer—Another name for a variable resistor. The value of a potentiometer can be changed without removing it from a circuit.

Resistor—A circuit component that controls current through a circuit.

Rotary switch—A switch that connects one center contact to several individual contacts. An antenna switch is one common use for a rotary switch.

Semiconductor—Material that has some properties of a conductor and some properties of an insulator.

Schematic symbol—A drawing used to represent a circuit component on a wiring diagram.

Single-pole, double-throw (SPDT) switch—A switch that connects one center contact to one of two other contacts.

Single-pole, single-throw (SPST) switch—A switch that only connects one center contact to another contact.

Solid-state devices—Circuit components that use semiconductor materials. Semiconductor diodes, transistors and integrated circuits are all solid-state devices.

Switch—A device used to connect or disconnect electrical contacts.

Triode—A vacuum tube with three active elements: cathode, plate and control grid.

Chapter 5

Circuit Components

Before we look at the operation of electronic circuits, let's discuss some basic information about the parts that make up those circuits. This chapter presents the information about circuit components that you need to know for your Novice exam. You will find descriptions of several types of fuses, switches, resistors and semiconductor devices. We combine these components with other devices to build practical electronic circuits.

Every circuit component has a **schematic symbol**. A schematic symbol is nothing more than a drawing used to represent a component. We use these symbols when we are making a circuit diagram, or wiring diagram, to show how the components connect for a specific purpose. You will learn the schematic symbols for the circuit components discussed in this chapter. As you discover more about electronics, you will learn how these symbols can be used to illustrate practical circuit connections.

RESISTORS

Resistors are important components in electronic circuits. We talked about the concept of resistance in Chapter 4. A resistor opposes the flow of electrons. We can control the electron flow (the current) by varying the resistance in a circuit.

Most resistors have standard fixed values, so they can be called fixed resistors. Variable resistors, also called **potentiometers**, allow us to change the value of the resistance without removing and changing the component. Potentiometers are used as the volume and tone controls in most stereo amplifiers. Figure 5-1 shows two types of fixed resistors, a potentiometer and their schematic symbols.

[Now turn to Chapter 12 and study those questions with numbers that begin 2F-1. Review this section if you have any problems.]

SWITCHES

How do you control the lights in your house? What turns on your car radio? A **switch**, of course.

The simplest kind of switch just connects or disconnects a single electrical contact. Two wires connect to the switch; when you turn the switch on, the two wires are connected. When you turn the switch off, the wires are disconnected. This is called a **single-pole, single-throw switch**. It connects a single pair of wires (single pole) and has only two positions,

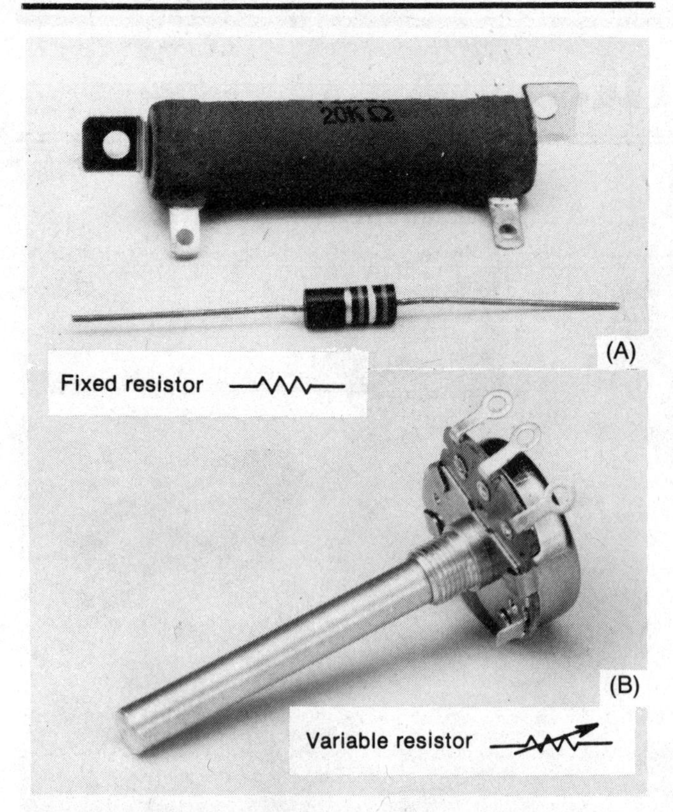

Figure 5-1—Fixed resistors come in many standard values. Most of them look something like the ones shown at A. Variable resistors (also called potentiometers) are used wherever the value of resistance must be adjusted after the circuit is complete. Part B shows a potentiometer.

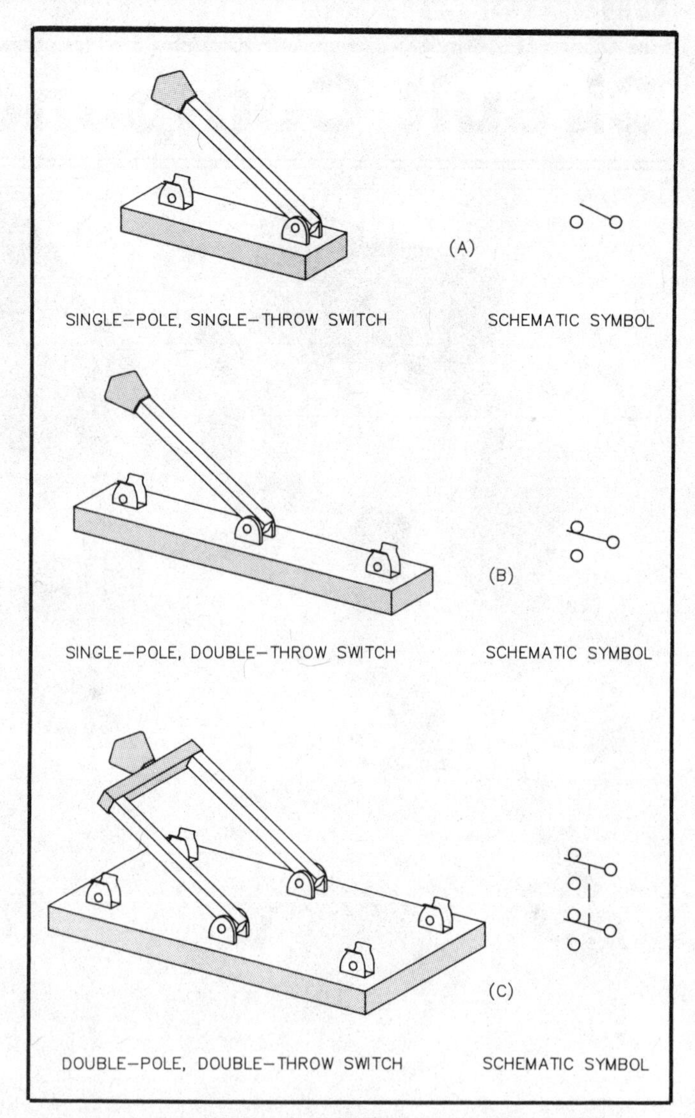

Figure 5-2—A single-pole, single-throw (SPST) switch can connect or disconnect one circuit. A single-pole, double-throw (SPDT) switch can connect one center contact to one of two other contacts. A double-pole, double-throw (DPDT) switch is like two SPDT switches in one package. Each half of the DPDT switch can connect one contact to two other contacts.

on or off (single throw). Sometimes we abbreviate single pole, single throw as **SPST**.

If we want to control more devices with a single switch, we need more contacts. If we add a second contact to a single-pole, single-throw switch we can control a second device or select between two devices. This kind of switch is called a **single-pole, double-throw switch**. Sometimes we abbreviate single pole, double throw as **SPDT**. An SPDT switch connects a single wire (single pole) to one of two other contacts (double throw). The switch connects a center wire to one contact when the switch is in one position. When you flip the switch to the other position, the switch connects the center wire to the other contact.

We can add even more contacts to the switch. A **double-pole, double-throw switch** has two sets of three contacts. We use the abbreviation **DPDT** for double pole, double throw. You can think of a DPDT switch as two SPDT switches in the same box with their handles connected together. A DPDT switch has two center contacts. The switch connects each of these two center contacts (double pole) to one of two other contacts (double throw). Figure 5-2 shows SPST, SPDT and DPDT switches and their schematic symbols.

All these switches are very useful, but they only connect a center contact to one or two other contacts. What if we want to connect a single contact to *several* other contacts? We might want to use one switch to connect our transmitter to several different antennas. We can do this with a **rotary switch**. As its name implies, a rotary switch turns around a central shaft to connect one center contact to several outer contacts.

Switches like this can have many contacts. They can also have more than one center contact. We specify the particular kind of switch by the number of contacts (positions) it has around the outside, and by the number of center contacts and switch arms (poles) it has. Figure 5-3 shows a photograph of a two-pole, five-position rotary switch; it has five separate contacts and two rotary arms.

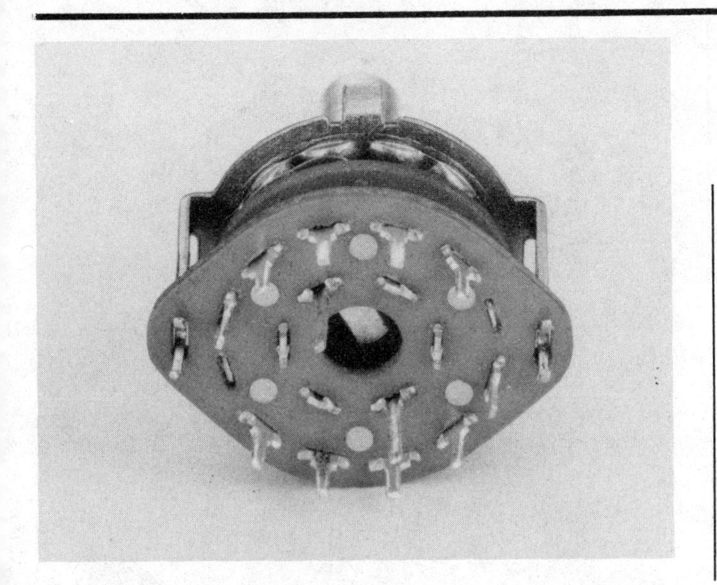

Figure 5-3—A rotary switch can connect one wire to several contacts. Most antenna switches use a rotary switch to connect a transceiver to several antennas. This photograph shows a two-pole, five-position switch; it can connect each center contact to five outside contacts.

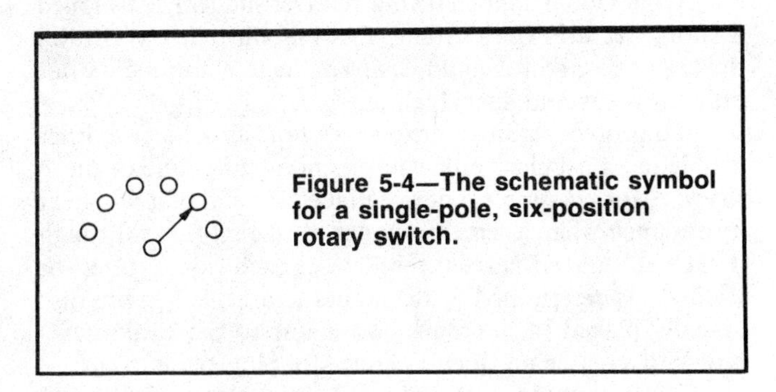

Figure 5-4—The schematic symbol for a single-pole, six-position rotary switch.

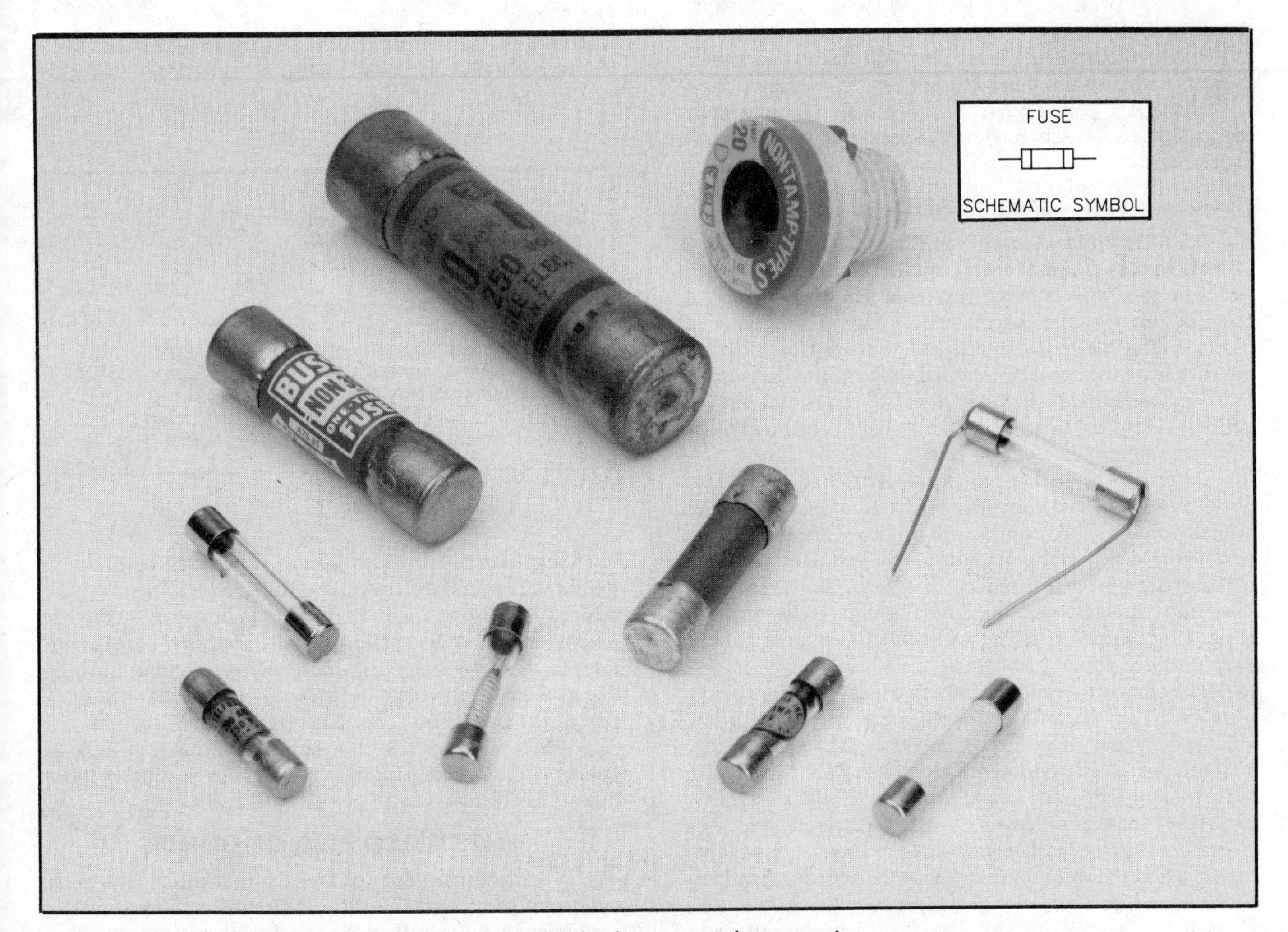

Figure 5-5—Some common fuses. Fuses protect circuits from excessive current.

The schematic symbol in Figure 5-4 is a single-pole, six-position rotary switch.

[You should now turn to Chapter 12. Study those questions that have numbers that begin 2F-2. Review this section if you have any problems.]

FUSES

What would happen if your receiver suddenly developed a short circuit? The current, flowing without opposition through the circuit, could easily damage components not built to withstand such high current.

To protect against unexpected short circuits and other problems, most electronic equipment includes one or more fuses. A **fuse** is simply a device made of metal that will heat up and melt when a certain amount of current flows through it. The amount of current that causes each fuse to melt (or "blow") is determined by the manufacturer. When the fuse (usually placed in the main power line to the equipment) blows, it creates an open circuit, stopping the current.

Fuses come in all shapes and sizes. Figure 5-5 shows some of the more common fuse types and the schematic symbol for a fuse. A fuse in a transistor radio using little power may be designed to blow at 500 mA. The fuses for your home's 120-V circuits may be designed to blow at 15 or 20 A. The principle is the same: When excessive current flows through the fuse it melts, creating an open circuit to protect your equipment. Remember that fuses are designed to protect against too much current, not too much voltage.

[To check your understanding of this section, study exam question 2F-3.1. Review this section if you have any problems.]

BATTERIES

We talked about batteries in Chapter 4. Simply put, a **battery** changes chemical energy into electrical energy. When we connect a wire between the terminals of a battery, a chemical reaction takes place inside the battery. This reaction produces free electrons, and these electrons flow through the wire from the negative terminal to the positive terminal. Batteries come in all shapes and sizes, from tiny ones for hearing aids to the large battery that starts your car. Figure 5-6 shows some different batteries.

Figure 5-6—Batteries come in all shapes and sizes. A battery changes chemical energy into electrical energy.

Figure 5-7—Some small batteries contain only one cell. We use the symbol at A for a single-cell battery. Manufacturers add several cells together in series to produce more voltage. At B, the schematic symbol for a multiple-cell battery.

+ −
(A)
SINGLE CELL

+ −
(B)
MULTIPLE—CELL BATTERY

Batteries are made up of small *cells*. Each cell has a positive electrode and a negative electrode. The cells produce a small voltage. The voltage a cell produces depends on the chemical process taking place inside the cell. Rechargeable nickel-cadmium cells produce about 1.2 volts per cell. Common zinc-acid and alkaline flashlight cells produce about 1.5 volts per cell. The lead-acid cells in a car battery each produce about 2 volts.

The number of cells in a battery depends on the voltage we want to get out of the battery. If we only need a low voltage the battery may contain only one cell. Small hearing-aid batteries usually only contain one cell. Part A of Figure 5-7 shows the schematic symbol for a single-cell battery. The two lines in the schematic symbol represent the two electrodes in the cell. The long line represents the positive terminal and the short line represents the negative terminal.

To produce a battery with a higher voltage, several cells must be connected in series so that their outputs add together. Each cell produces a small voltage, and the battery manufacturer connects several cells in series to produce the desired battery voltage. Part B of Figure 5-7 shows the schematic symbol for a multiple-cell battery. We use several lines to show the many cells in the battery. Again, the long line at one end represents the positive terminal, and the short line at the other end represents the negative terminal.

[Make a quick trip to Chapter 12 now and look at questions 2F-4.1 and 2F-4.2. Review this section if those questions confuse you.]

ANTENNAS AND GROUNDS

When you use your receiver or transmitter, you must connect it to an antenna. You should also connect all the equipment in your station to a good earth ground. We'll go over these connections later, in Chapter 9. There are really two kinds of ground connections: **chassis ground** and **earth ground**. The metal box that your radio is built on is called a *chassis*. Most manufacturers use the chassis as a common connection for all the places in the circuit that connect to the negative side of the power supply. This common connection is called the chassis ground. The chassis ground has a special schematic symbol.

To keep your station safe, you should also connect the chassis ground to a ground rod driven into the ground or to a cold water pipe. This connection is called an earth ground because it goes into the earth. An earth ground has a different schematic symbol. Figure 5-8 shows the symbol for a chassis ground and an earth ground.

Your receiver won't hear signals and no one will hear your transmitter without an antenna. The antenna is a very important part of any radio installation. We use the symbol shown in Figure 5-9 to represent the antenna on a schematic diagram.

[Now turn to Chapter 12 and study those questions with numbers that begin 2F-5 and 2F-6. If you have any trouble answering the questions, review this section.]

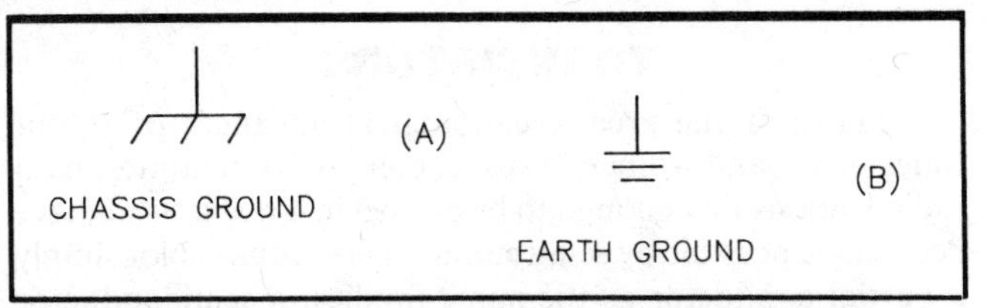

Figure 5-8—The metal chassis of a radio is sometimes used to make a common ground connection for all the circuit points that connect to the negative side of the power supply. We use the symbol at A to show those connections on a schematic. Your radio equipment should be connected to a ground rod or a cold water pipe for safety. We use the symbol shown at B to show an earth-ground connection.

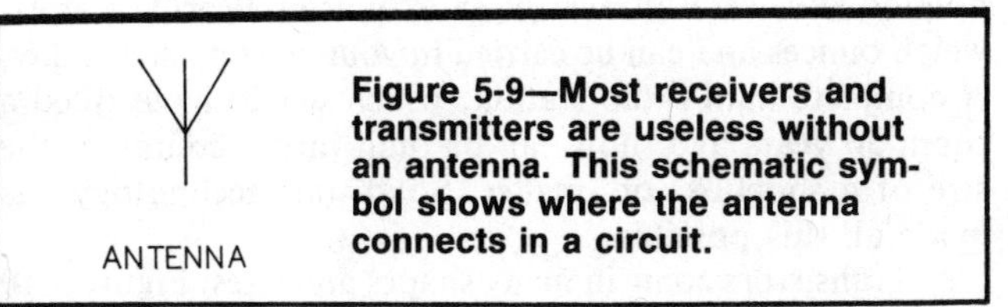

Figure 5-9—Most receivers and transmitters are useless without an antenna. This schematic symbol shows where the antenna connects in a circuit.

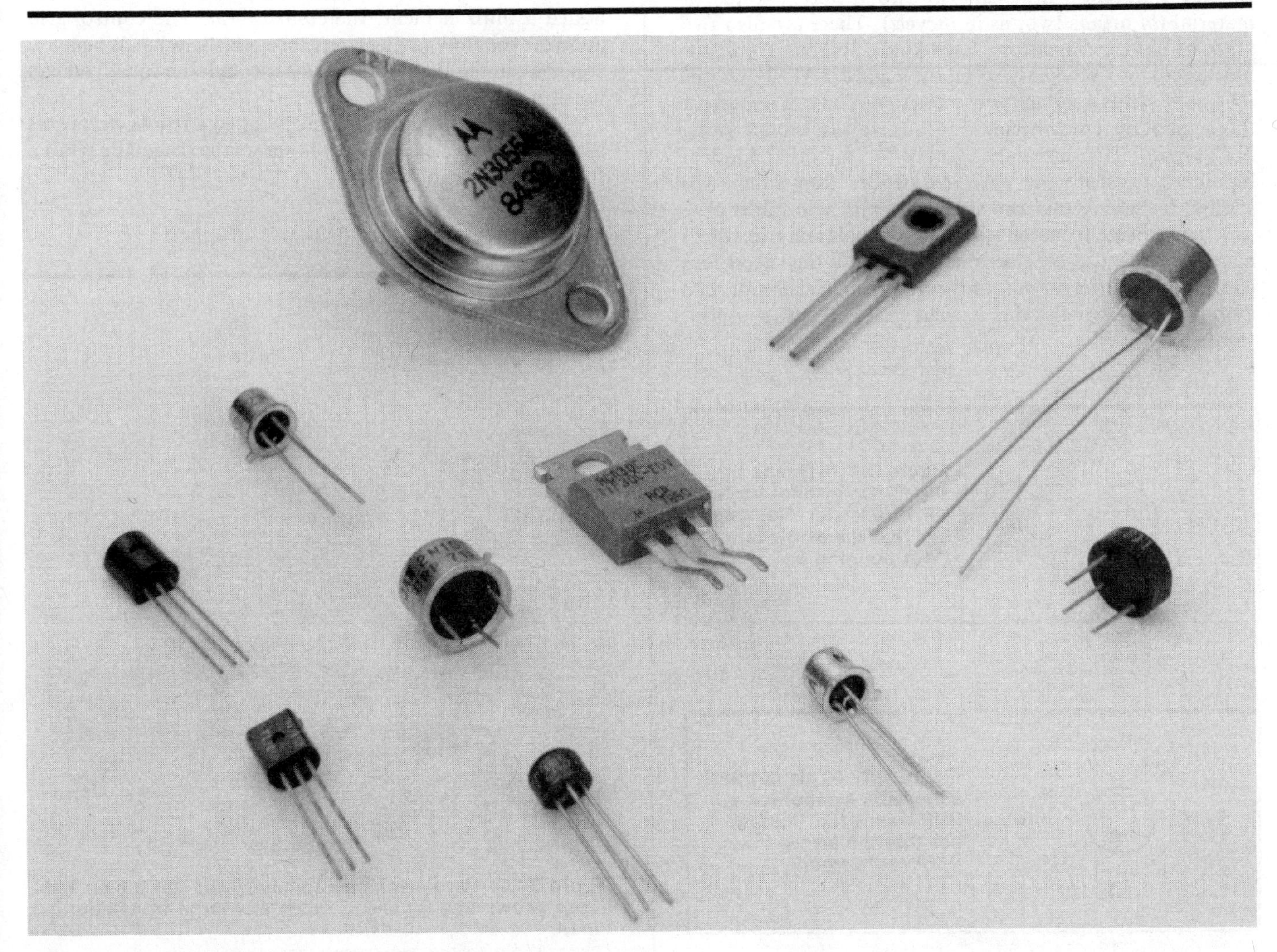

Figure 5-10—Transistors are packaged in many different cases.

TRANSISTORS

Many of the great technological advances of recent times—men and women in space, computers in homes, ham radio stations tiny enough to be carried in a pocket—all have been made possible by **semiconductor** electronics. Not simply a partial conductor as the name implies, a semiconductor has some of the properties of a conductor and some properties of an insulator.

Diodes and transistors are two types of semiconductors, the **solid-state devices** that have replaced vacuum tubes in most uses. Most semiconductor devices are much smaller than comparable tubes, and they produce less heat. Semiconductors are also usually less expensive than tubes.

Portable broadcast-band radios, which weighed several pounds and were an armful shortly after World War II, weigh ounces and can be carried in your shirt pocket today. A complete ham radio station, which would have filled a room 50 years ago, now can be built into a container the size of a shoebox, or smaller. Solid-state technology has made all this possible.

Transistors come in many shapes and sizes. Figure 5-10 shows some of the more common case styles for transistors. The most common type of transistor is the **bipolar transistor**. Bipolar transistors are made of two different kinds of material (*bi* means two, as in *bi*cycle). There are also two kinds of bipolar transistors. Each kind of bipolar transistor has a separate schematic symbol. Figure 5-11 shows the schematic symbol for an *NPN transistor*. You can remember this symbol by remembering that the arrow is "*n*ot *p*ointing i*n*." Figure 5-12 shows the symbol for the other kind of bipolar transistor, the *PNP transistor*. Remember this symbol by saying that the arrow "*p*oints i*n* *p*roudly."

You can see from the schematic symbols that transistors have three leads, or electrodes. Each of the electrodes connects to a different part of the transistor. Transistors can amplify small signals; this is what makes them so useful. Usually, we use a low-level signal applied to the *base* of the transistor to control the current through the *collector* and *emitter*.

[Time for another look at the question pool. Study the questions in Chapter 12 with numbers that begin 2F-7. Review this section if you have problems.]

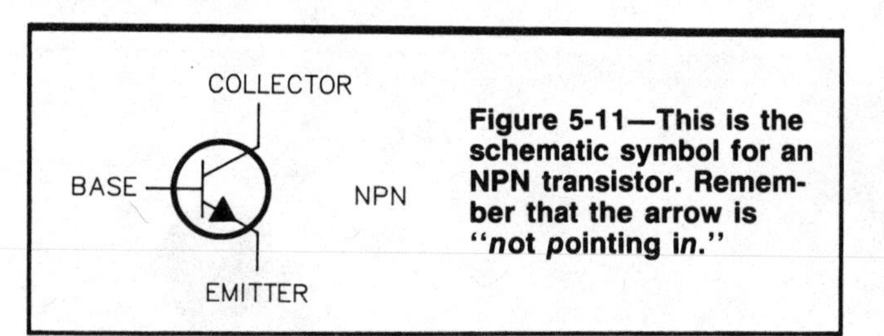

Figure 5-11—This is the schematic symbol for an NPN transistor. Remember that the arrow is "*n*ot *p*ointing i*n*."

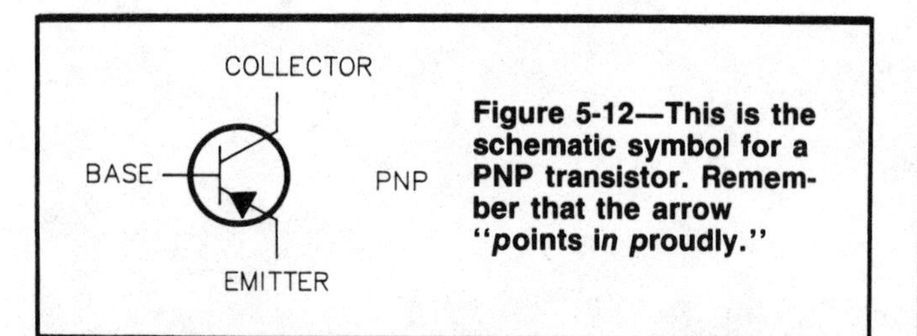

Figure 5-12—This is the schematic symbol for a PNP transistor. Remember that the arrow "*p*oints i*n* *p*roudly."

VACUUM TUBES

The development of the vacuum tube was an important milestone in the history of radio. The vacuum tube was the first active electronic device—that is, the vacuum tube can amplify, or produce an enlarged version of the input signal.

From the outside, a tube looks like a glass bulb with pins sticking out of the bottom. Sometimes there are also leads coming out of the top or sides. Some tubes have a metal collar or band around the base, and some tubes have ceramic or metal envelopes (outer shell). Tubes are quite fragile, and will break easily if mishandled. They usually plug into a socket wired into a circuit. Figure 5-13 shows some common tubes.

Tubes are named for the number of elements they have inside them. All tubes have at least two elements, the plate and the cathode. A tube with only two elements is not very useful, however. Tubes became really useful when inventors added a third element, the *control grid*. The control grid controls the flow of electrons through the tube. When you can control the flow of electrons through the tube, you can use it as an amplifier.

A tube with three elements is called a **triode** (*tri* means three, as in *tri*cycle). Figure 5-14 shows the schematic symbol for a triode vacuum tube.

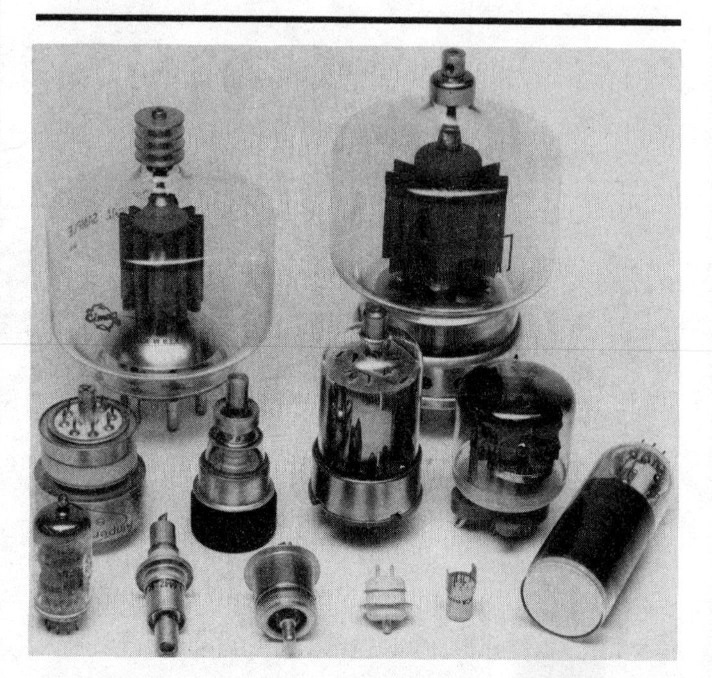

Figure 5-13—Here are some common vacuum tubes. This figure shows tiny receiving tubes and large transmitting tubes.

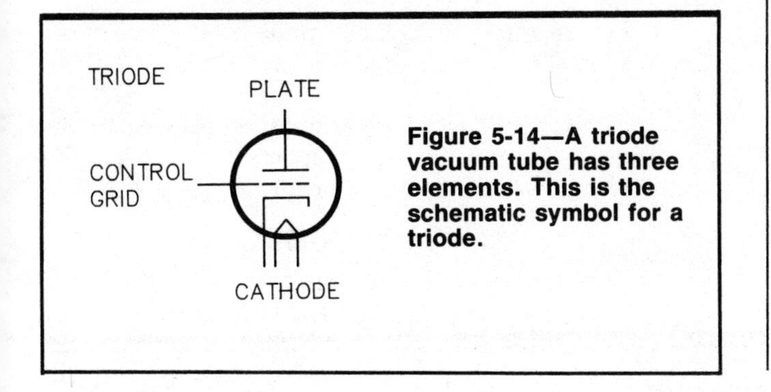

Figure 5-14—A triode vacuum tube has three elements. This is the schematic symbol for a triode.

While transistors have replaced tubes in many modern applications, tubes are still used in many types of electronic circuits. Radio and television sets using tubes are still in operation, and the ability of tubes to handle high power and high voltage makes them ideal for use in the output stage of an amateur transmitter.

[Turn to Chapter 12 and study question 2F-8.1. That's it for this chapter! By now, you should have a basic understanding of how all of the circuit components included on the Novice exam work. If you had trouble with any of the related questions in the question pool, review those sections before you proceed to the next chapter.]

KEY WORDS

Antenna switch—A switch used to connect one transmitter, receiver or transceiver to several different antennas.

Block diagram—A picture using boxes to represent sections of a complicated device or process. The block diagram shows the connections between sections.

Digital communications—Amateur Radio transmissions that are designed to be received and printed automatically. Also, transmissions used for the direct transfer of information from one computer to another.

Electronic keyer—A device that generates Morse code dots and dashes electronically.

Hand key—A simple switch used to send Morse code.

Impedance-matching network—A device that matches the impedance of an antenna system to the impedance of a transmitter or receiver. Also called an antenna-matching network or Transmatch.

Modem—Short for modulator/demodulator. A modem modulates a radio signal to transmit data and demodulates a received signal to recover transmitted data.

Microphone—A device that converts sound waves into electrical energy.

Packet radio—A system of digital communication where information is broken into short bursts. The bursts ("packets") also contain addressing and error-detection information.

Power supply—A circuit that provides a direct-current output at some desired voltage from an ac input voltage.

Radioteletype (RTTY)—Radio signals sent from one teleprinter machine to another machine. Anything that one operator types on his teleprinter will be printed on the other machine.

Receiver—A device that converts radio signals into audio signals.

SWR meter—A measuring instrument that can indicate when an antenna system is working well.

Teleprinter—A machine that can convert keystrokes (typing) into electrical impulses. The teleprinter can also convert the proper electrical impulses back into text. Hams use teleprinters for radioteletype work.

Terminal node controller—A TNC accepts information from a computer and converts the information into packets. The TNC also receives packets and extracts information to be displayed by a computer.

Transceiver—A radio transmitter and receiver combined in one unit.

Transmit-receive switch (TR switch)—A device that allows you to connect one antenna to a receiver and a transmitter. The switch connects the antenna to the receiver or transmitter as you operate the switch.

Transmitter—A device that produces radio-frequency signals.

Chapter 6

Practical Circuits

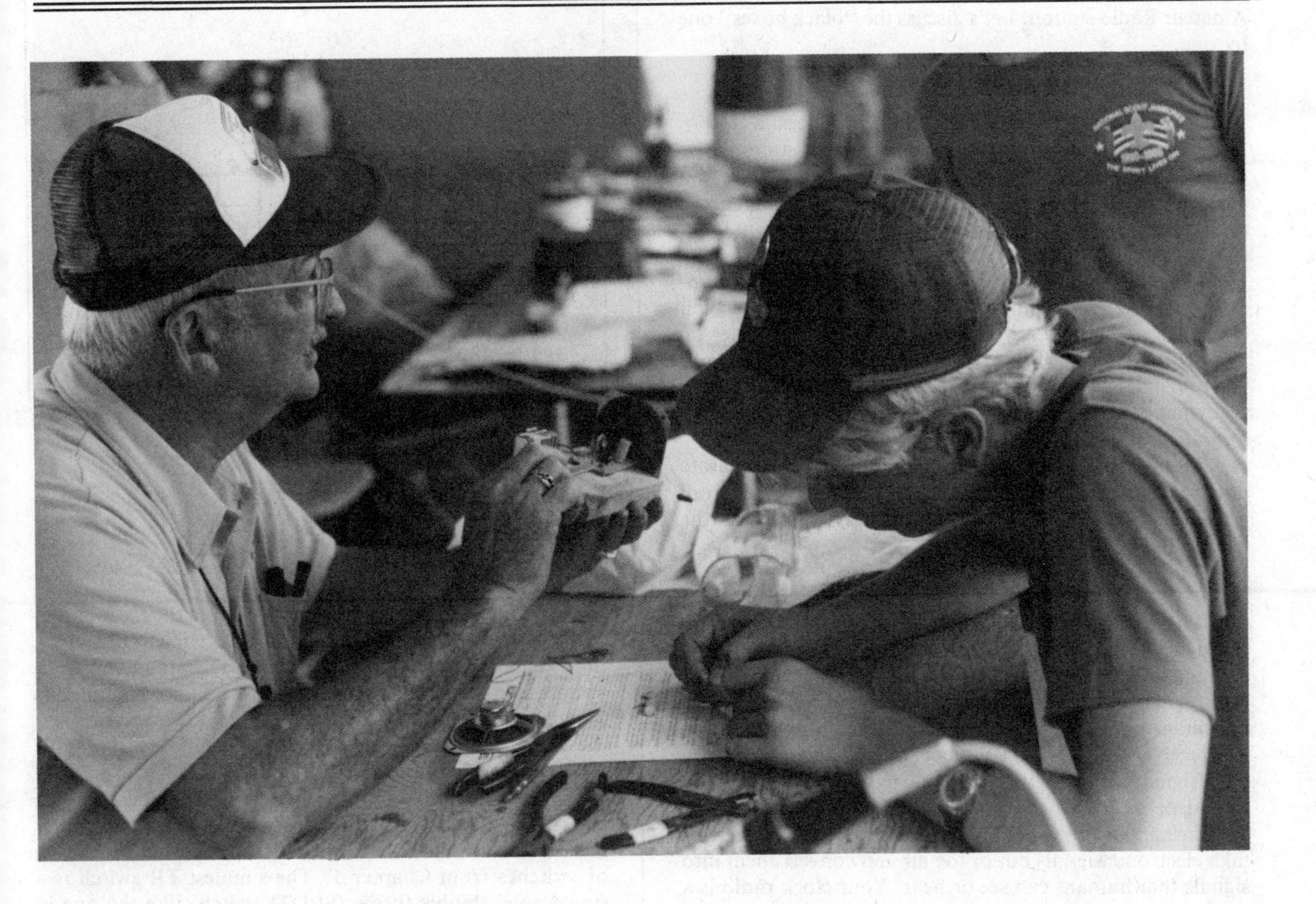

In the last chapter, you learned about some of the components that make up electronic circuits. In this chapter, we'll introduce some of the basic equipment that goes into an Amateur Radio station. We will briefly cover equipment like receivers, transmitters and antenna switches. We'll show you how to connect equipment to make a fully functional ham radio station.

Some of the terms may be confusing at first. Don't worry if you don't understand everything. Most of the ideas presented in this chapter are reinforced later in the book. We'll talk more about connecting your station equipment in Chapter 10.

Throughout this chapter we use **block diagrams**. In a block diagram, each part of a station is shown as a "black box." The diagram shows how all the boxes connect to each other. Study the block diagrams carefully and remember to turn to Chapter 12 when we send you there.

BASIC STATION LAYOUT

Figure 6-1 shows a block diagram of a very simple Amateur Radio station. Let's discuss the "black boxes" one by one.

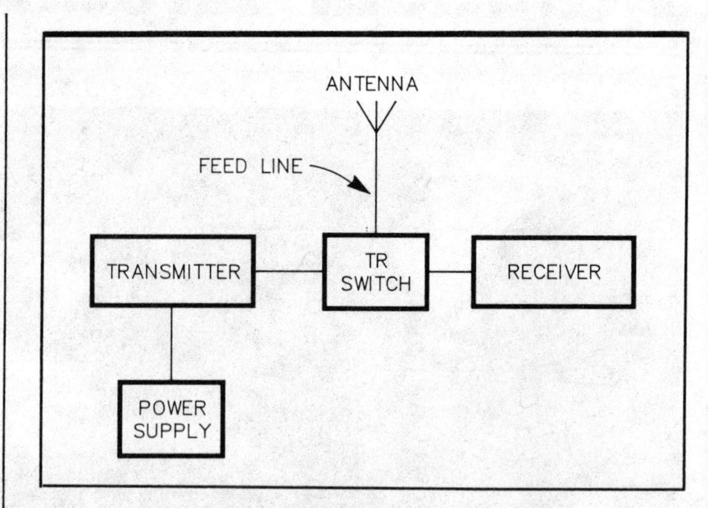

Figure 6-1—A TR switch connects between the transmitter, receiver and antenna.

TRANSMITTERS

A **transmitter** is a device that produces a radio-frequency (RF) signal. Television and radio stations use powerful transmitters to put their signals into the air. Radio amateurs use lower-powered transmitters to send signals to each other. A transmitter produces an electrical signal that can be sent to a distant receiver.

We call the signal from a transmitter the *radio-frequency carrier*. To transmit international Morse code, you could use a telegraph key as a switch to turn the carrier on and off in the proper code pattern. If you want to transmit voice signals, you need extra circuitry in the transmitter to add voice content to the carrier. We call this extra circuitry a *modulator*.

Many modern amateur transmitters have a separate **power supply**. The power supply converts the 120 V ac from your wall sockets into 12 V dc (usually) to power the transmitter.

RECEIVERS

The transmitter is a sending device. It sends a radio-frequency (RF) signal to a transmitting antenna, and the antenna radiates the signal into the air. Some distance away, the signal produces a voltage in a receiving antenna. That ac voltage goes from the receiving antenna into a **receiver**. The receiver converts the RF energy into an audio-frequency (AF) signal. You hear this AF signal in headphones or from a loudspeaker.

Just about everyone is familiar with receivers. Receivers take electronic signals out of the air and convert them into signals that humans can see or hear. Your clock radio is a receiver and so is your television set. If you look around the room you're in right now, you'll probably see at least one receiver. The receiver is a very important part of an Amateur Radio station.

TRANSCEIVERS

In many modern Amateur Radio stations, the transmitter and receiver are combined into one box. We call this combination a **transceiver**. It's really more than just a transmitter and receiver in one box, though. Some of the circuits in a transceiver are used for both transmitting and receiving. Transceivers generally take up less space than a separate transmitter and receiver.

SWITCHES

Okay. Now we have the basic parts of our amateur station. The guys at the radio club gave you an old receiver, transmitter and power supply. You bring them home and set them on your desk. Now what?

Well, you know that you need to connect an antenna to the receiver if you want to hear signals. You also have to connect the antenna to the transmitter when you send out a signal. But you only have one antenna. What should you do?

You could disconnect the antenna from the receiver and connect it to the transmitter. You'd have to do this every time you switch over from receive to transmit and again to switch from transmit to receive. Most radio gear has its antenna connectors on the rear panel, however. Besides, these connectors aren't designed to make it easy to remove and reconnect cables rapidly.

What you need is a **transmit-receive switch** (also known as a **TR switch**). We use a TR switch to connect one antenna to a receiver and a transmitter. Remember our discussion of switches from Chapter 5? The simplest TR switch is a single-pole, double-throw (SPDT) switch, like the one in Figure 6-2. The antenna connects to the center arm, and the receiver and transmitter connect to the outside contacts. If you throw the switch one way, the transmitter connects to

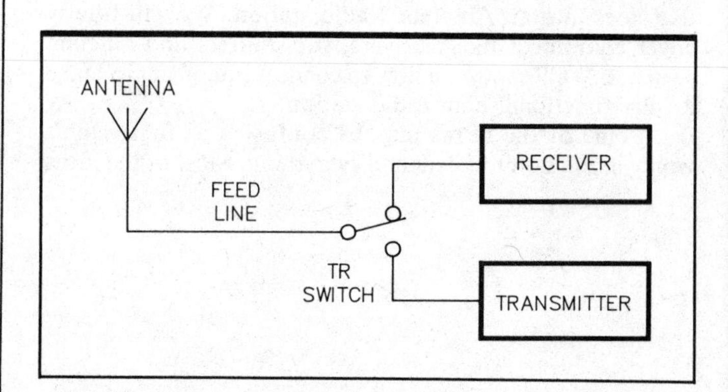

Figure 6-2—The simplest TR switch is an SPDT switch with the center arm connected to the antenna. One position connects the antenna to the receiver; the other position connects the antenna to the transmitter.

the antenna. The other switch position connects the receiver to the antenna.

Many TR switches use relays to switch the antenna. A relay uses a magnetic coil to make or break contacts. You can think of a relay as a remotely operated switch. When you throw a switch at your operating position, current passes through the relay coil and the relay arm switches from one set of contacts to another. A transmit-receive relay may have several contacts. Extra contacts can be used to quiet the receiver in the transmit mode or to switch accessory devices.

THE SIMPLEST STATION

The block diagram in Figure 6-1 shows how we connect all this equipment together. The power supply connects to the transmitter. The receiver and transmitter both connect to the TR switch. The TR switch connects the transmitter and receiver to the antenna, one at a time. This is just about all you need for the most basic Amateur Radio station.

CONNECTING MANY ANTENNAS

What if you have more than one antenna? Again, you could disconnect the antenna from your transmitter or receiver and reconnect another feed line. This can be very inconvenient. A simpler technique is to use an **antenna switch.** We mentioned antenna switches in Chapter 5. An antenna switch connects one transmitter, receiver or transceiver to several antennas. You can switch from one antenna to another with a simple flick of the switch.

The antenna switch connects at the point where the feed lines from all the antennas come into the station. See Figure 6-3. An antenna switch connects one receiver, transmitter or transceiver to one of several antennas.

IMPEDANCE-MATCHING NETWORKS

Another useful accessory that you will see in many ham shacks is an **impedance-matching network**. This device may let you use one antenna on several bands. The matching network may also allow you to use your antenna on a band it is not designed for. Sometimes we call the impedance-matching network an *antenna tuner* or *Transmatch*. All of these names indicate the impedance-matching network's main function. The network matches (tunes) the impedance of the load (the antenna and feed line) to the impedance of your transmitter. We usually connect the impedance-matching network right where the antenna comes into the radio station. See Figure 6-4.

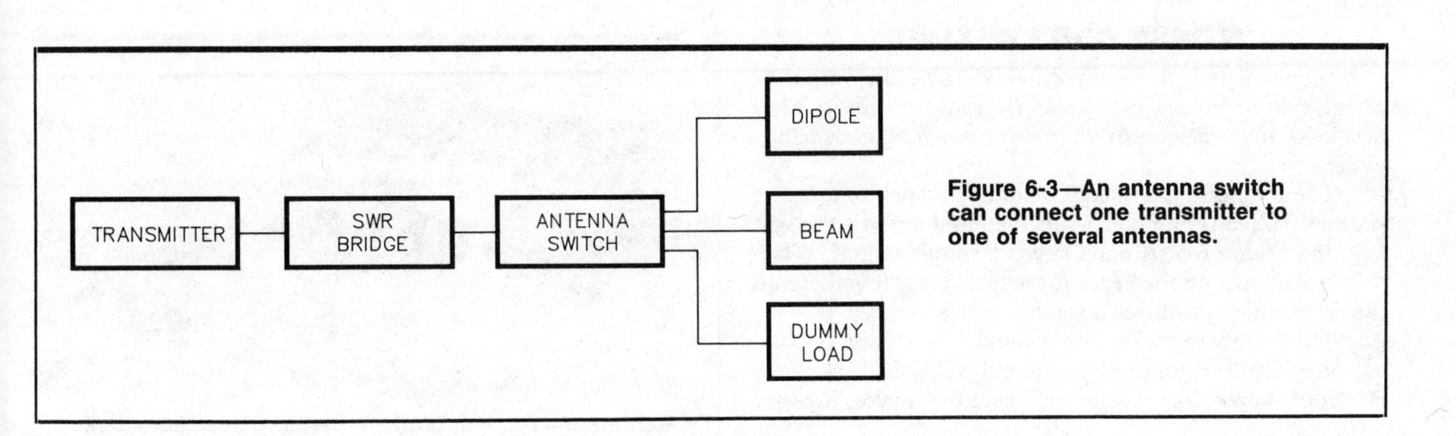

Figure 6-3—An antenna switch can connect one transmitter to one of several antennas.

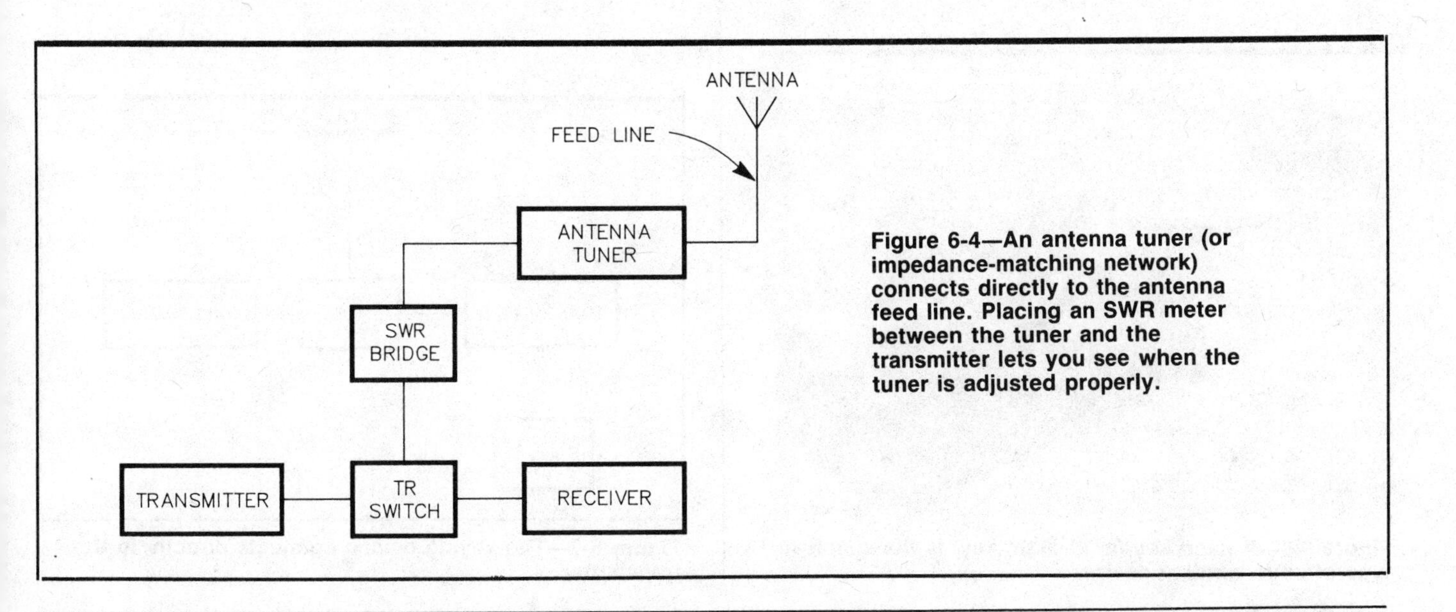

Figure 6-4—An antenna tuner (or impedance-matching network) connects directly to the antenna feed line. Placing an SWR meter between the tuner and the transmitter lets you see when the tuner is adjusted properly.

MONITORING THE SYSTEM

Another thing you may want to add to your station is an **SWR meter**. This device is also called an SWR bridge. The SWR meter measures something called the *standing-wave ratio*. You don't need to know too much about SWR in this chapter. We will go over it in more detail later in the book.

Standing-wave ratio is a good indicator of how well your antenna system is working. If you install an SWR meter in your station, you can keep an eye out for problems with your antenna. If you spot the problems early you can head them off before they damage your equipment.

The SWR meter can be connected at several points in your station. One good place to connect the meter is between the antenna switch and transceiver. See Figure 6-4. If you use a separate receiver and transmitter, you can connect the SWR meter between the TR switch and the rest of the antenna system. An SWR meter can also be used between an impedance-matching network and the transmitter. The SWR meter then indicates when the matching network is adjusted properly.

[Now turn to Chapter 12. Study those questions with numbers that begin 2G-1-1 and 2G-1-2. Review this section if you have any problems.]

STATION ACCESSORIES

So far, we have been talking about very basic station layout. We showed you how to connect a transmitter, receiver and antenna switch together to make a simple station. To communicate effectively, you will need a few simple accessories. Let's look at what you need.

MORSE CODE KEYING

Morse code is transmitted by switching the output of a transmitter on and off. Inventive radio operators have developed many devices over the years to make this switching easier.

The simplest kind of code-sending device is one you're probably already familiar with: the **hand key**, or straight key. See Figure 6-5. A hand key is a simple switch. When you press down on the key, the contacts press together and the transmitter produces a signal.

The code you make with a hand key is only as good as your "fist" or your ability to send well-timed code. An **electronic keyer**, like the one in Figure 6-6, makes it easier to send well-timed code. You must connect a *paddle* to the keyer. The paddle has two switches, one on each side. When you press one side of the paddle, one of the switches closes

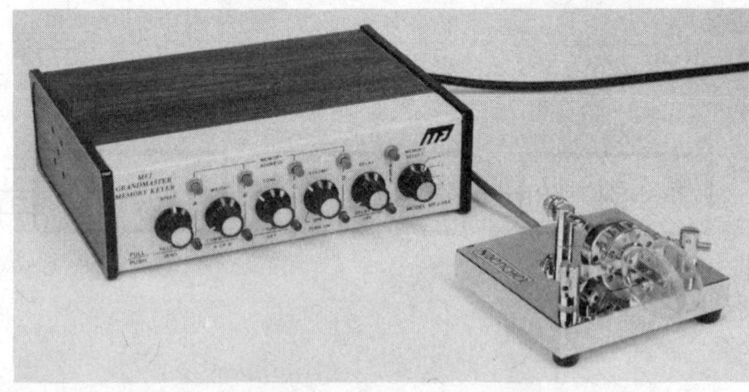

Figure 6-6—You can produce perfect code characters with an electronic keyer and a little practice.

Figure 6-5—A hand key (or straight key) is the simplest type of code-sending device.

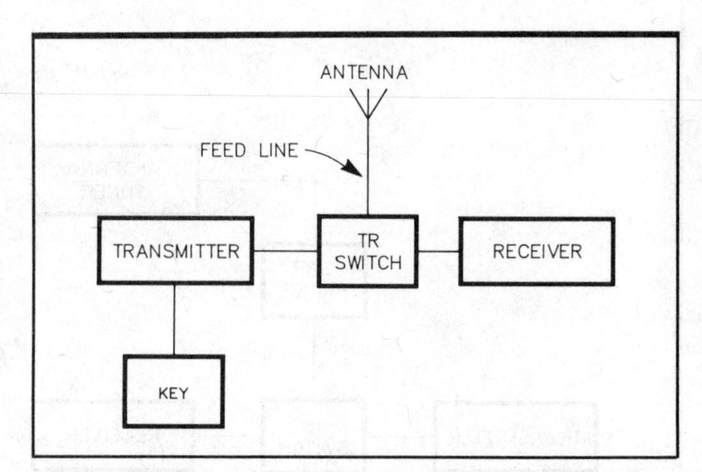

Figure 6-7—The keying device connects directly to the transmitter.

and the keyer sends a continuous string of dots. When you press the other side of the paddle, the keyer sends dashes. With a little practice and some rhythm, you can send perfectly timed code with a keyer. You may want to start out with a hand key, however. Using a hand key can help you develop the rhythm you need to send good code. When you can send good code with a hand key, you're ready to try a keyer.

Both the hand key and the electronic keyer connect directly to the transmitter. The key in the block diagram in Figure 6-7 is connected to the transmit section of the transceiver.

MICROPHONES

If you want to transmit voice, you'll need a **microphone**. A microphone converts sound waves into electrical signals that can be used by a transmitter. All voice transmitters require a microphone of some kind. Like a code key, the microphone connects directly to the transmitter. The microphone in Figure 6-8 is connected to the transmit section of the transceiver.

[Time for a trip to Chapter 12. Study those questions with numbers that begin 2G-2 and 2G-3. Review this section if you have problems.]

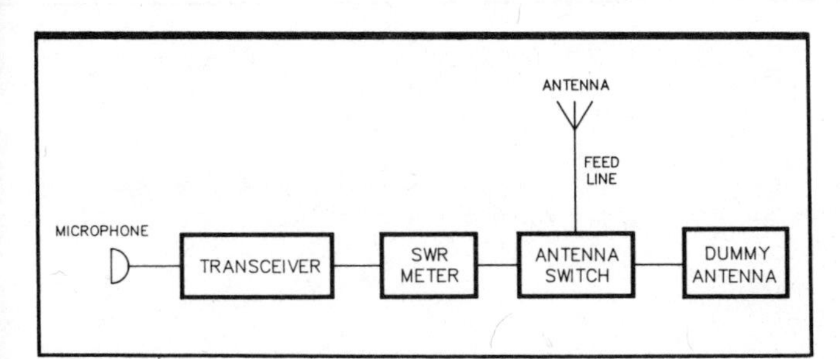

Figure 6-8—A microphone also connects directly to the transmitter. You will need a microphone to transmit voice.

DIGITAL COMMUNICATIONS

Digital communications are Amateur Radio transmissions that are designed to be received and printed automatically. Digital communications often involve direct transfer of information between computers. When you type information into your computer, the computer (with the help of some accessory equipment) processes the information. The computer then sends the information to your transmitter and the transmitter sends it out over the air. The station on the other end receives the signal, processes it and prints it out on a computer screen or printer. Digital communications have become a popular form of ham-radio communication. Here we will talk about setting up a station for two popular kinds of digital communications: **radioteletype** and **packet radio**.

Radioteletype

Radioteletype (RTTY) communications have been around for a long time. You may have seen big noisy **teleprinter** machines in old movies. A teleprinter is something like an electric typewriter. When you type on the teleprinter keys, the teleprinter sends out electrical codes that represent the letters you are typing. If we send these codes to another teleprinter machine, the second machine will reproduce everything you type. Hams have been converting this equipment and using it on the air for years. You can also send and receive radioteletype with a computer. These days computers are so cheap and readily available that they have just about replaced the old noisy teleprinter machines.

We use a **modem** for Amateur Radio digital communications. Modem is short for *mo*dulator-*dem*odulator. The modem accepts information from your computer and uses the information to modulate a transmitter. The modulated transmitter produces a signal that we send out over the air. When another station receives the signal, the other station uses a similar modem to demodulate the signal. The modem then passes the demodulated signal to a computer. The computer processes and displays the signal.

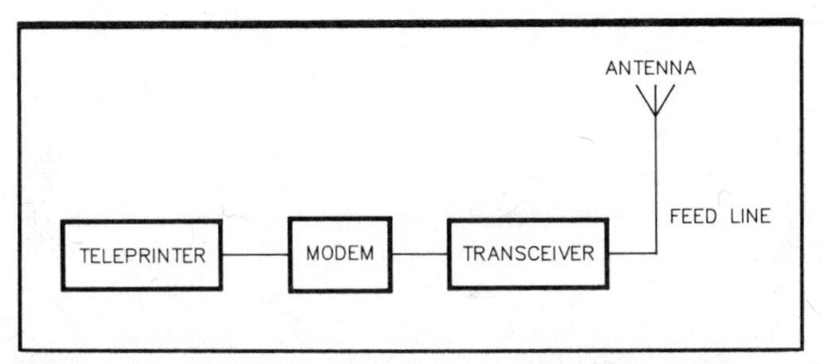

Figure 6-9—A typical radioteletype station. The modem connects between the transceiver and the teleprinter.

Sometimes hams use an older teleprinter instead of a computer. The teleprinter converts and displays information from the modem. A complete radioteletype station must have a computer or teleprinter, a modem and a transmitter. The modem connects between the computer and the transmitter, as shown in Figure 6-9.

Packet Radio

Packet radio is a very popular form of digital communications. This mode uses a **terminal node controller (TNC)** as an interface between your computer and transceiver. We might call a TNC an "intelligent" modem. The TNC accepts information from your computer and breaks the data into small pieces called "packets." Along with the information from your computer, each packet contains addressing, error-checking and control information.

The addressing information includes the call signs of the station sending the packet and the station the packet is being sent to. The address may also include call signs of stations that are being used to relay the packet. The receiving station uses the error-checking information to determine whether the received packets contain any errors. If the received packet contains errors, the receiving station asks for a retransmission. The retransmission and error checking continue until the receiving station gets the packet with no errors.

Breaking up the data into small parts allows several users to share a channel. Packets from one user are transmitted in the spaces between packets from other users. The addressing information allows each user's TNC to separate packets for that station from packets intended for other stations. The addresses also allow packets to be relayed through several stations before they reach their final destination. The error-checking information in each packet assures perfect copy.

A TNC connects to your station the same way a modem does. The TNC goes between the radio and the computer, as shown in Figure 6-10.

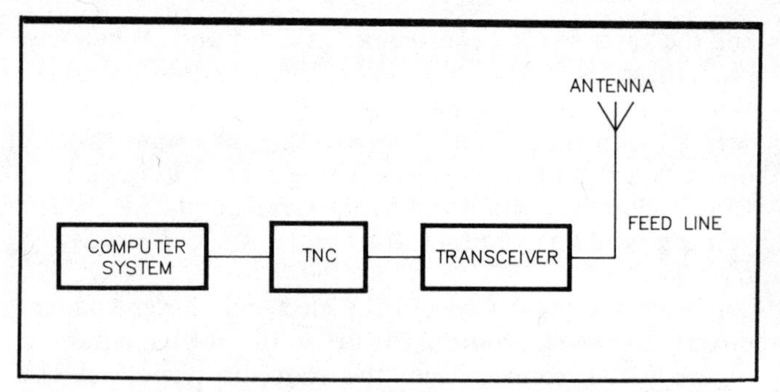

Figure 6-10—In a packet radio station, the terminal node controller (TNC) connects between the transceiver and a computer.

That's all for this chapter. Don't worry if you're a bit confused by some of the terms we used in this chapter. We covered a lot of information fairly quickly. We'll go over a lot of the operating information later in the book. For now you should be familiar with the terms, and study the block diagrams carefully.

[Before you go on to Chapter 7, turn to Chapter 12 and study the questions with numbers that begin 2G-4 and 2G-5. If you have any problems, review this last section.]

KEY WORDS

Amplitude modulation (AM)—A method of combining an information signal and an RF carrier. In double-sideband voice AM transmission, we use the voice information to vary (modulate) the amplitude of a radio-frequency signal. Shortwave broadcast stations use this type of AM, as do stations in the Standard Broadcast Band (510-1600 kHz). Amateurs seldom use double-sideband voice AM, but a variation, known as single sideband, is very popular.

Band spread—A receiver quality used to describe how far apart stations on different nearby frequencies will seem to be. We usually express band spread as the number of kilohertz that the frequency changes per tuning-knob rotation. Note that band spread affects frequency resolution.

Bandwidth—A range of frequencies (in hertz) that will pass through a filter with little or no loss.

Beat-frequency oscillator (BFO)—A circuit in a receiver that provides a signal to the detector. The BFO signal mixes with the incoming signal to produce an audio tone for CW reception.

Continuous wave (CW)—Morse code telegraphy.

DX—Distance; foreign countries.

Frequency resolution—The space between markings on a receiver dial. The greater the frequency resolution, the easier it is to separate signals that are close together. Note that frequency resolution affects band spread.

General-coverage receiver—A receiver used to listen to both the shortwave-broadcast frequencies and the amateur bands.

Ham-bands-only receiver—A receiver designed to cover only the bands used by amateurs.

Lower sideband (LSB)—The common mode of single-sideband transmission used on the 40, 80 and 160-meter amateur bands.

Multimode transceiver—A VHF or UHF transceiver capable of SSB, CW and FM operation.

Offset—The slight difference in transmitting and receiving frequencies in a transceiver.

Receiver—A device that converts radio waves into sound.

Receiver incremental tuning (RIT)—A transceiver control that allows for a slight change in the receiver frequency without changing the transmitter frequency. Some manufacturers call this a Clarifier (CLAR) control.

Rig—The radio amateur's term for a transmitter, receiver or transceiver.

Selectivity—The ability of a receiver to separate two closely spaced signals.

Sensitivity—The ability of a receiver to detect weak signals.

Shack—The room where an Amateur Radio operator keeps his or her station equipment.

Single sideband (SSB)—A common mode for voice operation on the amateur high-frequency bands. This is a variation of amplitude modulation.

Ticket—The radio amateur's term for an Amateur Radio license.

Transceiver—A radio transmitter and receiver combined in one unit.

Transmitter—A device that produces radio-frequency signals.

Transmit-receive switch (TR switch)—A device used for switching between transmit and receive operation. The TR switch allows you to connect one antenna to a receiver and a transmitter. As you operate the switch, it connects the antenna to the correct unit.

Upper sideband (USB)—The common single-sideband operating mode on the 20, 17, 15, 12 and 10-meter HF amateur bands. Hams also use upper sideband on all the VHF and UHF bands.

Variable-frequency oscillator (VFO)—A circuit used to control the frequency of an amateur transmitter.

Chapter 7

Selecting Your Equipment

Having an amateur license without a station is a little like having a driver's license without a car. Without a station, your license is just another piece of paper. You can probably arrange to use a friend's equipment or operate from a club station when your license first arrives. Eventually, though, you'll want to assemble your own gear.

Most hams look with pride at the operating position they've carefully built over the course of their years as an amateur. When hams meet, conversation always turns to the station. Whenever hams visit each other, the "shack" is usually the first stop on the tour. Many amateurs who develop friendships on the air eventually swap pictures of their equipment. It's no wonder, really. Hams devote many hours to their hobby, and they spend most of these hours in their shacks. The shack is where a ham operates on the air, repairs equipment, makes improvements and experiments with new projects.

As a newcomer to Amateur Radio, you may find the wide assortment of equipment, antennas and available accessories confusing at first. You'll wonder, "Why is this antenna better than that one?" or "What features does this receiver offer? Do I really need them all?"

You have to decide what your goals are. Take a look at your available resources (how much money you have to spend). Do a little research and then choose the equipment that best suits your needs. Actually, selecting your station equipment can be very easy, if you know what you want it to do. This chapter will help you select a radio that will provide you with many hours of enjoyable operating.

LOOK BEFORE YOU LEAP

There is one very important thing you should do before you go out and purchase a room full of equipment. Find out how big the room is! Try to get some idea about how much space you will be able to dedicate to your shack. Chapter 9 has more detailed information about where to locate your station.

You should have some idea if you will be able to use an entire room for your radio station. Some hams use a corner of a bedroom or den, and some use a fold-out shelf in a closet. Available space may be an important consideration in what you actually purchase.

After you decide how much space you have, you're ready to select your equipment. There are many factors to consider. "Specs" (specifications) are very important, but don't choose your equipment on that basis alone. Some rigs are technical marvels but very difficult and frustrating for a new ham to operate. Other gear may look beautiful but have a lot of technical problems. The most important consideration is what works for you. Choose equipment that will be enjoyable and comfortable to operate.

Where should you look for this equipment? There are many different places. Check with local hams to see what kind of equipment they are using. Find out what they like about certain pieces of gear and what problems they have had. Learn from their experience. They are as proud of their shacks as you will be of yours. When you are trying to decide what you want, there is no substitute for sitting down and listening to a rig. (For the name of a local radio club, contact ARRL Headquarters.)

Another place to look is your local radio store. Check the telephone book for ham radio retailers in your area. Most large metropolitan areas have at least one ham radio retail store. If you visit the store, you will be able to see and compare several of the newest pieces of equipment. In addition, most ham stores have antennas set up so you can listen to a receiver you might buy.

The popular ham magazines carry many ads in each issue. When you look through any issue of *QST*, you can find ads from many manufacturers. Local dealers and companies who do a large volume of mail-order business also advertise in *QST*. The back section of this book includes ads from most manufacturers and many dealers. Many of the ads list features and specifications, and will give you a rough idea of what the equipment costs.

QST even has an advertising acceptance policy that can help protect you. If a piece of equipment doesn't live up to its manufacturer's claims, you won't see it advertised in *QST*. It's a good idea to study the ads and have some equipment in mind when you visit the local ham store.

BUYING USED EQUIPMENT

Many newcomers to Amateur Radio simply cannot afford to purchase new equipment. Even if you don't have to worry about money, you might not want to spend a lot at first. You might be asking yourself, "What if I don't enjoy ham radio?" or "What if I lose interest?" Buying used equipment is a good alternative to purchasing all new equipment for your station. You can still get good-quality gear, and pay quite a bit less than if you purchased it new.

Even if you decide to buy a new transceiver, think about buying used accessories. You can probably locate all the station accessories that you will *need* to make your station a joy to operate.

There was a time when almost all hams built their own rigs from spare parts. The cost was small, and the operators took great pride in their work. Many hams still build their own equipment as well, either from junk parts or as kits. We hope you will try building at least some part of your station as well. The satisfaction of being able to say, "I built it myself!" is a joy you will never forget.

Most Novices won't want to build a modern transmitter, receiver or transceiver, however. The electronics have gotten much more complicated over the years. This complexity can frustrate an inexperienced builder. If you have a limited ham radio budget, buy used equipment.

WHAT SHOULD I BUY?

The answer to that question depends largely on available resources and personal preferences. It seems to be a rule of Murphy's Law ("if anything can go wrong, it will!") that whenever you're looking for a particular item it cannot be found; then, when you least expect it, it appears!

The older issues of amateur magazines can provide a lot of information on equipment you may see on the used market. The New Products and Product Review columns in *QST* are a good start. Many local libraries carry back issues of *QST*. The old-timers in your radio club may also have some *QST*s. If you can't find the issues you need, you can purchase them from ARRL HQ.

Don't just rush out and buy the first piece of equipment you can find, however. Buying used radio equipment is much like buying a used car. You really should "kick the tires" a bit. Examine the equipment closely (as well as you can). Don't get snowed by salesmanship. A couple of old adages apply: "You (generally) get what you pay for" and "caveat emptor"—buyer beware. Perhaps we should add a third: "All things come to those who wait..." Have patience!

Surplus Equipment

There's still some WW II government surplus equipment on the market, and you may also see some newer surplus gear. Generally speaking, a beginner who doesn't have an Elmer should stay away from this equipment.

Much of the military equipment is big and heavy. You might also have to find some special connectors or build a power supply. Some units produce large amounts of television interference (TVI), and that's one thing we can all do without!

Many units that appear to be bargains will require quite a bit of work to get them on the air. Surplus equipment is simply not cost-effective for the beginner. Many hams do enjoy using surplus gear, however. You might try a surplus rig later, when you have gained some experience and want to do some tinkering.

Tube Availability

You might wonder about the practicality of buying tube-type equipment. After all, most modern equipment uses transistors and integrated circuits, and many companies have stopped making tubes. You don't need to worry; tubes are still available. Most electronic supply houses still stock tubes. Many amateurs have plenty of tubes in their "junk boxes" (often free for the asking). Finally, there are tube discount outlets that may have just what you need.

WHERE CAN I FIND USED GEAR?

There are many different sources of used ham gear. You can buy gear from equipment retailers and local hams. You may see what you want listed in the pages of ham magazines (see the *QST* Ham Ads) or used-equipment flyers. Other sources are auctions, flea markets, hamfests and even garage sales.

If you belong to an Amateur Radio club, ask your instructor and other members of the club. They can often provide you with some leads and other useful information about buying used gear. The club members can help you select a particular unit or a complete station. They can also help you test it to ensure it works properly. They will certainly provide some helpful hints on connecting all the pieces together so they operate efficiently.

One of the safest ways to purchase used equipment is to buy it from a local retailer. Often the dealer sets aside a section of the ham shop for used equipment. When you buy from such a source, there's usually some sort of warranty (30 days or so) on the equipment. If a problem comes up within that time, you can bring it back to the retailer for repairs.

Many dealers route the used equipment they've received through their service shop. This ensures that the gear is working properly before they place it on the used-equipment shelf. Some dealers will even allow you to see the equipment operate in the service shop before you buy it. This is the best way to determine the rig's capabilities and condition.

When you buy used gear from a local ham you generally have no guarantees. Ask to see the unit in operation. You might even wish to take along a more knowledgeable ham who can help you make the decision. A local ham can be your best source of information on used equipment.

As a beginner, you are at a disadvantage when it comes to buying used gear. You may already know what you want or what you need. The hard part is deciding which of those used rigs provides the best value for your hard-earned dollars. An "old-timer" can be an invaluable source of information. Chances are, he or she may have at one time owned a similar piece of equipment. With the seller's permission, your ham friend could operate the equipment and note any potential problems—your future headaches!

Local hamfests, flea markets and auctions are some of the better opportunities to see a quantity of used equipment. For the careful buyer, they may also be the source of some excellent equipment. But you have to know what you're looking for. What are the capabilities of the equipment, its current market value, cost of repairs and availability of repair parts?

At a hamfest or flea market, you may not be able to test the equipment before you buy it. A very thorough visual inspection by an experienced eye will generally suffice if the price is right, however. Again, help from an experienced amateur is a wise choice. You can usually tell a lot from the external appearance of the equipment. If it looks physically abused, chances are it has been abused electronically as well.

There are some simple precautions you should take if you're buying gear from another ham through the mail. Try to be sure that you'll have the right to return the equipment if you don't like what you receive. Shipment by truck freight with the right of inspection permits you to examine the package contents before you accept delivery. If you don't like what you receive, simply refuse delivery. You may be asked to pay by bank check or money order, rather than personal check.

There is one other very important point to keep in mind when you buy used equipment from any source. Be sure there is an owner's manual to go with the radio! Some hams may even have the service or shop manual to go with equipment they have for sale. A shop manual can be a valuable addition, because it generally has more complete service procedures and troubleshooting guidelines.

If you are not at least getting an owner's manual with the radio, be cautious about making the purchase. Manuals for some pieces of equipment are available from the manufacturer. There are also several companies that specialize in manuals for used or surplus equipment. Table 7-1 lists sources for many of the manuals you may need.

Table 7-1
Sources of Equipment Manuals

Collins Telecommunications Products Division
Rockwell International
Cedar Rapids, IA 52406

R.L. Drake Co
540 Richard St
Miamisburg, OH 45342
513-866-2421

Hallicrafter Manuals
Ardco Electronics†
PO Box 95, Dept Q
Berwyn, IL 60402

Hammarlund Manuals
Irving J. Abend
PO Box 426
Bergenfield, NJ 07621
201-384-7589

Hammarlund Manuals
Pax Manufacturing Co
Attn: Peter Kjelson
100 East Montauk Hwy
Lindenhurst, NY 11757

HI Manuals, Inc††
PO Box Q-802
Council Bluffs, IA 51502

Howard Sams Publications
4300 W 62nd St
Indianapolis, IN 46205
317-291-3100

ICOM America, Inc
2380 116th Ave NE
Bellevue, WA 98004
206-454-7619

Swan Division of Cubic Communications
304 Airport Rd
Oceanside, CA 92054
714-757-7525

Ten-Tec, Inc
Hwy 411 East
Sevierville, TN 37862
615-453-7172

Kenwood USA Corporation
2201 E Dominguez St
PO Box 22745
Long Beach, CA 90801-5745
213-639-4200

US Government Surplus
General Services Administration
National Archives and Record Services
Washington, DC 20408

US Government Surplus
Slep Electronics Co†
PO Box 100
Otto, NC 28763

Yaesu USA
17210 Edwards Rd
Cerritos, CA 90701
213-404-4847

†Write for manual prices (specify model).

††HI Manuals will not answer requests for information unless you include a $5 research fee. Order their catalog ($1) to see if they have the manual you need.

GOING SHOPPING

Once you've faced the decision of whether to buy new gear or used, you must think about a few more details. Don't worry. After you finish this chapter you'll know just what to look for when you go out to buy that first rig.

TRANSCEIVERS OR SEPARATES?

One of the first decisions you'll face is whether to use a **transceiver** or a separate **receiver** and **transmitter**. When you make this decision, keep your space limitations in mind. A transceiver, as the name implies, combines both a transmitter and receiver in one package. That means you can fit a transceiver in a smaller space than that required by separates. In fact, some of the newer transceivers require less than one foot of desk space. See Figure 7-1.

If size is unimportant, you'll want to consider the relative advantages of both transceivers and separates. In general, transceivers are easier to set up. Usually, you just connect a telegraph key, attach an antenna, plug in the ac power cord and you're ready to go.

With separates, you'll probably have to do a little wiring between the transmitter and receiver. You'll also need a **transmit-receive switch (TR switch)** to switch the antenna between the two units. These connections are easy to make, however, so don't be afraid of separates if you find something you like.

Transceivers are generally easier to operate than a separate receiver and transmitter. A single control usually sets both the transmit and receive frequency. With one frequency control to set, you're sure to be transmitting and receiving on the same frequency. Some manufacturers do offer "twins"—separates interconnected so they can act as a transceiver. See Figure 7-2.

There was a time when separate transmitter-receiver combinations offered better performance and flexibility than transceivers. The separate receivers were generally superior to those in transceivers. In addition, separates offered the flexibility of being able to transmit on one frequency while receiving on another. You may never need this capability as a Novice. Later, as a General or higher class ham, you may want to operate "split-frequency." This means listening to foreign (DX) stations in their part of the band while you transmit in your sub-band.

Modern technology, however, makes it possible to purchase high-performance transceivers. These transceivers are at least equal to, and sometimes better than, a separate receiver and transmitter. Most new transceivers have "memories" that allow you to switch quickly from one frequency to another. Many manufacturers offer external VFOs. Either of these features allows split-frequency operation. Figure 7-3 shows two modern, full-featured transceivers.

After a little looking, you'll probably decide that either a transceiver or separates will work well for you. Some transceivers offer the flexibility of separates, and some separates offer the convenience of transceivers. What then?

Figure 7-1—The Yaesu FT-747 is a good example of a trend toward less complicated transceivers. You can work the world with this rig and a good antenna.

(A)

(B)

Figure 7-2—(A) The Heath HR-1680 and HX-1681 "twins." The transmitter is a CW-only rig designed to cover portions of the 80, 40, 20, 15 and 10-meter bands. At B, a pair of Kenwood "twins" produced in the 1970s. The receiver is all solid-state. The transmitter uses solid-state devices except for the driver and final-amplifier stages, which use tubes.

(A)

(B)

Figure 7-3—Two modern transceivers with lots of "bells and whistles." (A) The Kenwood TS-930S. (B) The ICOM IC-761. All the knobs and switches on these rigs can be intimidating to a Novice. If you read the instruction manual carefully you should have no trouble understanding everything, however.

PRICE

An important consideration for most of us is price—what else is new! Don't be dismayed if you can't afford a new transceiver right away. You may be better off buying a good receiver first and then purchasing a transmitter when you can afford it. This way, you can at least listen to the ham bands. This will let you find out what you will be doing once you get on the air.

When you decide what kind of gear you want, you'll have to examine specific features. The next few sections describe some of the things to look for. Remember that many of the features explained here apply to transceivers as well as separate transmitters and receivers.

RECEIVERS

In many ways, your choice of a receiver will be the most important decision you'll have to make. The statement, "You have to hear 'em to work 'em" is especially true on today's crowded ham bands.

When you decide what equipment you want, you should probably purchase the best receiver you can afford. Even if you are on a limited budget, it makes sense to buy a good receiver. Then buy an inexpensive CW-only or CW/AM transmitter. This way you can get on the air with minimum cost. You won't be able to operate in the voice portion of 10 meters, but you'll still have fun on CW.

If you choose a good receiver, chances are it will last you for many years. A good receiver will serve you well no matter what class of license you achieve. Likewise, when you consider transceivers, pay close attention to the receiver section.

There are two basic types of receivers to choose from: ham-bands-only and general-coverage units. As the name implies, a **ham-bands-only receiver** covers only those frequencies where hams operate. A **general-coverage receiver** operates over a large, continuous segment of the radio spectrum.

A typical general-coverage receiver spans 1.6 through 30 MHz. This wide range lets you listen to international broadcasts and commercial and military stations, as well as amateur communications. Some general-coverage receivers tune as low as 500 kHz to cover the Standard Broadcast AM band as well. Figure 7-4 shows some typical receivers.

Because manufacturers design ham-bands-only receivers specifically for the amateur bands, they offer several advantages. They are generally easier to use. The **band spread** is usually greater.

Band spread refers to how quickly the receiver changes frequency as you turn the tuning knob. One turn of the knob on a general-coverage receiver usually changes the frequency quite a bit. One turn of the knob on a ham-bands-only receiver will not change the received frequency nearly as much. The band will seem much larger and you will find more room between stations on a ham-bands-only receiver. The greater band spread also makes it easier to tell the exact frequency you are tuned to.

Moreover, manufacturers design amateur receivers specifically for use on the ham bands. The performance of these receivers will generally be superior (on the ham frequencies) to general-coverage receivers of comparable price. Modern general-coverage radios do not always have these limitations, however. You *can* get a first-rate receiver that includes the general-coverage feature. Expect to pay a bit more than you would for an older model ham-bands-only receiver, though.

When you make your decision, you must make some tradeoffs. Ham-bands-only receivers offer convenience, better performance and more-accurate tuning. A general-coverage receiver has a much greater frequency range. If your main interest is listening to the amateur bands, you should opt for the performance of the ham-bands-only receiver. If shortwave listening interests you, a general-coverage receiver is a good choice.

You may find two receivers of equal quality offering similar features. In this case, the general-coverage receiver will usually cost more because of the additional tuning range.

(A)

(B)

Figure 7-4—A modern general-coverage receiver from Heath (the SW-7800) is shown at A. This receiver, with digital frequency display, tunes the shortwave broadcast bands as well as the ham bands. B shows the R-5000, a full-featured general-coverage communications receiver from Kenwood.

Tuning Mechanism

Regardless of what type of receiver you choose, you should give some consideration to the tuning mechanism. This includes the knobs, dials, gears and readout. Make certain that the tuning mechanism works freely, that the dial markings are understandable, and that you feel comfortable operating it.

Many different types of tuning-dial mechanisms are in use today. It seems as though each manufacturer has developed a different idea of what is desirable. Some older receivers use a slide-rule dial (as shown in Figure 7-5). This is a linear scale marked on the radio, with a bar or pointer that moves along the scale.

Figure 7-5—The Hallicrafters SX-101 receiver, produced in 1958, weighed 75 pounds. It is shown here as an example of a radio with a slide-rule dial.

Figure 7-6—This Collins ham-bands-only receiver features a circular dial with 1 kHz frequency resolution.

Others use circular dials (Figure 7-6). This is a fixed line with a rotating disk behind it. Still others use a combination of both circular and slide rule dials.

Today, most manufacturers are using digital displays (Figure 7-7). This kind of dial mechanism is very accurate. When you use a receiver with a digital display, you *know* where you are tuned. This is helpful when you are tuning in a net or a shortwave broadcast station on a specific frequency. You can look up the operating frequency in your guide book or net directory and set the receiver to the exact frequency.

Why are there so many different types of tuning dials? The name of the game is **frequency resolution**. The farther apart the markings on the dial, the better the frequency resolution. The better the frequency resolution, the better you'll be able to tell precisely what frequency you're on. A receiver with markings every 1 kHz or less is best, although many adequate receivers have frequency markings only every 5 kHz.

Which type of dial mechanism is best? All the tuning mechanisms mentioned can be accurate. You should choose the dial mechanism you are most comfortable reading and tuning. Make sure the dial mechanism operates smoothly. Avoid individual mechanisms that feel sloppy and slip or skip—that's part of your comfort and accuracy.

Don't forget to consider the mechanical band spread of your receiver. This refers to the distance you must turn the knob to obtain a given change in dial setting. We usually specify band spread in terms of the number of kilohertz per knob revolution. The greater the band spread, the larger the band will seem. This makes it easier to set the dial to a specific frequency. A radio that tunes 10 kHz per revolution has a greater band spread than one that tunes 25 kHz per revolution.

Figure 7-7—The Yaesu FT-757GX is a marvel of modern miniaturization. The transceiver is all solid state and has a digital frequency display. The rig also includes a general-coverage receiver and 100-watt transmitter with a built-in electronic keyer. All this fits in a box the size of a portable typewriter.

Selectivity and Sensitivity

There are two very important electrical specifications you should know about: selectivity and sensitivity. **Selectivity** is the ability of a receiver to separate two closely spaced signals. This determines how well you can receive one signal in the presence of another signal that is very close in frequency.

Sensitivity is the ability of a receiver to detect very weak signals. While both of the specifications are important, as a Novice you'll find selectivity to be the more important of the two. That's because the Novice bands are rather narrow and filled with the signals of a growing number of operators. Isolating the signal you are receiving from all the others nearby is very important. The selectivity of your receiver will directly affect how much you enjoy your time on the air.

We use **bandwidth** to measure selectivity. Bandwidth is nothing more than how wide a range of frequencies you hear with the receiver tuned to one frequency. With a 6-kHz bandwidth, you can hear signals as much as 3 kHz above and below where you are tuned.

What if you can't hear signals that are more than 200 Hz above or below where your receiver is tuned? This means that your receiver has a bandwidth of 400 Hz. The narrower this bandwidth is, the greater the selectivity of the receiver. A narrow bandwidth makes it easy to copy one signal with another one close by in frequency.

Special filters built into the circuitry of a modern receiver determine the selectivity. Generally, these filters contain quartz crystals arranged to provide a specific selectivity. Some receivers have several filters built in to enable you to choose between different selectivity bandwidths.

As a general guideline, you should look for a selectivity of 600 Hz or less for CW operation. Receivers designed for single-sideband voice operation come with a standard filter selectivity of around 2.8 kHz. This is usable on CW, but you should try for a receiver with a selectable CW bandwidth of 600 Hz. If the receiver has provision for adding narrow-bandwidth accessory filters, you can add them later (but they will cost extra).

Another selectivity feature available on some receivers is a notch filter. This filter can be used to cut out, or notch, a specific frequency from within the received bandwidth. A notch filter is handy when you're trying to receive a signal that is very close in frequency to another signal. By adjusting the notch control, you can effectively eliminate the unwanted signal.

The BFO

When you are choosing a receiver, make sure it has a built-in **beat frequency oscillator (BFO)**. Some receivers intended strictly for listening to AM shortwave broadcasts do not contain a BFO. These receivers are unusable for CW or SSB.

You should have no trouble determining that a receiver has a BFO. Some older receivers, usually the general-coverage units, actually have a front-panel control labeled BFO. Other receivers have a built-in crystal oscillator that serves as a BFO. These receivers usually have a mode switch labeled LSB, USB, CW, RTTY or some combination of these labels. If your receiver has such a switch, you're in business.

Other Features

There are other features you may find on receivers. Most receivers have a meter that shows the strength of the incoming signal. These meters, called S-meters, are useful when you are comparing two antennas. You can also use the S-meter when you are trying to rotate a directional antenna for maximum signal strength.

Some receivers have a crystal calibrator. This is a low-power oscillator that will provide a signal every 100 kHz. You can use the calibrator to make sure that the dial mechanism is accurate.

Besides the filters already mentioned, some receivers have a "noise blanker." This can help filter out power-line noise and ignition noise from nearby vehicles. Some receivers have filters between the intermediate stages. These may be called variable IF filters, or the receiver may have a control labeled PASSBAND TUNING. An RF gain control can help if you have another ham in your neighborhood.

All these features can add to your operating enjoyment. They may not be necessities if you are on a tight budget, however. Frequency coverage, tuning mechanism, frequency resolution, selectivity and sensitivity are the most important things to look for.

BUYING USED RECEIVERS

Amateur equipment manufacturers have produced many different types of receivers over the years. Some of these receivers covered frequencies as low as the AM broadcast band to as high as 50 MHz (6 meters). These receivers were not as sensitive at the high frequencies as at lower frequencies. Sensitivity, at least up to the 20-meter band, is usually pretty good for most of the older receivers. On 15 and 10 meters, you may want to add a preamplifier to provide a little help.

A preamplifier is a unit placed ahead of the RF amplifier stage of the receiver. We can put the preamplifier ahead of the mixer stage if there is no RF amplifier. The preamp boosts the level of incoming signals.

Preamplifiers made by several manufacturers are on the used market today. They must be used with discretion. Under certain operating conditions, a preamp can provide *too* much signal at the receiver input. This overloads the receiver front end and, in effect, creates interference that doesn't really exist!

A simple preamplifier is easy to build. With a separate receiver you just connect the preamp in the antenna lead to the receiver. If you own a transceiver the installation will be a bit more complicated. (The preamplifier would be destroyed as soon as you began transmitting if it were in the antenna lead to the rig!)

A circuit for a simple preamp appeared in the Hints and Kinks column in the August 1984 issue of *QST*. See Figure 7-8. Circuit boards and complete parts kits for this

Figure 7-8—A simple one-transistor preamplifier is available in kit form. See text for details.

preamp are available at a modest cost. Write to A&A Engineering, 2521 W LaPalma Ave, Unit K, Anaheim, CA 92801.

Receiver stability, a function of both mechanical ruggedness and oscillator-circuit electrical stability, should be checked. Ensure that the receiver can withstand a reasonable amount of bumping without bouncing off frequency.

On today's crowded bands, selectivity is very important. It also comes with a price! The more filters a receiver has and the better the quality of the filters, the more the receiver will cost. Many older receivers obtained various degrees of selectivity with LC (inductance-capacitance) circuits. More expensive and modern receivers use either mechanical or crystal-lattice filters.

For CW communications, try to obtain a receiver that has a filter bandwidth of 500 or 600 Hz. For single-sideband, 2.8 kHz or so is an adequate bandwidth.

One way to use an older receiver with only an SSB filter bandwidth is to employ an audio filter. This type of filtering is somewhat after the fact. Nevertheless, an audio filter may mean the difference between maintaining a QSO or losing one.

Audio filters are simple to construct (see *The ARRL Handbook for the Radio Amateur*). If you don't want to build a filter, you can purchase one of the many available commercial units. Audio filters simply plug into the audio output jack of the receiver. Then your headphones or a speaker connect to the filter output.

You should also consider frequency resolution, as we mentioned earlier. The better the resolution, the easier it will be to zero in on a particular frequency. And, of course, you must ensure that you are operating within the band of frequencies allowed by your license class.

A final word about used receivers. Your receiver will probably make or break your amateur station. A good receiver will make your operating time enjoyable, while an inadequate receiver can be extremely frustrating. For this reason, you should choose your receiver with great care.

Try to avoid the "boat anchors." This is what hams call receivers that require two people simply to get them in the trunk of your car. Buy the best receiver you can possibly afford. If it means you have to wait a while before you can buy a transmitter, you'll still be better off. "Listen before you transmit" should be your motto throughout your ham career, and you might as well start now. You won't be sorry!

TRANSMITTERS

Your transmitter is the main reason you've been working toward your ham license. It provides the means to reach out to other hams. Just as with receivers, transmitters come in all shapes and sizes. What you'll have to pay for a transmitter depends more on what features you *want* than on what you actually *need*.

First, remember that your Novice license permits you to use 200 watts of output power in the HF bands. It allows you to operate CW in portions of the 80, 40, 15 and 10-meter amateur bands set aside for your use. To operate in these CW sub-bands, all you really need is a simple CW transmitter.

Most modern transmitters have an output power in the 100-watt range. It would take an external power amplifier to get up to the full 200-watt Novice limit! Very few Novices use this maximum output power, since 100 watts is more than enough for most contacts. Figure 7-9 shows a simple Novice operating position, with an inexpensive transceiver as the station rig.

What you might *want* in the way of a transmitter is another story. FCC rules permit Novices to use single-sideband voice and digital modes on the 10-meter band. When you upgrade to the General or higher class license, you can use much larger portions of the 80, 40, 15 and 10-meter bands. You can also use the 160, 30, 20, 17 and 12-meter bands when you upgrade. The final "carrot" for upgrading is that you will be able to run up to 1500 watts PEP output. Keep these ideas in mind. You might see something you like that is more than you can use right away. Remember that you will be able to use it someday.

A typical ham transmitter runs between 75- and 200-W PEP output from the final amplifier. It operates upper and lower sideband and CW, and has VFO control. The transmitter will cover the 80, 40, 20, 15 and 10-meter amateur bands. Some also include the 160-meter band. Newer transmitters cover the 30, 17 and 12-meter bands as well. Such a transmitter will be more than adequate for all the operating you're likely to do as a Novice. It will also serve you well as a General. It will even drive a high-power linear amplifier, if you decide to get one when you upgrade.

As with the receiver, you should choose a transmitter that is easy to operate. If the transmitter has a VFO, check the tuning mechanism. Look for the same features in the transmitter tuning dial as you would in a receiver dial. The dial should have easily understood markings, a smooth, freely operating mechanism and operating comfort.

Figure 7-9—The Ten-Tec Century 21 is a transceiver available on the used market. This solid-state, CW-only transceiver makes operation as uncomplicated as possible.

SELECTING AN OLDER TRANSMITTER

Used transmitters come in a variety of shapes, sizes, colors and capabilities. Finding one with a built-in VFO shouldn't be difficult. This will save you the trouble of buying or building a separate VFO. Many older transmitters are CW-only types, others have AM capability and more recent units incorporate single sideband (SSB).

Some transmitters cover the "top band"—160 meters. Others cover the MARS (Military Affiliate Radio Service) frequencies, which lie outside the amateur bands. When you upgrade to the General license you will be able to make use of these features. That may play a part in your ultimate decision.

Double-sideband AM is still in use, but on a very limited scale. CW and SSB are the most common modes. Many people like slow-scan television (SSTV), radioteletype (RTTY), packet radio and other specialized operating modes. Older rigs designed for AM operation have rugged power supplies, and are good candidates for use on RTTY. Though as a Novice you may plan to operate mainly on CW, you will want the added capabilities later.

Most transmitters with single-tube power-amplifier (PA) stages operate at power levels of 50 to 100 watts. Parallel-tube amplifiers (those having two or more tubes in the final amplifier) develop more than 100 watts PEP output. Most solid-state equipment runs at power levels in the 100-watt PEP output range. Just remember: You can work the world with 100 watts and a decent antenna.

The workhorse of the final-amplifier-tube family is the 6146. It's tough and can take a beating—such as prolonged key-down periods or antenna mismatches. Given a little time to cool down, the 6146 will bounce right back as good as new. The Kenwood TS-820S is a popular transceiver that uses 6146 tubes in its final amplifier. It should be available at reasonable prices on the used market. This transceiver is shown in Figure 7-10.

A few years ago, some manufacturers used TV sweep tubes (horizontal deflection amplifiers) as final amplifier tubes in SSB/CW transmitters. Sweep tubes (like the 6HF5, 6JS6C, 6KD6 and 6LQ6 among others) are less tolerant. Though they do have their advantages, they cannot sustain much abuse. For example, prolonged tune-up periods are sure to harm them. If you treat these tubes correctly, they perform well. In the hands of a beginner they might suffer an early death, however. You might consider these points when you are deciding which unit you wish to add to your station.

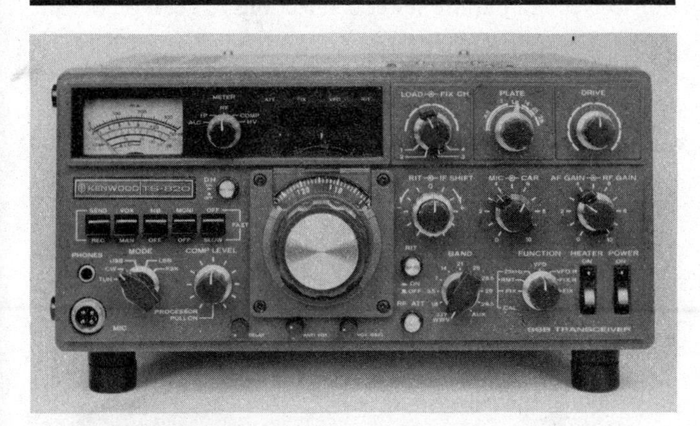

Figure 7-10—The Kenwood TS-820S transceiver uses 6146 tubes in the final-amplifier section.

Another veteran tube type, the 807, found extensive use in early transmitters. The 807 is rugged and practically indestructible unless it is severely abused. You will find these old "bottles" in many an old-timer's "junk box." This makes finding replacements easy.

Tube-type transmitters use two basic types of keying circuits: cathode keying and grid-block keying. If you use a hand key, the voltage polarity at the key jack is of little consequence. If you use an electronic keyer with a transistor switch in the output circuit, polarity must be considered.

Keyers that use relays with "floating" (ungrounded) contacts in the output circuit are more flexible since key-jack voltage polarity is unimportant. But relay contacts can stick and the trend is to use transistorized output stages in modern keyers.

Cathode keying presents a positive voltage to the transmitter key jack. For cathode keying, the key line opens and closes the cathode circuit of one or more tubes in the transmitter. When the key is up, there is an open circuit in the tube cathode. This effectively shuts off the tube so it draws no plate or screen current; thus, no radio-frequency output.

Grid-block keying uses a negative voltage applied to the grid(s) of the keyed tube(s). This voltage, usually at a reduced amplitude, is also present at the transmitter key jack. With the key open, the bias voltage cuts off the tube. Closing the key removes or reduces the bias and the tube conducts.

As a rule, older transmitters can be expected to use cathode keying while more modern tube equipment uses grid-block keying. Fully transistorized rigs may use either positive or negative polarity at the key jack.

Before you connect a keyer to any rig, you should check the key-jack voltage and polarity. Make sure your keyer will work with the rig. Check your equipment manual. It should tell you the type of keying circuit used in your transmitter and the key-jack voltage. The manual for your keyer should tell you what voltage and polarity it can handle. If you built the keyer, the construction article should tell you the appropriate keying voltage.

There are ways to change the keyer output circuit. You can use a transistor or optoisolator, but these techniques are beyond the scope of this book. A more-experienced amateur may be able to help you if you face this problem. Also, *The ARRL Handbook for the Radio Amateur* contains more information.

VFOs and Crystal Control

Quite a few older transmitters use crystal control and require that you add an external VFO for good flexibility. The VFO gives you the ability to move freely around the bands. You won't have to buy, store and constantly change a bunch of crystals.

Other transmitters have built-in VFOs as well as the option for crystal control. These transmitters are often good buys. You may find one for as much or less than the cost of a separate transmitter and VFO combination.

Crystal control might find its uses in net operations where you have to "hit" a particular frequency over and over again. You might also find crystals in homemade portable equipment of simple design. The days of being "rock-bound"—crystal controlled on specific frequencies—have faded into the past for most HF operations.

Crystals might have a bit of an edge in stability when compared with older VFOs. VFOs make operating much more enjoyable, however. Modern-day VFOs and frequency synthesizers (synthesizers are internally crystal controlled) have replaced the use of individual crystals. A handful of crystals can cost as much as a VFO. The crystals won't supply anywhere near as much frequency coverage, however.

Separate VFOs are sometimes more difficult to locate than the transmitters they mate with. They don't last long on the dealer's shelves or the used-equipment market in general. You're not the only one looking for them!

The most important quality to look for in a VFO is stability. An unstable VFO will result in unwanted frequency changes. This can be frustrating, especially on today's crowded bands, and will rapidly decrease your "QSO enjoyment ratio." Mechanical ruggedness is directly related to electrical stability, too. The VFO should be capable of absorbing slight shocks or bumps without warbling like a sick canary.

need BFO

TRANSCEIVERS

Transceivers combine the functions of both a receiver and transmitter in one package. This means that everything we said about separate units also applies to the receiver and transmitter portions of a transceiver. Generally, transceivers use VFO control and provide both CW and voice operation.

A primary advantage of using a transceiver is that the rig automatically transmits and receives on the same frequency. It takes some care and practice to be sure a separate receiver and transmitter are operating on the same frequency. Some simple transceivers suitable for Novice operation are shown in Figure 7-11. The transceiver at B in the Figure operates only on 10 meters, but it can be a good value for a new Novice.

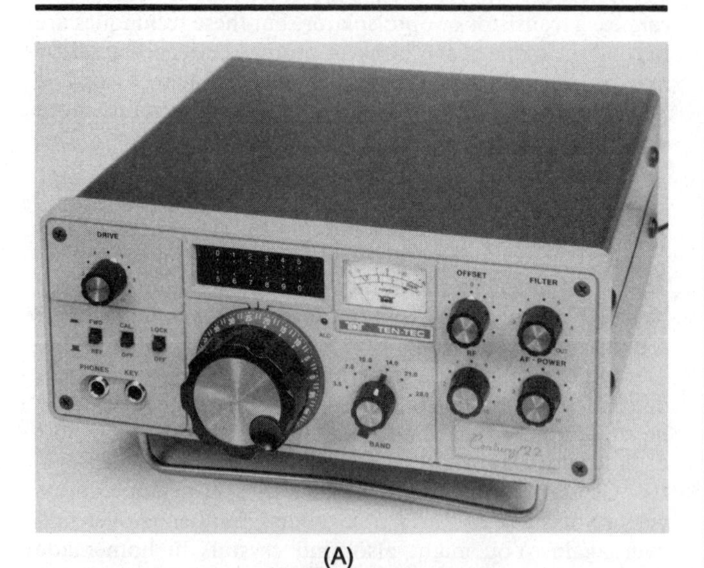

(A)

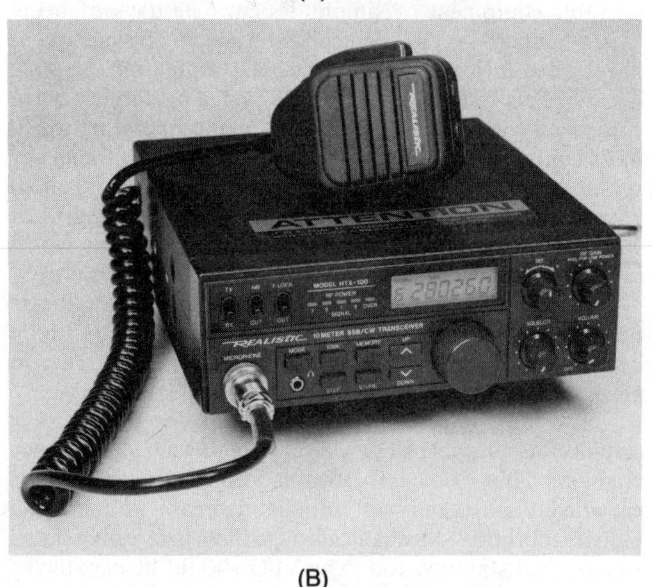

(B)

Figure 7-11—The successor to the Century 21, Ten-Tec's Century 22 is shown at A. B shows the Radio Shack HTX-100, a small, inexpensive 10-meter transceiver.

Receiver incremental tuning (RIT) is a useful transceiver feature. This feature allows you to shift the receiver frequency over a limited range without affecting the transmitter frequency.

Although transceivers theoretically transmit and receive on exactly the same frequency, there may actually be a slight frequency difference, or *offset*. This offset can vary from transceiver to transceiver. RIT lets you retune the receiver slightly. This enables you to compensate if the station you are working has a different offset.

There are many good transceivers on the used market. Most of the things we have said so far about receivers and transmitters apply when you are considering a used transceiver. See Figure 7-12. Look for signs of wear and tear. You may be able to find a good used transceiver for less than the cost of a separate receiver and transmitter. A transceiver will require fewer connections to get it on the air.

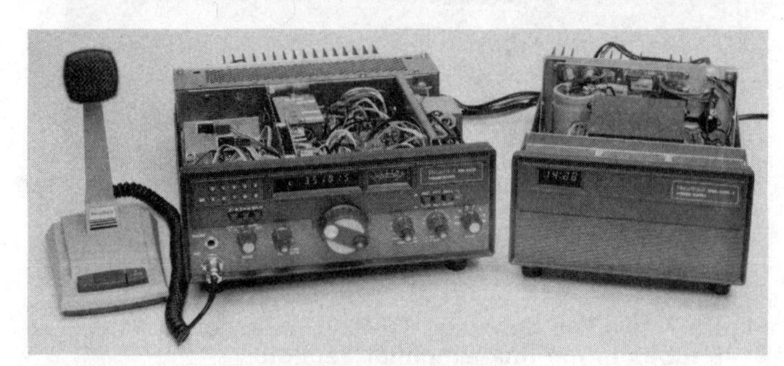

Figure 7-12—The Heath HW-5400 transceiver is a CW and SSB rig often available used. The covers have been removed from the rig and power supply to show some of the internal circuitry.

SINGLE-SIDEBAND EQUIPMENT

So far, we've been talking mostly about rigs for CW operation. Most of the things we've mentioned also apply when you are looking for a rig for single-sideband voice operation. There are a few features that directly apply to SSB operation however.

Sidebands

There are two kinds of single sideband: **upper sideband (USB)** and **lower sideband (LSB)**. Hams usually use LSB on 160, 80 and 40 meters, and USB on 20, 17, 15, 12 and 10 meters. Upper sideband is used on VHF and UHF. Some radios automatically switch to the correct sideband for the operating frequency. Others have separate USB and LSB positions on the mode switch. You will always use the USB position for your Novice 10-meter and VHF SSB operation.

Microphone

To talk on SSB, you will need to add a microphone to your transceiver. The microphone connects to the microphone jack on your radio.

Individual voices and microphones have different characteristics. Your rig should have a microphone gain control so that you can set everything up for a clean, nice-sounding signal. Most SSB transmitters have an automatic level control (ALC) meter to help you determine the correct microphone gain setting. See your equipment manual for instructions on adjusting everything to the right level.

Speech Processor

Speech processing increases the average power of a single-sideband signal. Used properly, a **speech processor** can greatly improve the readability of a signal. Misused, however, it can severely degrade the audio quality and make the signal more difficult to understand.

Virtually all SSB transmitters and transceivers built in the last 10 years have speech processors as standard equipment. There is also a variety of accessory speech processing equipment for transmitters without built-in processing.

On some rigs, there is no adjustment for the speech processor; there is just an on/off control. Other gear has one or more variable controls to set the speech processing level. Always check your equipment instruction manual for information on setup and operation. When you think you've got everything adjusted properly, ask other amateurs for on-the-air checks.

Voice-Operated Switch (VOX)

Virtually all transceivers manufactured for SSB operation have a **voice-operated switch (VOX)**. This feature switches the rig into transmit automatically when you speak into the microphone. The VOX switches the rig back to receive when you stop talking. VOX is handy because it allows you to listen during pauses and lets you keep both hands free for writing.

There are usually three VOX controls: gain, delay and anti-VOX. *VOX gain* sets the sensitivity. You should adjust this control so that the VOX keys the transmitter when you speak in a normal voice. Don't set the gain so high that the VOX keys the transmitter from background noises. For VOX operation, it's best to use a microphone that requires close talking, rather than one that picks up stray sounds.

VOX delay sets the interval between when you stop talking and when the transceiver switches back to receive. Most hams set this control so the rig does not switch back instantly between words. You should adjust the delay so the transceiver switches after a short pause.

Anti-VOX works with VOX gain to prevent audio from the receiver's speaker from keying the VOX. Without it, the speaker audio will key the transmitter. You could set the VOX gain so that speaker audio won't key the transmitter. If you did that, though, you'd have to shout into the mike to activate the VOX. Anti-VOX circuitry allows the VOX to ignore audio from the speaker, yet respond when you speak into the microphone.

VHF AND UHF EQUIPMENT

There is a wide variety of VHF and UHF equipment for you to consider. Think about the type of operation you might like to try before you choose a specific piece of gear. Unlike five- or six-band HF transceivers, most VHF and UHF equipment operates on only one band. You'll need separate rigs to work 222 and 1270 MHz.

VHF and UHF equipment can be categorized by the mode of operation. Many transceivers for 222 and 1270 MHz operate only in the FM mode. Other gear can be used only for SSB and CW work. For both FM and SSB/CW operation, you will need a third type, the **multimode transceiver**, like the one in Figure 7-13.

Figure 7-13—The ICOM IC-375A 220-MHz transceiver. This rig can be used on FM, SSB and CW.

You should also consider where the equipment will be used. Some VHF/UHF equipment is best for home station operation. Other, more compact, units work well for mobile operation in a car. Portable, hand-held transceivers can be used anywhere.

Although manufacturers usually optimize equipment for a certain application, you can use it wherever you want. If you like, you can use a hand-held transceiver in a car or a mobile rig at your home station. There are, however, some trade-offs that you must consider when selecting a rig with more than one application in mind.

Base-Station and Mobile Equipment

For VHF and UHF FM work, base-station equipment usually consists of a multimode transceiver or a mobile transceiver. Multimode base-station transceivers often have built-in power supplies and are physically larger than their FM-only counterparts.

Mobile rigs are often smaller than base-station transceivers, and they operate from a 13.8-V automobile electrical system. See Figure 7-14. You'll need an external, accessory power supply if you plan to use one of these at home.

FM mobile transceivers usually have all the features found in base-station equipment. Power output for mobile and base-station equipment is generally in the 10- to 25-watt range. The rig's physical size and the power supply are the most important things to consider. Except for these factors, there is no particular advantage to purchasing a base-station

Figure 7-14—Modern VHF FM rigs are usually very small, like the Kenwood TM321A shown here.

unit or a mobile unit for FM work. Many amateurs simply purchase a mobile transceiver and external power supply and use the same transceiver both at home and in the car.

There is a limited selection of equipment available for 222-MHz SSB and CW work. Most SSB and CW operation is done with a *linear transverter* used with a regular 10-meter HF transceiver. A transverter is a small box that converts signals from one band to another. For example, a 222-MHz-to-28-MHz transverter receives a signal at 222.1 MHz and converts it to 28.1 MHz for reception on an HF receiver. Likewise, the transverter accepts a 28.1-MHz signal from your HF transmitter and converts it for transmission on 222.1 MHz.

The transverter has no external controls. The transceiver operates exactly like it does on HF; the only difference is that you're transceiving on 222 MHz instead of 28. Most transverters have 10 to 25 watts RF output.

Hand-Held Transceivers

A very popular piece of equipment is the hand-held VHF or UHF FM transceiver. Such equipment is the most compact of the lot. See Figure 7-15. Hand-held transceivers are self-contained stations. They use battery power and have an antenna, speaker and microphone built in.

Most hand-held units have provisions for using an external 12-V supply and an external antenna. Some offer accessory external speakers and microphones, so you can use them in mobile or home-station installations as well.

To conserve physical size and battery power, hand-held transceivers usually have lower power output than their mobile counterparts. An average hand-held unit may have 1 or 2 watts output; a "high-power" hand-held may produce 5 watts.

Figure 7-15—The Kenwood TH31BT and the Yaesu FT-109H 220-MHz FM hand-held transceivers. This type of rig is a complete portable station, with an autopatch tone pad and a flexible vertical antenna.

FM TRANSCEIVER FEATURES

VHF and UHF FM transceivers have many features not usually found on HF rigs. While many of these features are not essential, they are useful. All FM transceivers have a squelch control that quiets the receiver when no signal is present. Squelch lets you leave your transceiver on without having to listen to noise when no one is around.

When you evaluate your equipment choices, you'll notice a few things. Virtually all newer transceivers use a synthesizer to control the operating frequency, while many older ones use crystal control. Transceivers that use synthesizers have circuitry built in that allows them to operate on any frequency in the band. This is similar to a VFO on an HF transceiver. Crystal-controlled radios have a front-panel control to allow you to switch between 10 or so repeater or simplex channels.

WHAT KIND OF RIG TO BUY?

There is a variety of equipment to choose from for your new bands and modes. Listed here are transceivers and antennas that work on 10-meter SSB, 1.25 meters and 23 cm. Some of this equipment is available new; some is out of production but available on the used equipment market. Multiband HF transceivers that can be used for 10-meter SSB and digital modes are not listed. This is by no means a complete list of all available equipment. ARRL does not endorse specific products. Check with local amateurs or look through the ads at the back of this book for ideas on what is currently available. Advertisements and product evaluations in *QST* and other amateur magazines are another source of help.

28-MHz Single-Band Transceivers

Manufacturer	*Model*	*Power*	*Features*
Clear Channel	AR3500	30 W	CW, SSB, 12 V
Kantronics	KT110	20 W	CW, SSB, 12 V
Radio Shack	HTX-100	25 W	CW, SSB, 12 V
Uniden President	HR2510	24 W	CW, SSB, 12 V

220-MHz Transceivers

Manufacturer	*Model*	*Power*	*Type*	*Features*
Clegg†	FM-76	10 W	Mobile	Crystal
Drake†	UV3	10 W	Base	Multi-band
ICOM†	IC-37A	25 W	Mobile	
ICOM	IC-38A	25 W	Mobile	
ICOM	IC-3A/AT	2.5 W	Hand-held	
ICOM	IC-03AT	2.5 W	Hand-held	
ICOM	IC-375A	25 W	Base	Multimode
KDK	FM-4033	25 W	Mobile	
Kenwood†	TH-31AT	1 W	Hand-held	
Kenwood	TH-31BT	1 W	Hand-held	
Kenwood	TH-315A	2.5 W	Hand-held	
Kenwood	TM-321A	25 W	Mobile	
Kenwood	TM-3530A	25 W	Mobile	
Kenwood	TM-621A	25 W	Mobile	Dual-band (2 meters)
Midland†	13-509	10 W	Mobile	Crystal
Midland†	13-513	20 W	Mobile	Crystal
Santec†	ST-220	2.5 W	Hand-held	
Tempo†	S-2		Hand-held	
Yaesu	FT-33R	5 W	Hand-held	
Yaesu†	FT-103R	2.5 W	Hand-held	
Yaesu	FT-109RH	5 W	Hand-held	
Yaesu†	FT-311RM	25 W	Mobile	
Yaesu†	FT-127A	10 W	Mobile	Crystal

†Out of production.

1270-MHz Transceivers

Manufacturer	*Model*	*Power*	*Type*	*Features*
ICOM	IC-12AT	1 W	Hand-held	
ICOM	IC-12GAT	1 W	Hand-held	
ICOM†	IC-120	1 W	Mobile	
ICOM	IC-1200A	10 W	Mobile	
ICOM	IC-1271A	10 W	Base	Multimode
Kenwood	TH-55AT	1 W	Hand-held	
Kenwood	TR-50	1 W	Mobile	
Kenwood†	TM-521A	10 W	Mobile	
Kenwood	TM-531A	10 W	Mobile	
Yaesu	FT-2311R	10 W	Mobile	

†Out of production.

Linear Transverters

Manufacturer	*Model*	*Power*	*Type*
Microwave Modules	MMT220/28	15 W	220-28 MHz
Microwave Modules	MMT1296/144	2 W	1296-144 MHz
SSB Electronics	LT 23S	10 W	1296-144 MHz

28 MHz Antennas

Manufacturer	*Model*	*Description*
Cushcraft	10-3CD	3-element Yagi
Cushcraft	TEN-3	3-element Yagi
Cushcraft	10-4CD	4-element Yagi
Cushcraft	AR10	Vertical (base)
Hustler	MO1,2,3	Vertical (mobile)
Hustler	10-MB4	4-element Yagi
Hy-Gain	105BAS	5-element Yagi
Hy-Gain	103BAS	3-element Yagi
KLM	10M-4	4-element Yagi
KLM	10M-6	6-element Yagi
Telrex	10M313	3-element Yagi
Telrex	10M523	5-element Yagi
Telrex	10M636	6-element Yagi

220 MHz Antennas

Manufacturer	*Model*	*Description*
AEA	HR-2	1/2 wave vertical (hand-held)
AEA	IsoPole 220	Vertical (base)
Antenna Specialists	AP-220.3G	On-glass mount (mobile)
Antenna Specialists	HMR-223	5/8-wave vertical (mobile)
Antenna Specialists	HMR-229	For portable operation
Butternut	220-CV-5	Vertical collinear (base)
Cushcraft	220B	17-element Yagi
Cushcraft	225WB	15-element Yagi
Cushcraft	A220-11	10-element Yagi
Cushcraft	A220-7	7-element Yagi
Cushcraft	224WB	3-element Yagi
Cushcraft	ARX-220B	Vertical (base)
Cushcraft	BN-220	For hand-held rig
Cushcraft	CS-220	5/8-wave vertical (mobile)
Hustler	CG-220	Collinear (mobile)
Hustler	G7-220	Collinear vertical (base)
Hustler	HMD220	For hand-held rig
Hustler	SF-220	Vertical (mobile)
Hy-Gain	V3-S	Vertical collinear (base)
KLM	220-7	7-element Yagi
KLM	220-14X	14-element Yagi
KLM	220-22LBX	22-element Yagi
KLM	JV-220X	Vertical (base)
Larsen	BSA-220-K	Vertical (base)
Larsen	FB1-36	Collinear (base)
Larsen	KD4-220	For hand-held rig (3 sizes)
Larsen	NMO/NLA 220	Vertical (mobile)
Mosley	MY-220-5	5-element Yagi
Mosley	MY-220-9	9-element Yagi
Telrex	220M15	15-element Yagi

1270 MHz Antennas

Manufacturer	*Model*	*Description*
Down East Microwave	2345LY	Yagi
KLM	1.2-44LBX	Yagi
Larsen	FB3-1290	Collinear vertical (base)
Larsen	NMO/NLA 1290	Collinear vertical (mobile)
Spectrum International	1296-LY	28-element Yagi
Tonna		23-element Yagi
Tonna		55-element Yagi

If you operate on only a few repeater or simplex frequencies, a crystal-controlled rig is fine. There are many of them on the used marked, often in the $100-$150 price range. See Figure 7-16. If you buy one locally, it will probably include crystals for the local repeaters and popular simplex frequencies.

If you live in an area with many repeaters, you'll probably want the convenience of a synthesizer-controlled transceiver. You should also consider this if you plan to travel a lot and use repeaters in other areas. Remember that you may have to add the cost of crystals to the cost of a second-hand crystal-controlled transceiver. If you have to buy many crystals, the cost may approach that of a synthesizer-controlled transceiver.

Crystal-controlled transceivers usually require two crystals for each repeater or simplex channel. One crystal is for the transmitter (repeater input frequency) and one is for the receiver (repeater output frequency). For simplex operation, you generally need two crystals for the same frequency.

Synthesizer-controlled transceivers have a lot of frequency flexibility. They usually have a switch to select between simplex and duplex (repeater) operation. The transceiver comes from the factory programmed for the proper frequency split for the band. You simply dial up the repeater output frequency on your transceiver. The transceiver will automatically switch to the repeater input frequency when you transmit. If it's necessary, you can usually set most synthesizer-controlled transceivers for non-standard frequency splits.

It may be inconvenient to have to tune a synthesizer-controlled radio all over the band to change frequency. This is true if you normally operate on only a few repeater or simplex frequencies, perhaps separated by hundreds of kilohertz. To make synthesizer-controlled transceivers more convenient, they often have memories to store your favorite channels. You simply program often used frequencies into the memory and recall them as you wish. For example, memory 1 may be for the local repeater input on 222.32 MHz, memory 2 for the 223.5 MHz simplex frequency, and so on.

Synthesizer-controlled transceivers often offer the ability to automatically scan up and down the band or among repeater channels. This feature is handy if you're traveling and want to find an active repeater.

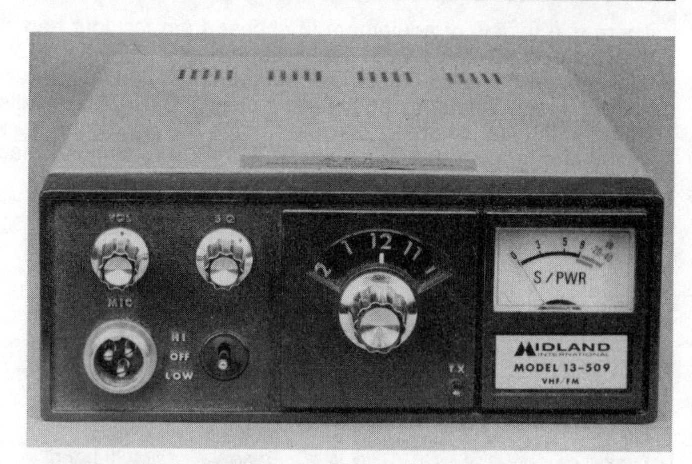

Figure 7-16—The Midland 13-509 is a crystal-controlled 220-MHz FM rig available on the used market.

There are other FM transceiver features, sometimes standard and sometimes available as accessories. These include tone pads for autopatch use and tone-burst or subaudible-tone generators for repeaters that require such tones for access.

FOR MORE INFORMATION

We've tried to give you a few guidelines in this chapter. There are lots of rigs out there, both new and used. Take your time and try to get as much information as you can. Talk to other hams and read the Product Reviews in *QST*. Remember that everyone has an opinion, though. What they like might not be what you like. Most of all, don't worry! If you buy a used rig and you don't like it, you can probably sell it for almost as much as you paid for it. Start small and simple. Ham radio is fun; part of the fun is dreaming about new equipment and upgrading your station as you upgrade your license.

KEY WORDS

Antenna—A device made from wire or metal tubing. It picks up or sends out radio waves.

Balun—Contraction for *bal*anced to *un*balanced. A device to couple a balanced load to an unbalanced source, or vice versa.

Beam antenna—A directional antenna. A beam antenna must be rotated to provide coverage in different directions.

Characteristic impedance—The opposition to electric current that an antenna feed line presents. Impedance includes factors other than resistance, and applies to alternating currents. Ideally, the characteristic impedance of a feed line will be the same as the transmitter output impedance and the antenna input impedance.

Coaxial cable—coax (pronounced kó-aks). This is a type of feed line with one conductor inside the other. Insulation surrounds the inner conductor, and in turn, the insulation is surrounded by a braided shielding conductor. A plastic covering protects the shield. Sometimes the shielding conductor is solid.

Dipole antenna—See **Half-wave dipole**. A dipole may have lengths other than ½ wavelength.

Directivity—The ability of an antenna to focus transmitter power into a beam. Also its ability to enhance received signals from specific directions.

Director—An element in "front" of the driven element in a Yagi antenna.

Driven element—The element of an antenna that connects directly to the feed line.

Feed line (feeder)—See **Transmission line**.

Gain—A comparison of how much signal two different antennas will pick up. Also a comparison of the transmitter signal enhancement from two different antennas.

Half-wave dipole—A basic antenna used by radio amateurs. It consists of a length of wire or tubing, opened and fed at the center. The entire antenna is ½ wavelength long at the desired operating frequency.

Inverted-V dipole—A half-wave dipole antenna with its center elevated and the ends drooping toward the ground. Amateurs sometimes call this antenna an "inverted V."

Ladder line—Parallel-conductor feeder with insulating spacer rods every few inches.

Matching network—A device that matches one impedance level to another. For example, it may match the impedance of an antenna system to the impedance of a transmitter or receiver. Amateurs also call such devices a Transmatch, impedance-matching network, or match box.

Multiband antenna—An antenna that will operate well on more than one frequency band.

Omnidirectional—Antenna characteristic meaning it radiates equal power in all compass directions.

Open-wire feed line—Parallel-conductor feeder with air as its primary insulation material.

Parallel-conductor feed line—Feed line with two conductors spaced uniformly.

Polarization—Describes the characteristic of a radio wave. An antenna that is parallel to the surface of the earth, such as a dipole, produces horizontal polarization. One that is perpendicular to the earth's surface, such as a quarter-wave vertical, produces vertical polarization.

Radiate—To convert electric energy into electromagnetic (radio) waves. Radio waves radiate from an antenna.

Random-length wire antenna—An antenna having a length that is not necessarily related to the wavelength of a desired signal.

Reflector—An element in "back" of the driven element in a Yagi antenna.

Resonant frequency—The desired operating frequency of a tuned circuit. In an antenna, the resonant frequency is one where the feed-point impedance contains only resistance.

RF burn—A flesh burn caused by exposure to a strong field of RF energy.

Sloper—A ½-wave dipole antenna that has one end elevated and one end nearer the ground.

Standing-wave ratio (SWR)—Sometimes denoted as VSWR. A measure of the impedance match between the feed line and the antenna. Also, with a Transmatch in use, a measure of the match between the feeder from the transmitter and the antenna *system*. The system includes the Transmatch and the line to the antenna.

SWR meter—A device used to measure SWR.

Transmatch—See **Matching network**.

Transmission line—The wires or cable used to connect a transmitter or receiver to an antenna.

Twin lead—Parallel-conductor feeder with wires encased in insulation.

Vertical antenna—A common amateur antenna, usually made of metal tubing. The radiating element is vertical. There are usually four or more radial conductors parallel to or on the ground.

Wavelength—Often abbreviated λ. The distance a radio wave travels in one RF cycle. The wavelength relates to frequency. Higher frequencies have shorter wavelengths.

Yagi antenna—The most popular type of amateur directional (beam) antenna. It has one driven element and one or more additional elements.

Chapter 8

Choosing an Antenna

The "antenna farm" at W1AW (ARRL Headquarters) is a bit bigger than you'll need as a Novice. Large beam antennas on tall towers almost always assure the best communications.

We know that a transmitter generates radio-frequency energy. How do we convert this electrical energy into radio waves? We do it with an **antenna**. This is simply a piece of wire or other conductor designed to **radiate** the energy. An antenna converts current into an electromagnetic field (radio waves). The radio waves move away or *propagate* from the antenna. You might relate their travel to the ever-expanding waves you get when you drop a pebble in water. Waves from an antenna radiate in all directions, though, not just in a flat plane. It also works the other way. When a radio wave passes across an antenna, it generates a voltage in the antenna. That voltage isn't very strong, but it's enough to create a small current. That current travels through the **transmission line** to the receiver. The receiver detects the radio signal. In short, the antenna converts electrical energy to radio waves and radio waves to electrical energy. This process makes two-way radio communication possible with just one antenna.

Your success as a ham will depend heavily on the antenna you use. Although often the least expensive part of a station, the antenna is the most important part! A good antenna can make a fair receiver work like a champ. It can also make your 200 watts sound like a whole lot more. Remember, you'll normally use the same antenna both to transmit and receive. Clearly, any improvement you make to your antenna improves your transmitted signal. It also improves the strength of the signals you receive.

When you assemble your antenna system, you'll get a chance to use your creativity. You may discover, for example, that property size or landlord restrictions keep you from setting up an antenna "by the book." If so, you can innovate. There are some rather creative ways to get around many antenna problems. *The ARRL Antenna Book*, *The ARRL Antenna Compendium*, and similar publications offer suggestions. Later paragraphs give suggestions, too.

Some antennas work better than others. Antenna design and construction have been prime avocations of radio amateurs since Marconi. Experimentation continues today. You'll probably experiment with different antenna types over the years. Putting up a better antenna is a comparatively inexpensive and yet very rewarding way to improve your station.

WAVELENGTHS

Sometimes we talk about antenna lengths in wavelengths. A **wavelength** relates to the operating frequency. When you construct an antenna for one particular amateur band, you cut it to the proper wavelength. Often you'll see the lower case Greek letter lambda (λ) used as an abbreviation for wavelength. For example, ½ λ means "one-half wavelength."

Most popular ham antennas are less than 1 λ long. (A very popular antenna is a ½-λ dipole. You'll learn how to build one later in this chapter.) There is a simple relationship between operating frequency and wavelength. The wavelength is shorter at the higher frequencies. Wavelength is longer at the lower frequencies.

If numbers interest you, use this equation to find the wavelength for a specific frequency.

$$\lambda \text{ (in feet)} = \frac{984}{f \text{ (in MHz)}} \qquad \text{(Equation 8-1)}$$

With this equation the wavelength comes out in feet. The frequency is given in megahertz (MHz) for this equation. Let's say we wanted to know the wavelength for 7.15 MHz. We divide 984 by 7.15, and the answer is about 137.6 feet.

Whenever we talk about an antenna, we specify its design frequency. Or often we might refer instead to the amateur band it covers. We could talk about "an 80-meter dipole," for example, one intended for operation in the 80-meter band. Antennas are specially tuned circuits. A simple antenna such as a dipole or a ¼-λ vertical has a **resonant frequency**. Such antennas do best at their resonant frequency, just as most other tuned circuits do.

To change the resonant frequency of a tuned circuit, you would change the capacitor value or the inductor value. Since the antenna is often nothing more than a length of wire, you change its length to change its resonant frequency.

FEED LINES

To get the RF energy from your transmitter to an antenna you must use a **transmission line**. A transmission line is a special cable or arrangement of wires. Such lines commonly go by the name **feed line**. Sometimes we just call them **feeders** for short. They feed power to the antenna, or feed a received signal from the antenna to the receiver. In this section we'll talk about lines that have two conductors separated by insulating material.

CHARACTERISTIC IMPEDANCE

One electrical property of a feed line is its **characteristic impedance**. In Chapter 4 you learned that resistance is an opposition to electric current. Impedance is another term that describes the opposition to electric current that a circuit presents. Impedance includes factors other than resistance, however.

The spacing between line conductors and the type of insulating material pretty much determine the characteristic impedance. Characteristic impedance is important because we want the feed line to take all the transmitter power and feed it to the antenna. For this to occur, the transmitter (source) must have the same impedance as its load (the feed line). In turn, the feed line must have the same impedance as its load (the antenna).

We can use special circuits called **matching networks** if any of these impedances are different. Still, careful selection of a feed line can minimize such matching problems.

COAXIAL CABLE

Several different feeder types are available for amateur use. The most common is **coaxial cable**. Called "coax" (pronounced kṓ-aks) for short, this feed line has one conductor inside the other. It's like a wire inside a flexible tube. The center conductor is surrounded by insulation, and the insulation is surrounded by a wire braid called the shield. The whole cable is then encased in a tough vinyl outer coating, which makes the cable weatherproof. See Figure 8-1. Coax comes in different sizes, with different electrical properties.

The most common types of coax have either a 50-ohm

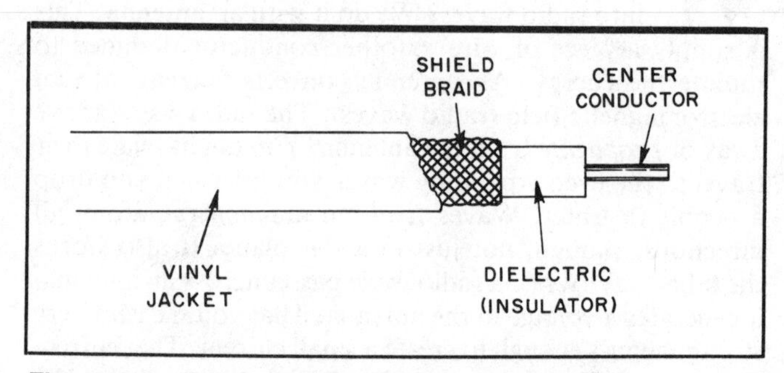

Figure 8-1—Coaxial cable, or coax, has a center conductor surrounded by insulation. The second conductor, called the shield, goes around that. Plastic insulation covers the entire cable.

or 72-ohm impedance. Coax designated RG-58 and RG-8 are 50-ohm cables. Some coax designations may also include a suffix such as /U, A/U or B/U, or bear the label "polyfoam." Feed line of this type may be used with most antennas. Cables labeled RG-59 or RG-11 are 72-ohm lines. Many hams use these types to feed dipole antennas.

The impedance of a ½-λ dipole in free space is about 73 ohms. Practical dipoles placed close to the earth, trees, buildings, etc, have an input impedance closer to 50 ohms. In any case, the small impedance mismatch caused by using 50- or 72-ohm cable as an antenna feeder is negligible.

In choosing the feed line for your installation, you'll have a trade-off between electrical characteristics and physical properties. The RG-58 and RG-59 types of cable are about ¼ inch in diameter, comparatively lightweight, and reasonably flexible. RG-8 and RG-11 are about ½ inch in diameter, nearly three times heavier, and considerably less flexible. As far as operation goes, RG-8 and RG-11 will handle much more power than RG-58 and RG-59. RG-58 and RG-59 will handle Novice power levels with ease, however.

Any line that feeds an antenna uses up a small amount of transmitter power. We consider this power lost, because it serves no useful purpose. (The lost power does nothing more than warm the feed line slightly.) The loss occurs because the wires are not perfect conductors, and the insulating material is not a perfect insulator. The larger coax types, RG-8 and RG-11, have less signal loss than the smaller types. If you use the smaller cable, it pays to keep the line length below 100 feet or so. Then you probably won't notice the small amount of additional signal loss, at least on the HF bands. This, combined with light weight and flexibility, is why most Novice operators find the smaller coax better suited to their needs. Also, the smaller feed line costs about half as much per foot as the larger types.

Coaxial cable has several advantages as a feed line. It is readily available, and is quite resistant to weather. It has a characteristic impedance near that of most common amateur antennas. Coax can be buried in the ground if necessary. It can be bent, coiled, and run next to metal with little effect. Its major drawback is that it is somewhat expensive. It is also somewhat cumbersome to attach to antennas and rigs because of its construction.

PARALLEL-CONDUCTOR FEED LINE

Parallel-conductor feed line is another popular line type. The most familiar example of this feeder is the 300-ohm ribbon used for TV antennas. It has two parallel conductors encased along the edges of a strip of plastic insulation. We often call this kind of line **twin lead**. See Figure 8-2A.

There is a type of twin lead designed for transmitting. This line uses heavier wires than the TV-receiving type. It has a characteristic impedance of 72 ohms. (This feed line is available from *QST* advertisers.) See Figure 8-2B.

Open wire feed line is a third type of parallel-conductor line. It contains two wires separated by plastic spacer rods. There is a rod every few inches along the feeder to maintain a uniform wire separation. The primary insulation is air. See Figure 8-2C. Often called **ladder line**, this type

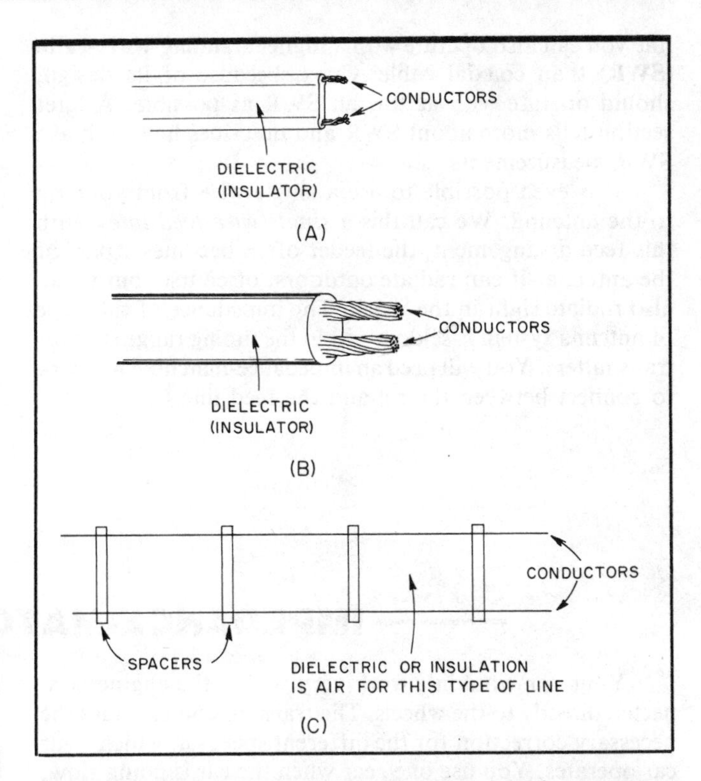

Figure 8-2—Parallel-conductor feed lines. At A, common 300-ohm TV twin lead. B shows 72-ohm transmitting twin lead, designed to carry higher power levels. C shows twin lead that uses air as the dielectric, with insulating spacers every few inches. Ladder line is its common name.

usually has a characteristic impedance between 450 and 600 ohms. The conductors can be bare wire, or they might be insulated with plastic. Ladder line can handle much higher power than twin lead.

You can make ladder line yourself. It is also available commercially, but may be difficult to find. A near equivalent coming into popularity is a cross between twin lead and ladder line. The wires are encased along the edges of a plastic ribbon, but the ribbon is not solid. Instead, it has punched rectangular holes along its length, leaving air as the primary insulation material. The common variety of this line has a 450-ohm impedance. For the same conductor spacing, this line has slightly more loss than ladder line. For most amateur work, the difference is negligible, however.

Parallel-conductor lines have some disadvantages. For example, they cannot be coiled or run next to metal drain pipes and gutters without adverse effects. Another drawback to this line is its characteristic impedance of 300 ohms or higher. It cannot be connected directly to most transmitters. Thus, you need an impedance **matching network** if you use any type of parallel-conductor feed line. Connect such a network between your transmitter and your feed line. If you use a matching circuit, even inexpensive TV twin lead can be used as your feed line.

The major advantage of ladder line is its very low loss. This means that for the same feeder length, more of your transmitter power will get to your antenna. With this feed line you can also operate with a higher **standing-wave-ratio (SWR)** than coaxial cable. Coax, because of its design, should operate with as low an SWR as possible. A later section tells more about SWR and describes how to make SWR measurements.

It is even possible to use a single wire from your rig to the antenna. We call this a *single-wire feed line*. With this feed arrangement, the feeder often becomes a part of the antenna. It can radiate outdoors, of course, but it can also radiate right in the shack. The impedance of this type of antenna system is seldom within the tuning range of most transmitters. You will need an impedance-matching network to connect between the rig and the feed line.

Many new amateurs worry needlessly about feed-line length. If you use coaxial cable between a rig and a dipole or vertical antenna, the cable can be any reasonable length. A good length to start with is simply the distance between the rig and the antenna! The feed line cannot be any shorter, and any extra length is probably unnecessary.

If you want to use your feed line to "tune" an unusual antenna system, consult an advanced text on the subject. Otherwise, cut the feed line to the length needed to get the RF to the antenna.

[Now study the questions in Chapter 12 with numbers that begin 2I-6 and 2I-7. Review this section if you have difficulty with any of the questions.]

IMPEDANCE-MATCHING NETWORKS

Your car wouldn't work very well if the engine connected directly to the wheels. The transmission provides the necessary correction for the different speeds at which your car operates. You use one gear when the car is going slow, another at moderate speeds, and still another at higher speeds.

An impedance **matching network** serves a function similar to the transmission in your car. Your transmitter won't operate very well if connected to a mismatched feed line. Let's say you want to use 50-ohm coax to feed a 35-ohm antenna. Your transmitter probably has an output circuit designed for a 50-ohm load. But that's not what the load will be with this antenna system. An impedance matching device will provide the proper impedance correction. A suitable network might contain only an inductor and a capacitor.

A **Transmatch** is a special type of matching network. These devices contain variable matching components (induc-

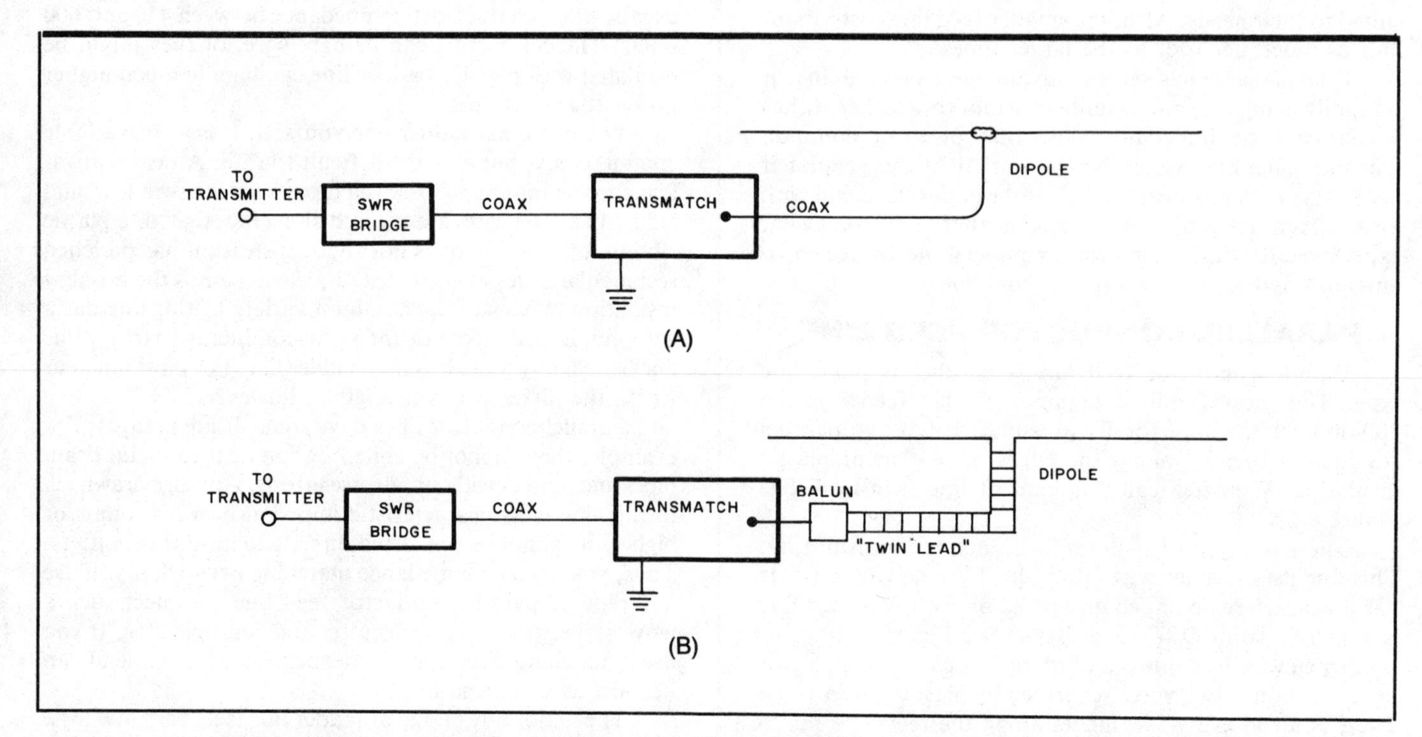

Figure 8-3—With a Transmatch you have flexibility in designing your antenna system. You can use the antenna on several bands, and the length isn't critical. You can use a dipole fed by coax (A) or twin lead (B). With a Transmatch, the dipole legs can be any length, although they should be as long as possible.

tors and capacitors), and often a band switch. They offer the flexibility of matching a wide range of impedances over a wide frequency range. With a Transmatch, it is possible to use one antenna on several bands. For example, you might use a center-fed wire antenna with inexpensive 300-ohm twin lead. Each band will use its own Transmatch settings; one combination for 80, one for 40, one for 15, and one for 10 meters.

Connect the Transmatch between the antenna and the **SWR meter**, as Figure 8-3 shows. (An SWR meter measures the standing-wave ratio—SWR—on an antenna feed line.) Tune the controls on your Transmatch for minimum SWR. Don't worry if you can't achieve a perfect match (1:1). Anything lower than 2:1 will work just fine.

Figure 8-4A shows the schematic diagram for a versatile Transmatch circuit. Part B shows a homemade Transmatch that includes an SWR meter, built from this circuit.

[Now study the questions in Chapter 12 numbered 2I-8.1 and 2I-8.2. Review this section if you have difficulty with the questions.]

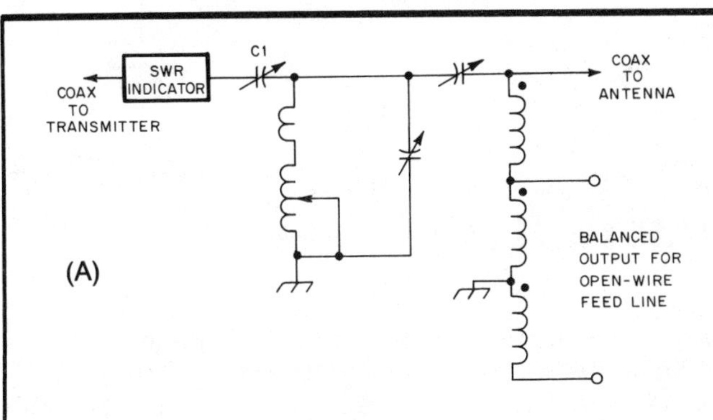

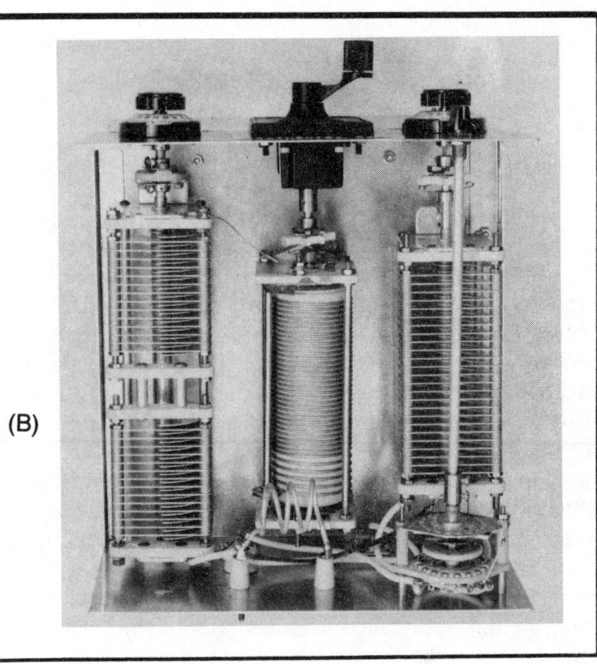

Figure 8-4—At A, the schematic diagram for a versatile Transmatch circuit. B shows a homemade Transmatch constructed from the circuit of A.

BALUNS

A center-fed wire with open ends, such as a dipole, is a *balanced* antenna. A balanced center-fed antenna has the following properties. The current flowing into one half of the antenna has the same level as that in the other half. The two currents are also opposite in phase. You can think of the term *balanced* as meaning that neither side connects to ground. A balanced antenna is balanced with respect to ground.

If we feed the antenna with a parallel-conductor line, we keep balance throughout the system. Parallel-conductor feed line is balanced. Neither side connects to ground. A 72-ohm twin-lead transmitting type of line is available. You can connect this feed line directly to a dipole antenna. Some amateurs feed their dipoles with such line to preserve system balance.

If we feed a dipole at the center with coax, we upset the system balance. One side of the antenna connects to the coax inner, or center, conductor. The other side connects to the coax shield. The shield connects to ground, so coax is an unbalanced line.

You don't have to go searching for 72-ohm twin lead to have a balanced dipole, however. You could feed it with ordinary 300-ohm twin lead. (To do this, you'd also need a Transmatch at the transmitter end of the line.) Another way is to feed your antenna with coaxial line and use a device called a **balun**. Balun is a contraction for *bal*anced to *un*-balanced. You'd install the balun at the antenna feed point.

Different types of baluns are available commercially. You can also make your own. A common type is a balun transformer, one with wires wound on a toroidal core. Besides providing a balance, these baluns can transform an impedance, such as from 50 to 75 ohms. Another type is a bead balun. Several ferrite beads go over the outside of the coax, in a stacked fashion. The beads tend to choke off any RF current that might otherwise flow on the outside of the shield.

You can easily make another type of choke balun from the coax itself. At the antenna feed point, coil up 10 turns of coax into a roll about 6 inches in diameter. Tape the coax turns together. The coiled turns will have inductance, which tends to choke off RF currents.

Feeding a dipole with parallel-conductor line doesn't assure a balanced antenna. You must also consider how the line connects to the transmitter. Most transmitters have a coaxial style connector. Suppose you are using 450-ohm parallel-conductor feed line. To feed such a balanced antenna system, you would need to install a balun at the transmitter end of the line. You'll need a balun at the transmitter end of the line if you are using any type of parallel-conductor feed line, including 72-ohm line or TV-type twin lead.

[Study questions in Chapter 12 that begin with 2I-9. Review this section if you have difficulty answering these questions.]

STANDING-WAVE RATIO (SWR)

If an antenna does not match the characteristic impedance of the feed line, there is a mismatch. Some of the transmitter energy reaching the antenna is reflected back down the line toward the transmitter. This reflected energy causes voltage standing waves on the line. When this happens the RF voltage is not uniform along the line. An **SWR meter** measures the **standing-wave ratio**. This is the ratio of the maximum voltage on the line to the minimum voltage. (These two points will always be ¼ λ apart.) Lower SWR values mean a better match exists between the antenna and the feed line. If a perfect match exists, the SWR will be 1:1. Your SWR meter thus gives a relative measure of how well the antenna impedance matches that of the feed line.

SWR METERS

The most common **SWR meter** application is tuning an antenna to the **resonant frequency** you want to operate on. (This discussion applies if you connect the feeder directly to the transmitter output, with no Transmatch.)

An SWR reading of 1.5:1 or less is quite acceptable. A reading of 4:1 or more is unacceptable. This means there is a mismatch between your antenna and your feeder.

To use the SWR meter, you transmit through it. You *must* have a license to operate a transmitter! What if your license has not arrived by the time you are ready to test your antenna? Just invite a licensed ham over to operate the transmitter.

How you measure the SWR depends on your type of meter. Some SWR meters have a SENSITIVITY control and a FORWARD-REFLECTED switch. If so, the meter scale usually gives you a direct SWR reading. To use the meter, first put the switch in the FORWARD position. Then adjust the SENSITIVITY control and the transmitter power output until the meter reads full scale. Some meters have a mark on the meter face labeled SET or CAL. The meter pointer should rest on this mark. Next, set the selector switch to the REFLECTED position. Do this *without* readjusting the transmitter power or the meter SENSITIVITY control. Now the meter pointer shows you the SWR value.

You can also measure SWR with a wattmeter, one showing RF power in watts. Your wattmeter may have meters to read both forward and reflected power. If not, it should have a switch or another way to change from forward to reflected power readings. To compute the SWR with a wattmeter, first note both the forward and reflected power for a given transmitter setting. Then consult a graph provided with the meter to find the corresponding SWR.

Find the resonant frequency of an antenna by connecting the meter between your transmitter and the feed line going to the antenna. Then measure the SWR at different frequencies across the band. Ideally, you will measure the lowest SWR at the center of the band, with higher readings at each end.

Sometimes your antenna may resonate far off frequency. See Figure 8-5. In this situation you will not get

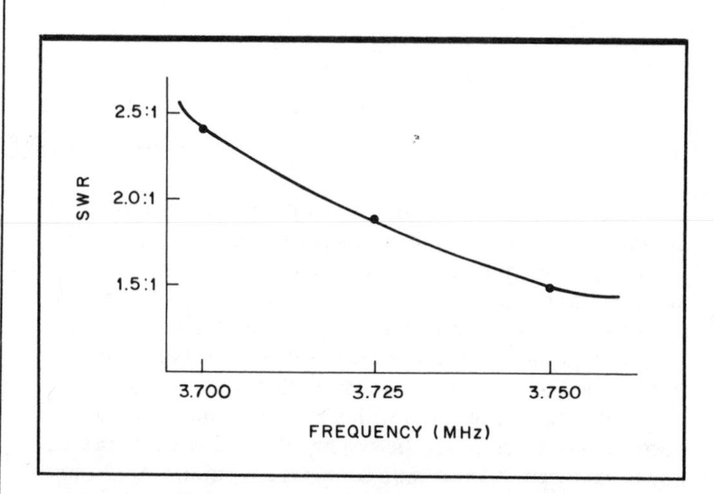

Figure 8-5—To determine if your antenna is cut to the right length, measure the SWR at different points in the band. Plot these values and draw a curve. Use the curve to see if the antenna is too long or too short. Adjusting the length will bring the lowest SWR to the desired frequency. Here the SWR is higher at the low end of the band. This antenna needs to be longer.

a "dip" in SWR readings with frequency. Instead, your readings will increase as you change frequency from one end of the band to the other. For example, you might read 2.5:1 at the low-freqency band edge, and 5:1 at the high-frequency end. This means antenna resonance is closer to the low-frequency end of the band than the high. It also means that resonance is below the low-frequency band edge. For a dipole or vertical antenna, this condition exists when the antenna is too long. Trimming the length will correct the problem.

Suppose the readings were 5:1 at the low-frequency end of the band and 2.5:1 at the high end. Here, the antenna is too short. Adding to the antenna length will correct this problem. Adjusting the antenna length for resonance in this way is what we call **tuning the antenna**.

The method described here to find resonance works for dipoles or vertical antennas. It does not show antenna resonance if you have a matching network between the SWR meter and the antenna. Nor does it show resonance for antenna systems that include a matching network at the antenna. An SWR meter *will* show when you have adjusted the matching network properly, however. Use the settings that give you the lowest SWR at your preferred operating frequency.

Finding Antenna Problems With An SWR Meter

At some point in your amateur career, it's likely that you'll run into antenna problems. Sometimes these problems occur when you first install an antenna. Or sometimes problems arise after the elements batter your antenna for weeks, months or years.

It's handy to have an SWR meter or power meter to help you diagnose antenna difficulties. This section tells how to interpret SWR meter readings in solving specific problems. This information applies to any type of antenna. We assume the antenna you're using normally provides a good match to your feed line at the measurement frequency.

One common problem is a loose connection where the feed line from your station attaches to the antenna wire. Splices or joints are another possible failure point. Your SWR meter will tell if the problem you're experiencing is a poor connection somewhere in the antenna. Observe the SWR reading. It should remain constant. If it is erratic, fluctuating markedly, chances are you have a loose connection. This problem is especially apparent on windy days.

If your SWR reading is unusually high, greater than 10:1 or so, you probably have a worse problem. *Caution:* Do not operate your transmitter with a very high SWR any longer than it takes to read the SWR! The problem could be an open connection or a short circuit. The most likely failure point is at the antenna feed point. The problem might be at the connector attaching your feed line to your transmitter. Carefully check your connections and your feed line for damage. You can also get unusually high SWR readings if the antenna is far from the correct length. This would happen if you try to operate your antenna on the wrong amateur band!

Most hams leave a meter in the line all the time, to monitor the SWR. Any sudden changes in the SWR may mean you have a problem, such as a broken wire or bad connection.

[Now study the questions in Chapter 12 with numbers that begin 2D-7-1, 2D-7-2 and 2D-7-3. Review this section if you have difficulty with any of these questions.]

PRACTICAL ANTENNAS

Hams use many different kinds of antennas. There is no one "correct" kind. Beginners usually prefer simpler, less expensive types. Some hams with more experience have antenna systems that cost thousands of dollars. Others have antennas that use several acres of property!

THE HALF-WAVE DIPOLE ANTENNA

A common Novice antenna is a wire cut to ½-λ at the operating frequency. The feed line attaches across an insulator at the center of the wire. This is the **half-wave dipole**. We often refer to an antenna like this as a **dipole antenna**. (*Di* means two, so a dipole has two equal parts. A dipole could be a length other than ½ λ.) The total length of a half-wavelength dipole is ½ λ. The feed line connects to the center. This means that each side of this dipole is ¼ λ long.

Use Equation 8-2 to find the total length of a ½-λ dipole for a specific frequency. Notice that the frequency is given in megahertz and the antenna length is in feet for this equation.

$$\text{Length (in feet)} = \frac{468}{\text{f (in MHz)}} \qquad \text{(Equation 8-2)}$$

For the center of the Novice bands, Equation 8-2 gives us the following approximate lengths for ½-λ dipoles.

Wavelength	*Frequency*	*Length*
80 meters	3.725 MHz	126 feet
40 meters	7.125 MHz	66 feet
15 meters	21.125 MHz	22 feet
10 meters	28.125 MHz	17 feet
10 meters	28.4 MHz	16.5 feet

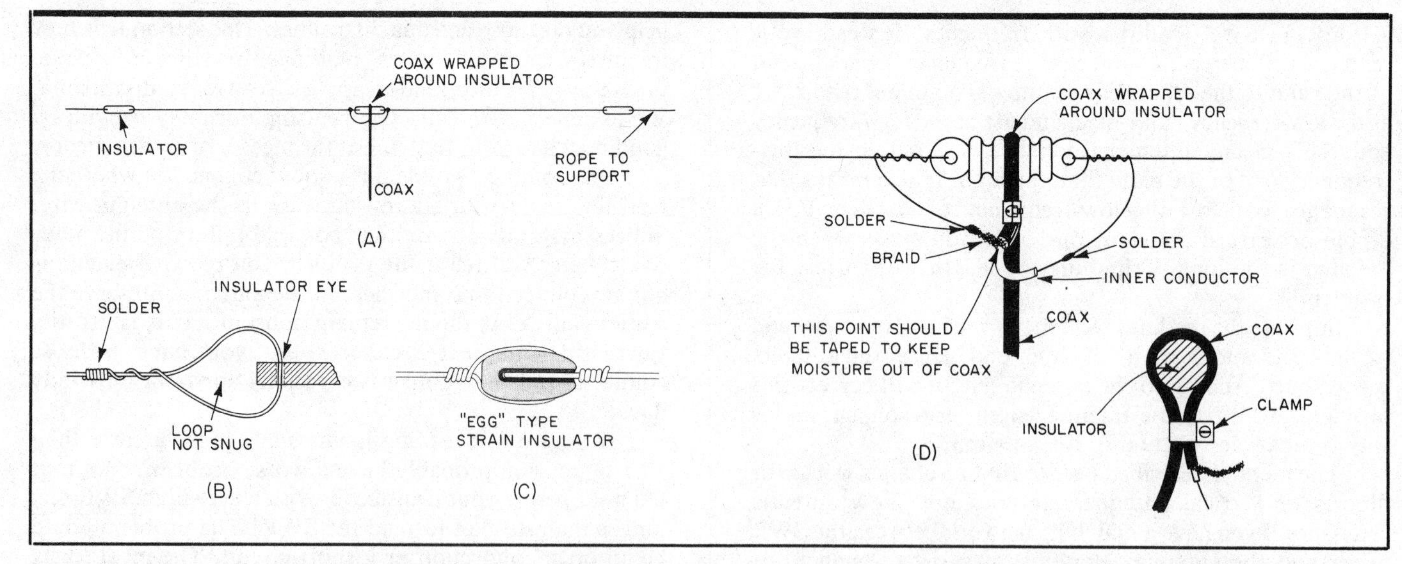

Figure 8-6—Simple half-wave dipole antenna construction. B and C show how to connect the wire ends to various insulator types. D shows the feed-line connection at the center.

Figure 8-6A shows the construction of a basic ½-λ dipole antenna. Parts B through D show enlarged views of how to attach the insulators. You can use just about any kind of wire for your dipole. It can range from regular electrical "zip cord" to strong copper-clad steel wire. If possible, use copper wire instead of steel or iron. Copper conducts electricity better. Most hardware or electrical supply stores carry suitable wire. Ordinary house wire, stripped of its insulation, works well. You can usually find copper-clad steel wire at a radio store. House wire and stranded wire will stretch with time, so a heavy gauge copper-clad steel wire is best. This wire consists of a copper jacket over a steel core. Such construction provides the strength of steel combined with the excellent conducting properties of copper.

Remember, you want a good conductor for the antenna, but the wire must also be strong. The wire must support itself *and* the weight of the feed line connected at the center.

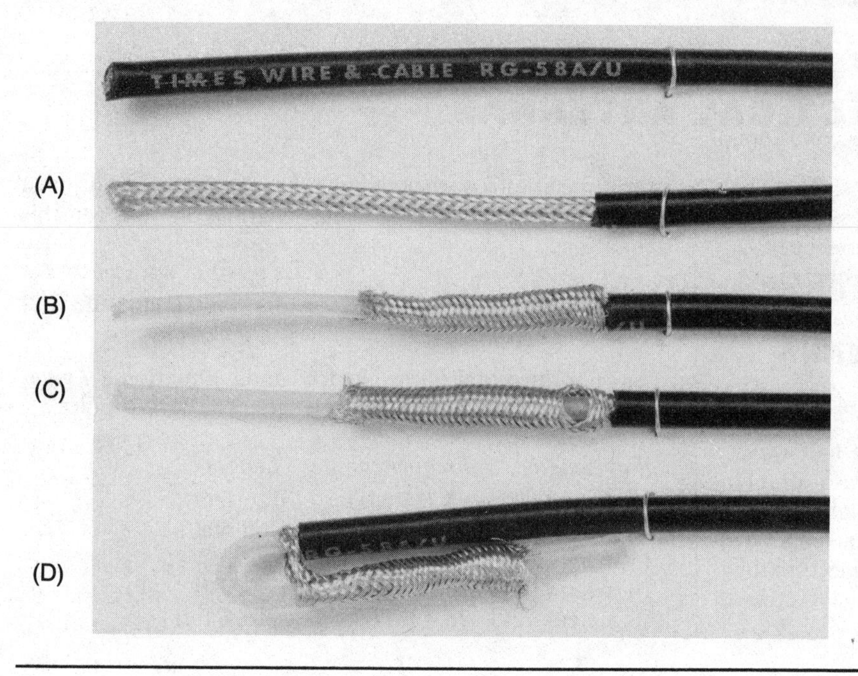

Figure 8-7—Preparing coaxial cable for connection to antenna wire. A—Remove the outer insulation with a sharp knife or wire stripper. If you nick the braid, start over. B—Push the braid in accordion fashion against the outer jacket. C—Spread the shield strands at the point where the outer insulation ends. D—Fish the center conductor through the opening in the braid. Now strip the center conductor insulation back far enough to make the connection and tin both center conductor and shield. Be careful not to use too much solder, which will make the conductors inflexible. Also be careful not to apply too much heat, or you will melt the insulation. A pair of pliers used as a heat sink will help. The outer jacket removed in step A can be slipped over the braid as an insulator, if necessary. Be sure to slide it onto the braid before soldering the leads to the antenna wires.

We use wire gauge to rate wire size. Larger gauge numbers represent smaller wire diameters. Conversely, smaller gauge numbers represent larger wire diameters. Although you can use almost any size wire for your dipole antenna, 12 or 14 gauge is usually best.

Cut your dipole to length (according to the dimension found by Equation 8-2). You'll need a feed line to connect it to your transmitter. For the reasons mentioned earlier, the most popular feed line for use with dipole antennas is coaxial cable. When you shop for coax, look for some with a heavy braided shield. If possible, get good quality cable that has at least 95 percent shielding. If you stick with name brand cable, chances are you'll get a good quality feed line. Figure 8-7 shows the steps required to prepare the cable end for attachment to the antenna wires at the center insulator.

The final items you'll need for your dipole are three insulators. You can purchase them from your local radio or hardware store (Figure 8-8). You can also make your own insulators from plastic or Teflon blocks. See Figure 8-9. One insulator goes on each end and another holds the two wires together in the center. Figure 8-10 shows some examples of how the feed line can attach to the antenna wires at a center insulator.

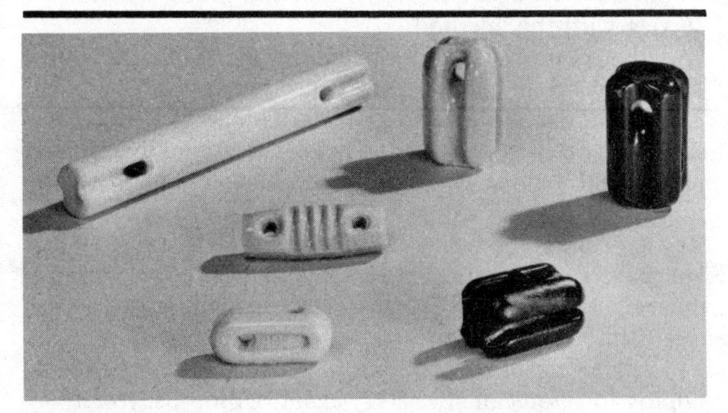

Figure 8-8—Various commercially made antenna insulators.

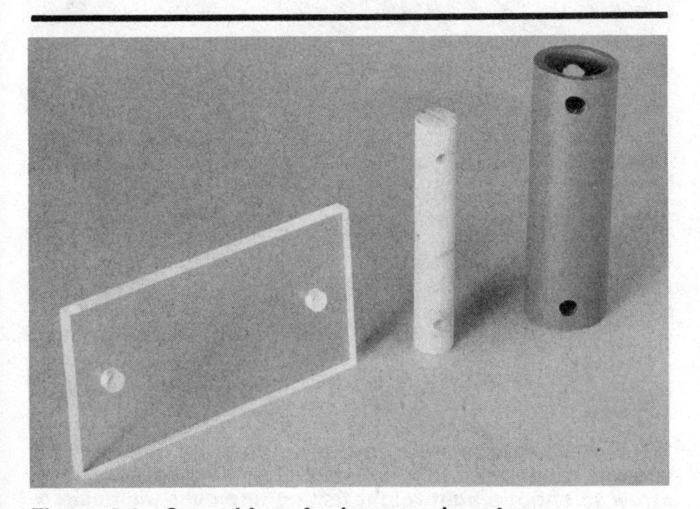

Figure 8-9—Some ideas for homemade antenna insulators.

Figure 8-10—Some dipole center insulators have connectors for easy feeder removal. Others have a direct solder connection to the feed line.

[Now turn to the questions in Chapter 12 with numbers that begin 2I-1 and study them. Review this section if you need to.]

ANTENNA LOCATION

Once you have assembled your dipole, find a good place to put it. *Never* put your antenna under, or over the top of electrical power lines. If they ever come into contact with your antenna, you could be electrocuted. Avoid running your antenna parallel to power lines that come close to your station. Otherwise you may receive unwanted electrical noise. Sometimes power-line noise can cover up all but the strongest signals your receiver hears. You'll also want to avoid running your antenna too close to metal objects. These could be rain gutters, metal beams, metal siding, or even electrical wiring in the attic of your house. Metal objects tend to shield your antenna, reducing its capability.

The key to good dipole operation is height. How high? One wavelength above ground is good, and this ranges from about 35 feet on 10 meters to about 240 feet on 80 meters. Of course very few people can get their antennas 240 feet in the air, so 40 to 60 feet is a good average height. Don't despair if you can only get your antenna up 20 feet or so, though. Comparatively low antennas can work well. Generally, the higher above ground and surrounding objects you can get your dipole, the greater the success you'll have. You'll find this to be true even if you can get only part of your antenna up high.

Normally you will support the dipole at both ends. The supports can be trees, buildings, poles or anything else high enough. Sometimes, however, there is just no way you can put your dipole high in the air at both ends. If you're faced with this problem, you have two reasonably good alternatives. You can support your dipole in the middle or at one end.

If you choose to support the antenna in the middle, both ends will droop toward the ground. This antenna, known as an **inverted-V dipole**, works best when the angle between the wires is no less than 90°. See Figure 8-11. If you use an inverted-V dipole, make sure the ends are high enough that no one can touch them. When you transmit, the high voltages present at the ends of a dipole can cause an **RF burn**.

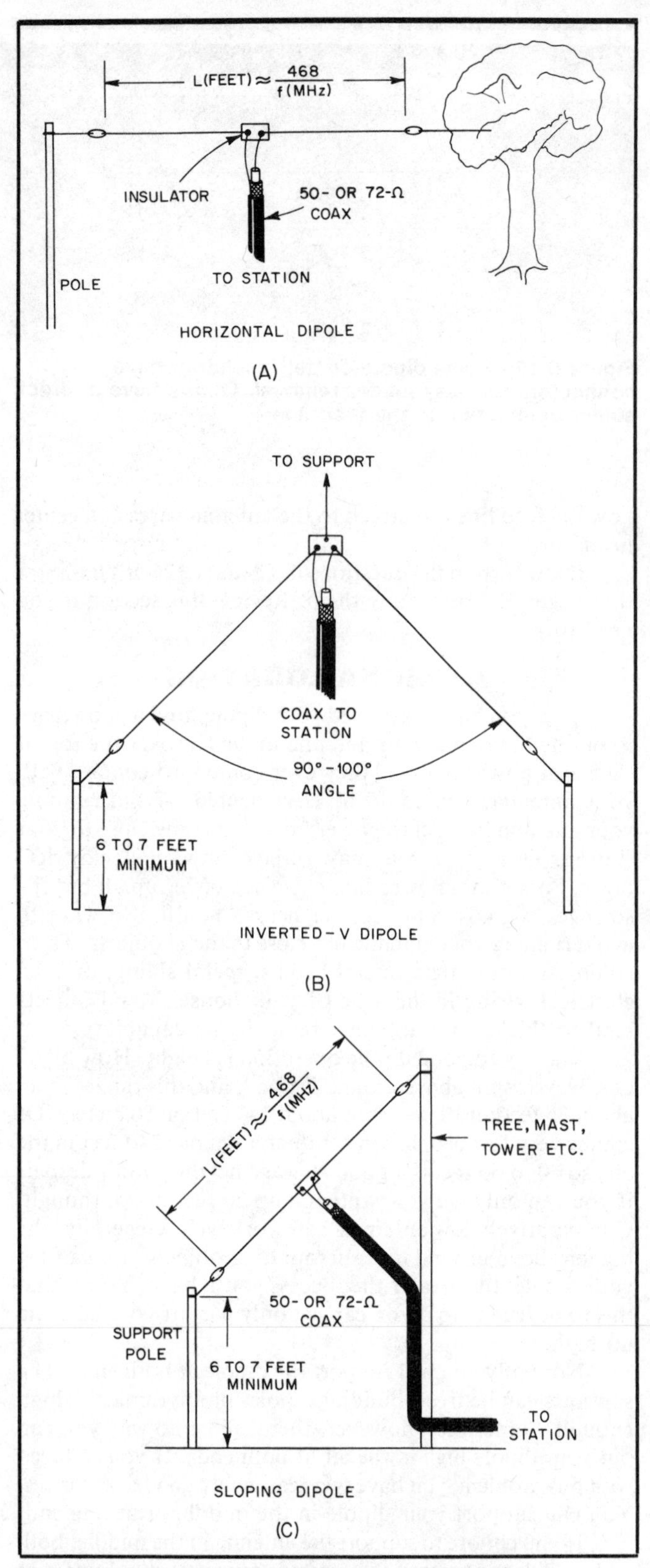

Figure 8-11—These are simple but effective wire antennas. A shows a horizontal dipole. The legs can be drooped to form an inverted-V dipole, as at B. C illustrates a sloping dipole (sloper). The feed line should come away from the sloper at a 90° angle for best results. If the supporting mast is metal, the antenna will have some directivity in the direction of the slope.

If you support your antenna at only one end, you'll have what is known as a *sloper*. This antenna also works well. As with the inverted-V dipole, be sure the low end is high enough to prevent anyone from touching it.

If you don't have the room to install a dipole in the standard form, don't be afraid to experiment a little. You can get away with bending the ends to fit your property, or even making a horizontal V-shaped antenna. Many hams have enjoyed countless hours of successful operating with antennas bent in a variety of shapes and angles.

ANTENNA INSTALLATION

After you've built your antenna and chosen its location, how do you get it up? There are many schools of thought on putting up antennas. Can you support at least one end of your antenna on a mast, tower, building or in an easily climbed tree? If so, you have solved some of your problems. Unfortunately, this is not always the case. Hams have used several methods over the years to get an antenna support rope into a tree. Most methods involve a projectile attached to a rope or line. You might be able to tie a rope around a rock and throw it over the intended support. This will work for low antennas. Even a major league pitcher, however, would have trouble getting an antenna much higher than 40 feet with this method.

The preferred method involves using a mechanical device to send the projectile over the highest possible branch. You could use a bow and arrow, a fishing rod, or even a slingshot. See Figures 8-12 and 8-13. You'll find that strong, lightweight fishing line is the best line to attach to the projectile. (Lead fishing weights make good projectiles.) Regular rope is too heavy to shoot any great distance. When you have successfully cleared the supporting tree, remove the weight. Then tie the support rope to the fishing line and reel it in.

If your first attempt doesn't go over the limb you were hoping for, you can always try again. Don't just reel in the line, however. Let the weight down to the ground first and

Figure 8-12—There are many ways to get an antenna support rope into a tree. These hams use a bow and arrow to shoot a lightweight fishing line over the desired branch. Then they attach the support rope to the fishing line and pull it up into the tree.

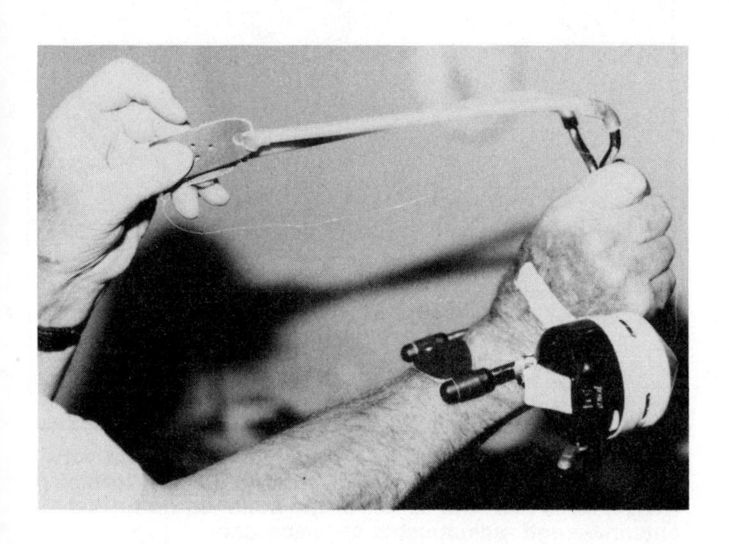

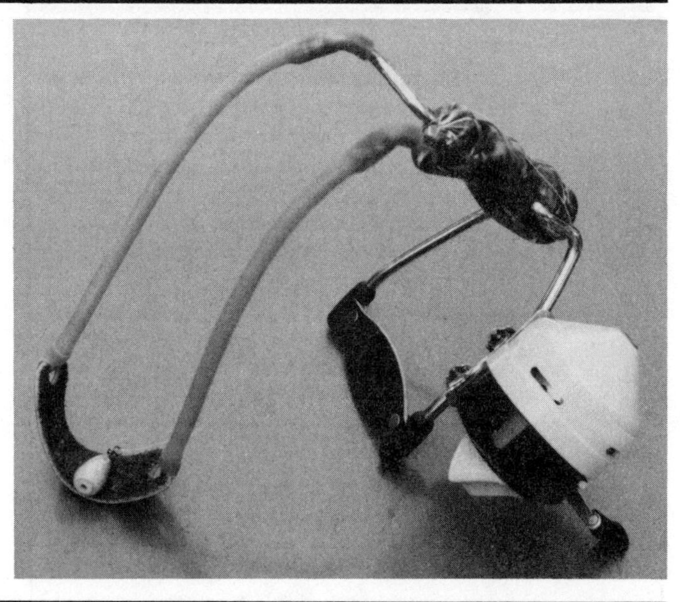

Figure 8-13—Another method for getting an antenna support into a tree. Small hose clamps attach a casting reel to the wrist bracket of a slingshot. Monofilament fishing line attached to a 1-ounce sinker is easily shot over almost any tree. Remove the sinker and rewind the line for repeated shots. When you find a suitable path through the tree, use the fishing line to pull a heavier line over the tree.

take it off the line. Then you can reel the line in without fear of getting the weight tangled in the branches.

You can put antenna supports in trees 120 feet and higher with this method. As with any type of marksmanship, make sure all is clear downrange before shooting. Your neighbors will not appreciate errant arrows, sinkers or rocks.

When your support ropes are in place, attach them to the ends of the dipole and haul it up. Pull the dipole reasonably tight, but not so tight that it is under a lot of strain. Tie the ends off so they are out of reach of passersby. Be sure to leave enough rope so you can let the dipole down temporarily if necessary.

Just one more step and your antenna installation will be complete. After routing the coaxial cable to your station, cut it to length and install the proper connector for your rig. Usually this connector will be a PL-259, sometimes called a UHF connector. Figure 8-14 shows how to attach one of these fittings to RG-8 or RG-11 cable. Follow the step-by-step instructions exactly as illustrated and you should not have any difficulty. Be sure to place the coupling ring on the cable *before* you install the connector body! If you are using RG-58 or RG-59 cable, use an adapter to fit the cable to the connector. Figure 8-15 illustrates the steps for installing the connector with adapter. The PL-259 is standard on most

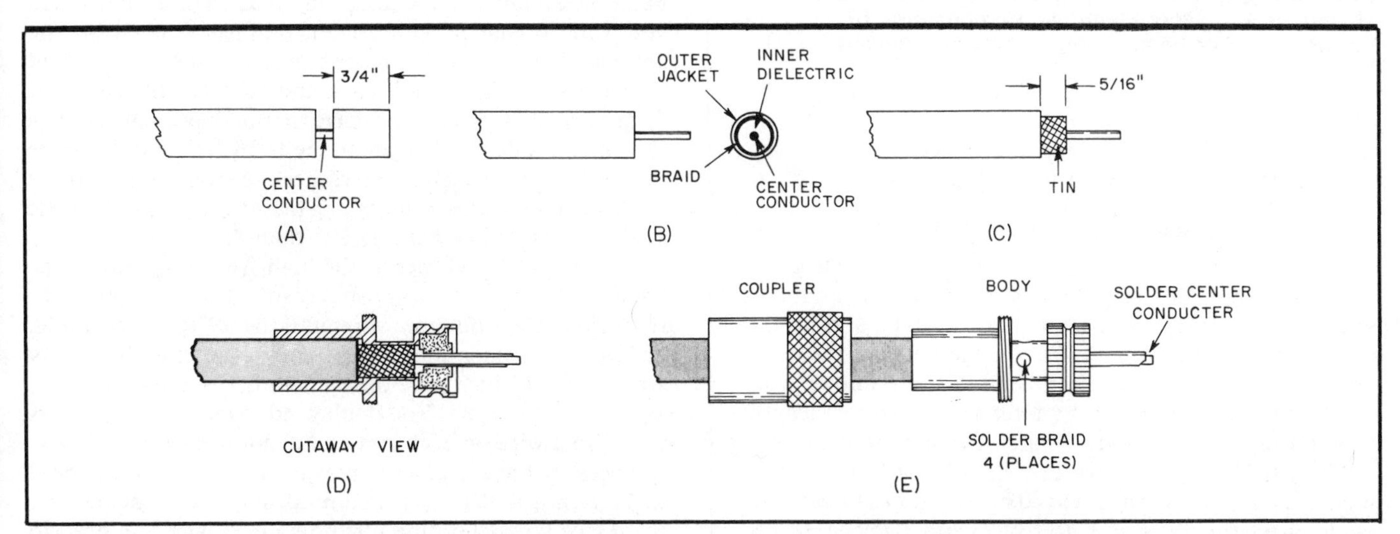

Figure 8-14—The PL-259 or UHF connector is almost universal for amateur HF work. It is also popular for equipment operating in the VHF range. Steps A through E illustrate how to install the connector properly.

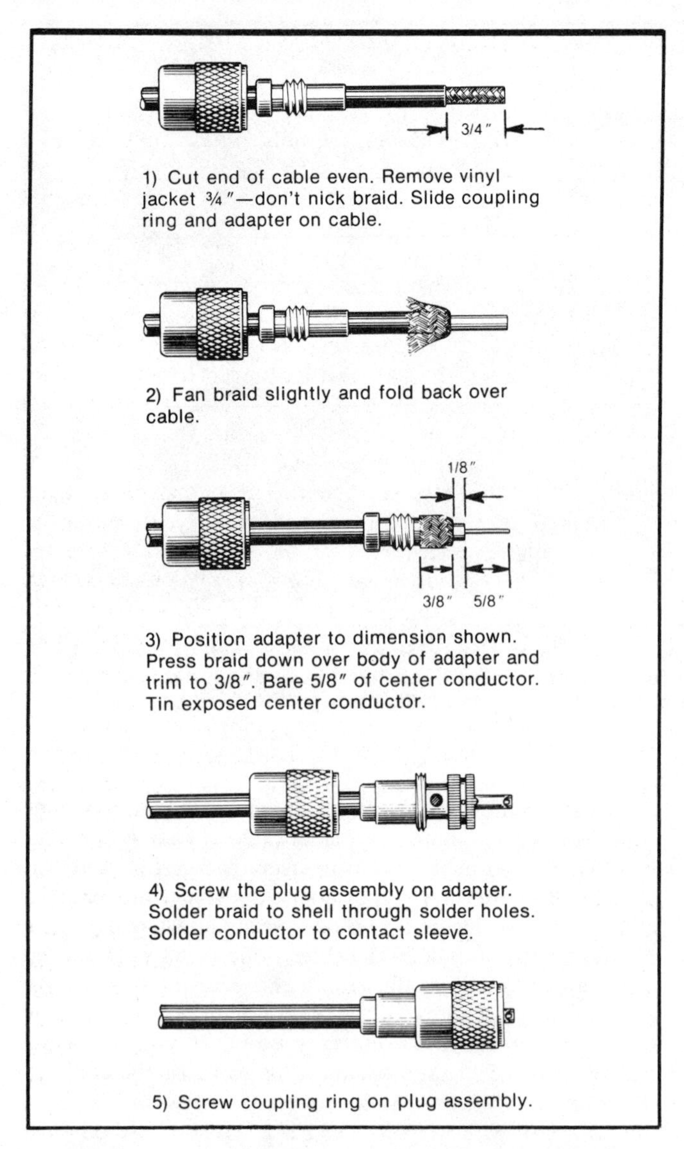

Figure 8-15—If you use RG-58 or RG-59 with a PL-259 connector, you should use an adapter, as shown here. Thanks to Amphenol Electronic Components, RF Division, Bunker Ramo Corp, for this information.

rigs. If you require another kind of connector, consult your radio instruction manual for installation information.

Tuning the Antenna

When you build an antenna, you cut it to the length given by an equation. This length is just a first approximation. Nearby trees, buildings or large metal objects, and height above ground all have an effect on the antenna resonant frequency. An SWR meter can help you determine if you should shorten or lengthen the antenna. The correct length provides the best impedance match for your transmitter.

The first step is to measure the SWR at the bottom, middle and top of the band. On 80 meters, for example, you would check the SWR at 3.701, 3.725 and 3.749 MHz. (A friend with a higher license class can help you check the SWR over a wider frequency range.) Graph the readings, as shown in Figure 8-16. You could be lucky—no further antenna adjustments may be necessary, depending on your transmitter.

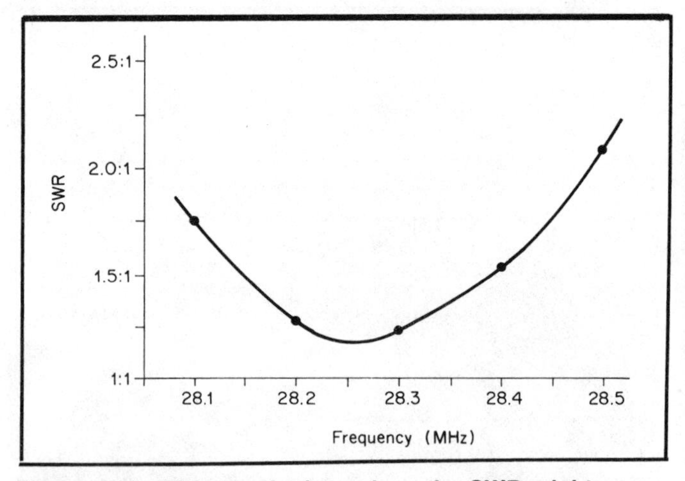

Figure 8-16—This graph shows how the SWR might vary across an amateur frequency band. The point of lowest SWR here is near the center of the band, so no further antenna-length adjustments are necessary.

Many tube-type transmitters include an output tuning network. These will usually operate fine with an SWR of 3:1 or less. Most solid-state transmitters (using all transistors and integrated circuits) do not include such an output tuning network. These *no-tune* radios begin to shut down—the power output drops off—with an SWR much higher than 1.5:1.

In any event, most hams like to prune their antennas for the lowest SWR they can get at the center of the band. With a full-size dipole 30 or 40 feet high, your SWR should be less than 2:1. If you can get the SWR down to 1.5:1, great! It's not worth the time and effort to do any better than that.

If the SWR is lower at the low-frequency end of the band, your antenna is probably too long. Disconnect the transmitter and try shortening your antenna at each end. The amount to trim off depends on two things. First is which band the antenna is operating on, and, second, how much you want to change the resonant frequency. Let's say the antenna is cut for the 80-meter band. You'll probably need to cut 8 or 10 inches off each end to move the resonant frequency 50 kHz. You may have to trim only an inch or less for small frequency changes on the 10-meter band. Measure the SWR again (remember to recheck the calibration). If the SWR went down, keep shortening the antenna until the SWR at the center of the band is less than 2:1.

If the SWR is lower at the high-frequency end of the band, your antenna was probably too short to begin with. If so, you must add more wire until the SWR is acceptable. Before you solder more wire on the antenna ends, try attaching a 12-inch wire on each end. Use alligator clips, as Figure 8-17 shows. You don't need to move the insulators yet. Clip a wire on each end and again measure the SWR. Chances are the antenna will now be too long. You will need to shorten it a little at a time until the SWR is below 2:1.

Once you determine just how much wire you need to add, cut two pieces and solder them to the ends. When you add wire, be sure to make a sound mechanical connection

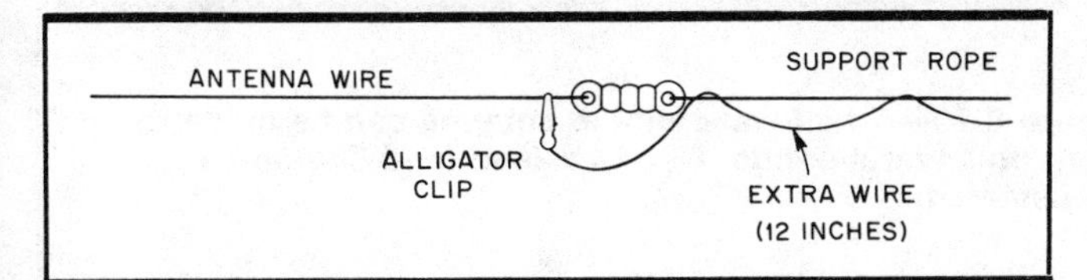

Figure 8-17—If your antenna is too short, attach an extra length of wire to each end with an alligator clip. Then shorten the extra length a little at a time until you get the correct length for the antenna. Finally, extend the length inside the insulators with a soldered connection. (See Figure 8-18.)

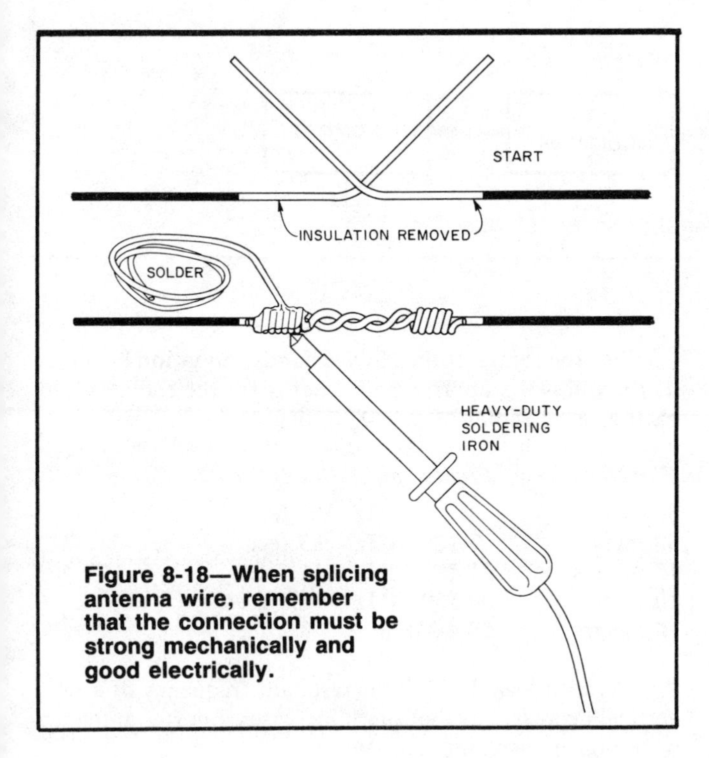

Figure 8-18—When splicing antenna wire, remember that the connection must be strong mechanically and good electrically.

before soldering. Figure 8-18 shows how. Remember that these joints must bear the weight of the antenna and the feed line. After you solder the wire, reinstall the insulators at the antenna ends, past the solder connections.

If the SWR is very high, you may have a problem that can't be cured by simple tuning. A very high SWR may mean that your feed line is open or shorted. Perhaps a connection isn't making good electrical contact. It could also be that your antenna is touching metal. A metal mast, the rain gutter on your house or some other conductor would add considerable length to the antenna. If the SWR is very high, check all connections and feed lines, and be sure the antenna clears surrounding objects.

That should give you enough information to construct, install and adjust your dipole antenna. You'll need a separate dipole antenna for each band you expect to operate on. Sometimes a 40-meter dipole will also work on 15 meters, though. Check the SWR on 15 meters—you may have a two-for-one dipole!

MULTIBAND DIPOLE OPERATION

The single-band dipole is fine if you operate on only one band. If you want to operate on more than one band, however, you could build and install a dipole for each band. What if you don't have supports for all these dipoles? Or what if you don't want to spend the money on several coaxial cable runs? The popular and inexpensive multiband dipole described here will solve your problems nicely.

A single-band dipole can be converted into a **multiband antenna** without too much difficulty. All you need to do is connect two additional ¼-λ wires for each additional band you want to use. Each added wire connects to the same feed line as the original dipole. The result is a single antenna system, fed with a single coaxial cable, that works on several different bands without adjustment. There is one potential problem with this antenna, though. The antenna will radiate signals on two or more bands simultaneously, so make sure your transmitter is adjusted properly. A poorly adjusted transmitter may produce harmonics of the desired output. If so, energy from your transmitter may show up on more than one band. The FCC takes a dim view of such operation!

Three-Band Dipole

You can build a three-band dipole for 80, 40 and 15 meters from ladder line. To build this antenna, you'll need a 100-foot roll of this line, three insulators and a coax feed line.

This antenna construction is similar to that for a regular dipole. Carefully remove the line from the spool and lay it on the ground. Take care to avoid twists and kinks in the wire. See Figure 8-19. At 33 feet, 6 inches from one end (X), cut *one* of the two wires. At 63 feet, 8 inches from the same end, cut the *other* wire. Remove the plastic spacers between these cuts, separating the open-wire line into two pieces. Measuring from the other end (Y), cut the shorter wire at 33 feet, 6 inches and the longer one at 63 feet, 8 inches. Figure 8-19A shows how the two antenna halves should look at this point.

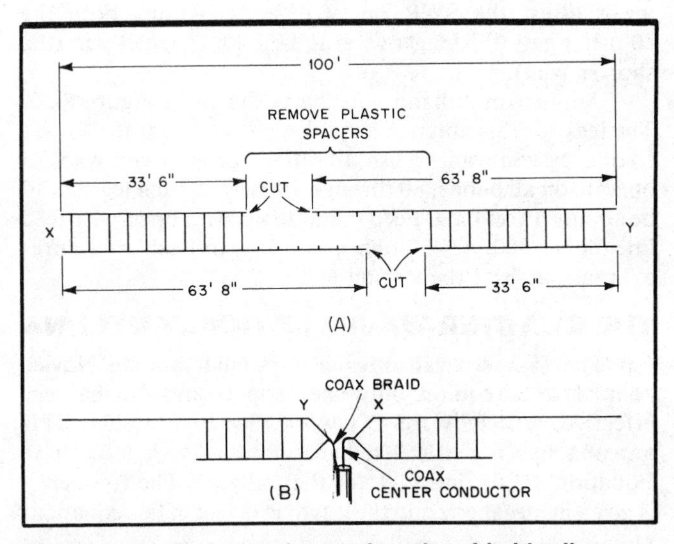

Figure 8-19—At A, cut the two lengths of ladder line as shown. Reverse the two sections to make the three-band dipole antenna, as shown at B.

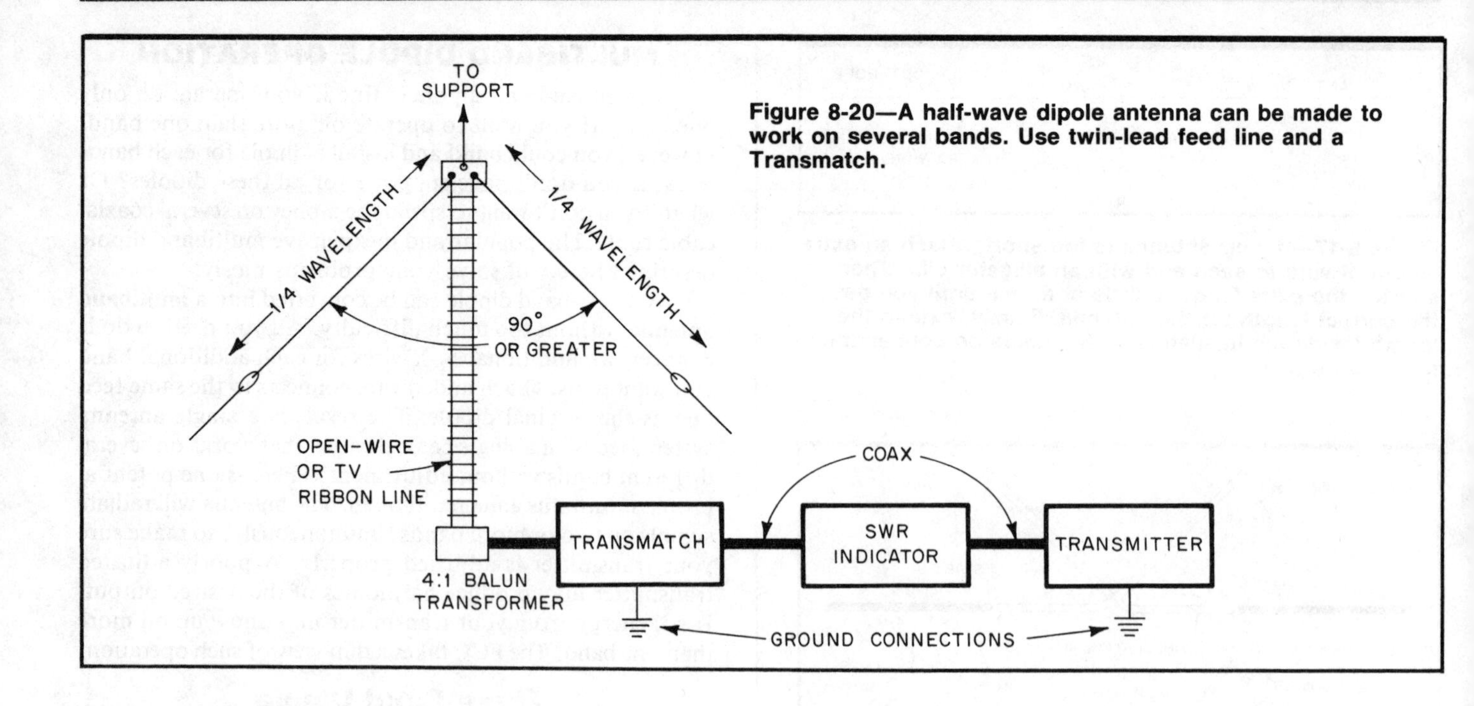

Figure 8-20—A half-wave dipole antenna can be made to work on several bands. Use twin-lead feed line and a Transmatch.

Reverse the position of the two halves, as Figure 8-19B shows. Now prepare the wire ends for connection to the feed line. Sandpaper the protective coating off both wires at the ends. (X and Y identify these ends in the drawing.) Connect the wire and a piece of coaxial cable in the same manner as described for a single-band dipole. Waterproof the connection with tape. Spray the taped connection with clear lacquer or coat it with rubber or silicone sealant for added protection against weather. Attach insulators to the ends, and your antenna is complete.

All information on antenna location and installation for a single-band dipole also applies to this multiband antenna. Use the same procedure described earlier to tune the antenna for the lowest SWR. Just remember that you must adjust the SWR on two bands. Adjust both the 80-meter part (the longer wire) and the 40/15-meter part (the shorter wire).

Another multiband antenna is shown in Figure 8-20. The legs for this antenna should each be ¼ λ at the lowest frequency you want to use. In other words, if you want to operate on all bands, 80 through 10 meters, each leg should be about 63 feet long. Feed this antenna with open-wire feed line (either 300- or 450-ohm), or TV twin lead. It requires a Transmatch at the station end.

THE QUARTER-WAVE VERTICAL ANTENNA

The ¼-λ **vertical antenna** is popular among Novice operators. It requires only one support and can be very effective, especially for DX work. See Figure 8-21. This antenna has a vertical radiator that is ¼-λ long. Use Equation 8-3 to find ¼ λ for the radiator. The frequency is given in megahertz and the length is in feet in this equation.

$$\text{Length (in feet)} = \frac{234}{\text{f (in MHz)}} \qquad \text{(Equation 8-3)}$$

For the center of the Novice bands, Equation 8-3 gives us the following approximate lengths for the radiator and each ground radial of a ¼-λ vertical.

Wavelength	*Frequency*	*Length*
80 meters	3.725 MHz	63 feet
40 meters	7.125 MHz	33 feet
15 meters	21.125 MHz	11 feet
10 meters	28.150 MHz	8.3 feet
10 meters	28.4 MHz	8.2 feet

As with ½-λ dipoles, the resonant frequency of a ¼-λ vertical decreases as the length increases. Shorter antennas have higher resonant frequencies.

The ¼-λ vertical also has wires laid on the ground, called ground radials. The key to successful operation with a ground-mounted vertical antenna is a good radial system. The best radial system uses *many* ground radials. Lay them out like the spokes of a wheel, with the vertical at the center. Some hams have buried ground radial systems containing over 100 individual wires. Ideally, these wires would be ¼ λ long or more at the lowest operating frequency. With such a system, earth or ground losses will be negligible. Radial length is not very critical, however. Studies show that with fewer radials you can use shorter lengths, but with a corresponding loss in antenna efficiency.[1,2] With 24 radials, there is no point in making them longer than about 1/8 λ. With 16 radials, a length greater than 0.1 λ is unwarranted. Four radials should be considered an absolute minimum.

[1]J. O. Stanley, "Optimum Ground Systems for Vertical Antennas," *QST*, December 1976, pp 14-15.

[2]B. Edward, "Radial Systems for Ground-Mounted Vertical Antennas," *QST*, June 1985, pp 28-30.

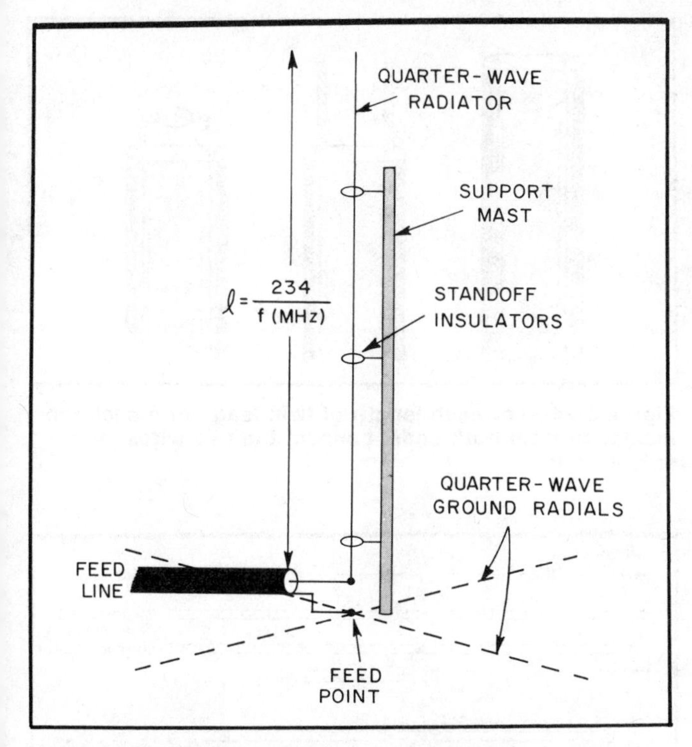

Figure 8-21—The quarter-wave vertical antenna has a center radiator and four or more radials spread out from the base. These radials form a ground plane.

Don't put the radials more than about an inch below the ground surface.

Earth losses are significant with ground-mounted verticals having only a few radials. These losses reduce antenna efficiency. Some of your transmitter power does no more than warm the earth beneath your antenna. Compared with 120 radials of 0.4 λ, antenna efficiency with 24 radials is roughly 63%. For 16 radials, the efficiency is roughly 50%. So it pays to put in as many radials as you can.

If you place the vertical above ground, you reduce earth losses drastically. Here, the wires should be cut to ¼ λ for the band you plan to use. Above ground, you need only a few radials—two to five. If you install a multiband vertical antenna above ground, use separate ground radials for each band you plan to use. These lengths are more critical than for a ground-mounted vertical. For elevated verticals, you should have two radials for each band, minimum. You can mount a vertical on a pipe driven into the ground, on the house chimney, or on a tower.

Once a vertical antenna is several feet above ground, there is little advantage in more height. (This assumes your antenna is above nearby obstructions.) For sky-wave signals, a height of 15 feet for the base is almost as good as 50. This is contrary to the case for a horizontal antenna, where height is important for working DX. Only if you can get the vertical up 2 or 3 wavelengths does the low-angle radiation begin to improve. Even then the improvement is only slight. At VHF and UHF it pays to get the antenna up high. At these frequencies you want the antenna to be above even distant obstructions.

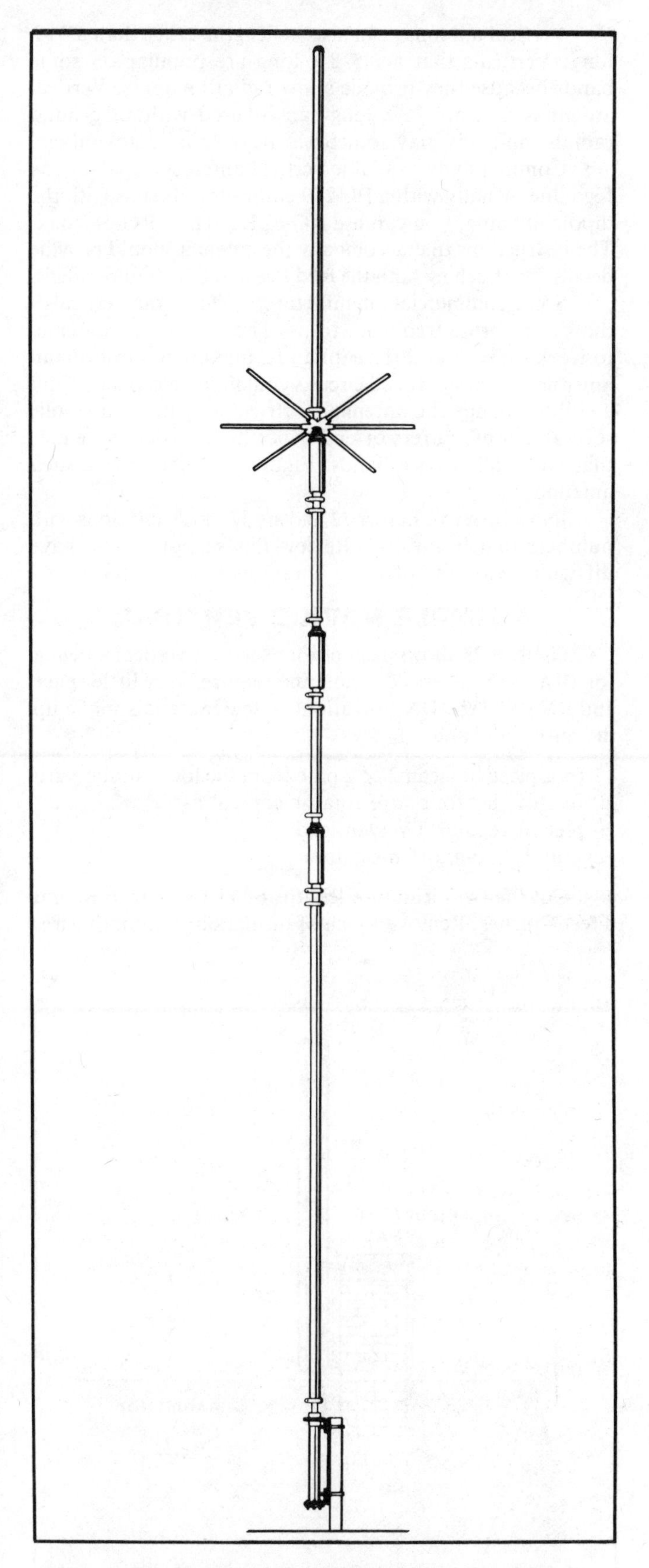

Figure 8-22—Commercially constructed "trap vertical" antennas generally look something like this. These antennas operate on several bands.

Vertical antennas can also be lengths other than a ¼ λ long. Verticals that are 5/8-λ long are popular on some bands because they provide a low radiation angle. Vertical antennas that are ½-λ long can be used without ground radials, and this may sometimes be a definite advantage.

Commercially available vertical antennas need a coax feed line, usually with a PL-259 connector. Just as with the dipole antenna, you can use RG-8, RG-11 or RG-58 coax. The instructions that accompany the antenna should provide details for attaching both the feed line and the ground radials.

Some commercial manufacturers offer "trap verticals" that incorporate frequency traps. These allow the antenna to work on several different bands, making it a **multiband antenna**. Traps are tuned circuits containing a capacitor and a coil to change the antenna electrical length. As a result, several manufacturers offer 20-foot-high vertical antennas that cover all Novice bands. Figure 8-22 shows one such antenna.

[Now turn to Chapter 12 and study those questions with numbers that begin 2I-2. Review this section if you have difficulty with any of these questions.]

A SIMPLE NOVICE VERTICAL

Figure 8-23 shows a simple, inexpensive vertical antenna for 10 and 15 meters. The antenna requires very little space and it's great for DX operation. A few materials make up the entire antenna.

12-foot piece of clean 2 × 2 pine from the local lumberyard
20 feet of flat four-wire rotator control cable
20 feet of regular TV twin lead
Several TV standoff insulators

Cut the twin lead into lengths of 11 feet 3 inches, and 8 feet 6 inches. Remove 1 inch of insulation from both wires

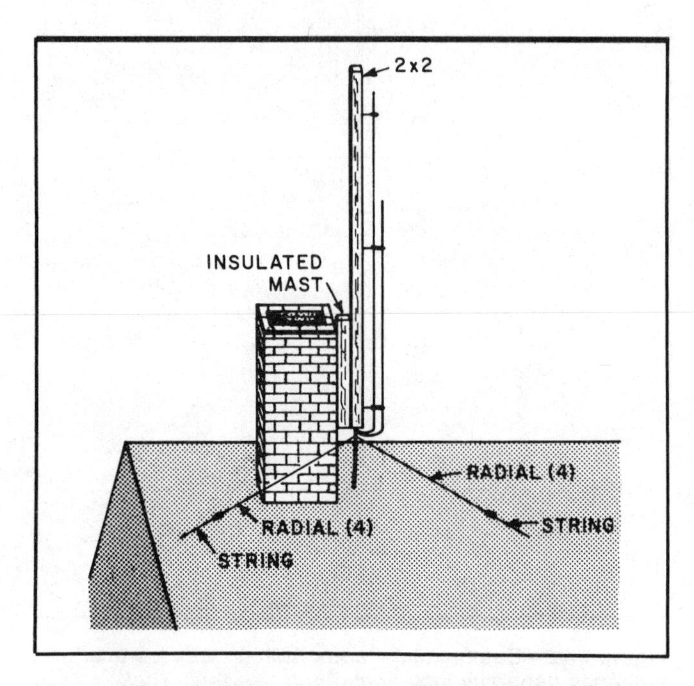

Figure 8-23—A simple Novice vertical antenna for use on 10 and 15 meters.

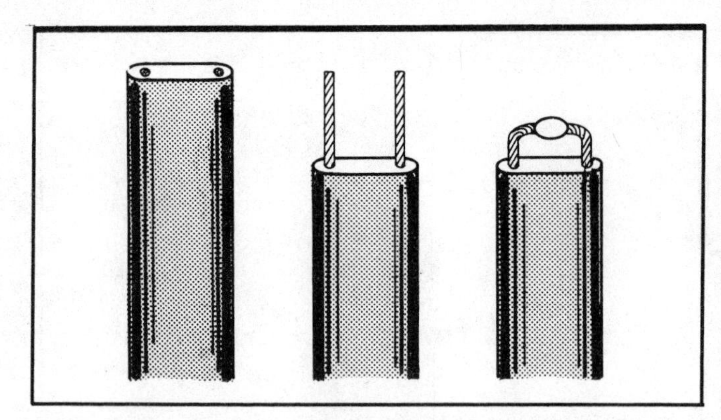

Figure 8-24—For each length of twin lead, strip back the insulation from both ends, connect the two wires, and solder them.

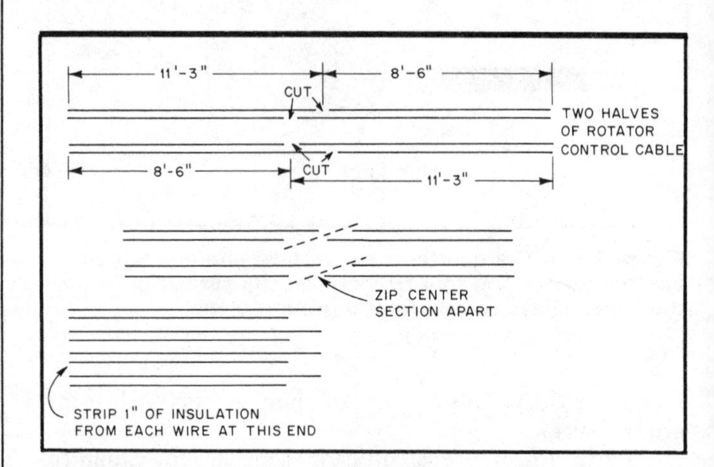

Figure 8-25—Cut some flat TV rotator-control cable to make the radials.

at both ends on each twin-lead length. Wrap the two wires together securely at each end, as Figure 8-24 shows, and solder. Cover one end of each length of twin lead with electrical tape.

Mount the TV standoff insulators at regular intervals on opposite sides of the 2 × 2. These will support the two pieces of twin lead. Separate the control cable into two pieces of two-conductor cable. Do this by slitting the cable a small amount at one end, then pulling the two pieces apart like a zipper.

Carefully cut the rotator cable as Figure 8-25 shows. Separate the pieces between the center cuts to make four identical sets of two-conductor cable. These make up the radials for the antenna system. Strip about 1 inch of insulation from each wire on the evenly cut end. The unevenly cut ends will be away from the antenna.

Attach a suitable length of RG-58 coaxial cable to the antenna. Connect the coax center conductor to the two wires on the 2 × 2. Then attach the cable braid to all the radial ends soldered together. Figure 8-26 shows the antenna construction. Make all the connections waterproof.

Now install the 2 × 2. You can do this in different ways. You could clamp it with U bolts to a TV mast, or hang it with a hook over a high branch. Or you might mount it

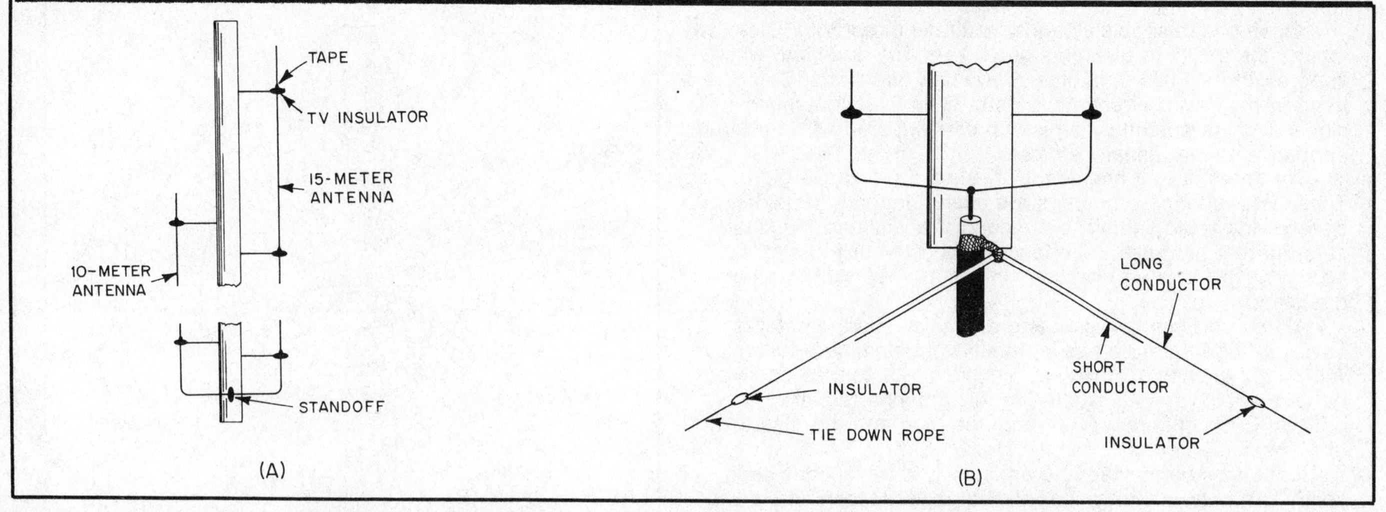

Figure 8-26—Attach the twin lead to the sides of the 2 × 2 with TV standoff insulators. A illustrates. At B, connect the feed line to both twin-lead lengths at the base.

on your house, at the side, near the roof peak. Mount the antenna as high as possible. Hang the radials at about a 45° angle away from the antenna base. For example, you could clamp the wood to your chimney with TV chimney-mount hardware. Let the radials follow the roof slope. Tie them off at the four corners. Unless you are surrounded by buildings or high hills, this should be a top DX antenna on 10 and 15 meters.

RANDOM-LENGTH WIRE ANTENNAS

If you can't install either a dipole or a vertical, you can still get on the air. Try a **random-length wire antenna**. See Figure 8-27. As the name implies, the antenna requires no specific length. As a rule of thumb, you should make your random-length antenna as long as possible. If you live in an apartment, you might use an antenna running along the ceiling in a few rooms. On the other hand, you may be able to string up a long length of wire outdoors. Use small wire (no. 22 to no. 28) if you want an antenna with low visibility. The only disadvantage of small wire is that it breaks easily. You may have to replace it often.

Random-length antennas have the advantage of versatility. They can be used anywhere. They do have one major disadvantage. They are not like the dipole and vertical, which can be fed directly from the transmitter through coaxial cable. The random-length antenna must be matched to the transmitter. That is because it has an impedance that is not 50 ohms.

The chances of a random-length wire working out to be exactly ¼ λ are very poor. Since electrical length determines resonant frequency, a random-length antenna probably won't resonate on any band you're likely to use. What results is an unacceptably high SWR. For this antenna you need a matching network or Transmatch.

BEAM ANTENNAS

Although generally impractical on 80 meters, and very large and expensive even on 40 meters, directional antennas often see use on 15 and 10 meters. The most common directional antenna that amateurs use is the **Yagi antenna**, but there are other types in use.

Generally called **beam antennas**, these directional antennas have two important advantages over dipole and vertical systems. First, the **directivity** of the antenna suppresses signals coming from directions other than where you point the antenna. This reduces the interference effects from stations in other directions, and increases your operating enjoyment. Second, a beam antenna concentrates the transmitted signal more in one direction than in others. The antenna provides **gain** in its pointed direction. Gain makes your signal sound stronger to other operators, and their signals sound stronger to you. Figure 8-28 shows some of the Yagi beams at W1AW, the station located at ARRL Headquarters in Newington, Connecticut. Figure 8-29 shows the typical radiation pattern of a Yagi beam.

A Yagi beam antenna has several elements attached to

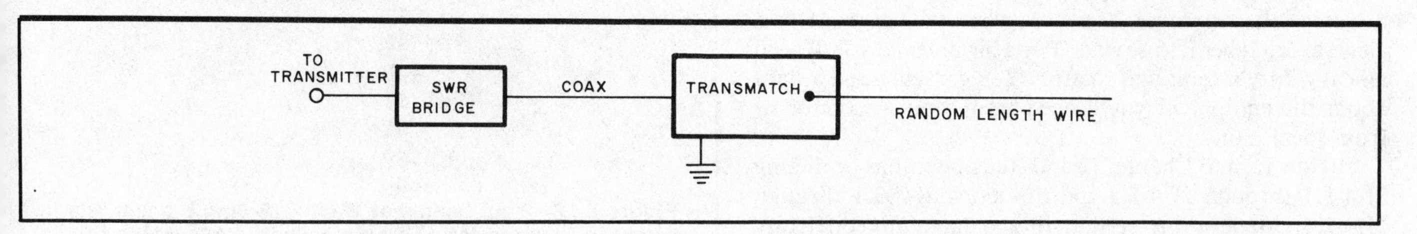

Figure 8-27—With a Transmatch, you can use a wire cut to a convenient random length as an antenna.

ANTENNA RADIATION PATTERNS

Often one desirable antenna feature is **directivity**. Directivity means the ability to pick up signals from one direction, while suppressing signals from other unwanted directions. Going hand in hand with directivity is **gain**. Gain tells how much signal a given antenna will pick up as compared with that from another antenna, usually a dipole.

An antenna that has directivity should also have gain. These two antenna properties are useful not only for picking up or receiving radio signals, but also for transmitting them. An antenna that has gain will effectively boost your transmitted energy in the favored direction while suppressing it in other directions.

When you mention gain and directivity, most amateurs envision large antenna arrays, made from aluminum tubing, with many elements. However, simple wire antennas can also be very effective, as illustrated in this antenna radiation pattern. Such patterns reveal both the gain and the directivity of a specific antenna.

Let's say we connect the antenna to a transmitter and send. The pattern shows the relative power received at a fixed distance from the antenna, in various compass directions. If you connect the antenna to a receiver, the pattern shows how the antenna responds to signals from various directions. In the direction where the antenna has gain, the incoming signals will be enhanced. The incoming signals will be suppressed in other directions.

Here is an important point to remember. You can never have antenna gain in one direction without a loss (signal suppression) in one or more other directions. Never! Another way to think of this is that an antenna cannot create power. It can only focus or beam the power supplied by the transmitter.

We call the long, thin lobes in a pattern the major lobes. The smaller lobes in a pattern are minor lobes. One or more major lobes mean directivity. An antenna with less directivity than this one would have fatter lobes. An antenna with no directivity at all would have a pattern that is a perfect circle. A theoretical antenna called an *isotropic radiator* has such a pattern. Radiation patterns are a very useful tool in measuring antenna capability.

The calculated or theoretical radiation pattern for an extended double Zepp antenna. In its favored directions, this antenna exhibits roughly 2 decibels gain over a half-wave dipole. This would make a 200-watt signal as strong as 317 watts into a dipole. An extended double Zepp antenna may be made with a horizontal wire hanging between two supports. (The wire is 1.28 λ long at the operating frequency and should be fed at the center with open-wire line.) The wire axis is along the 90/270-degree line shown in the chart.

a central *boom*, as Figure 8-30 shows. The feed line connects to only one element. We call this element the **driven element**. On a three-element Yagi like the one shown in Figure 8-30, the driven element is in the middle. The element at the front of the antenna (toward the favored direction) is a **director**. Behind the driven element is the **reflector** element. The driven element is about ½ λ long at the antenna design frequency. The director is a bit shorter than ½ λ, and the reflector a bit longer.

Yagi beams can have more than three elements. Seldom is there more than one reflector. Instead, the added elements are directors. A standard 4-element Yagi would have a reflector, a driven element, and two directors.

Because beams are directional, you'll need something to turn them. A single-band beam for 10 or 15 meters can be mounted in the same manner as a TV antenna. You can use a TV mast, hardware and rotator. You could plan to buy a large triband beam for use on 10 and 15 meters as a Novice. It will cover 20 meters when you upgrade to a General or higher license class. For a big antenna you'll need a heavy-duty mount and rotator. You can get good advice about the equipment you'll need from your instructor or from local hams.

[Now turn to Chapter 12 and study questions beginning 2I-4-1.1 through 2I-4-1.3 and questions 2I-4-2.1 through 2I-4-2.4. Review this section if you have any difficulty answering these questions.]

Figure 8-28—Yagi beams at W1AW. A single beam sits atop the tower at left. Three stacked beams adorn the taller tower at the right.

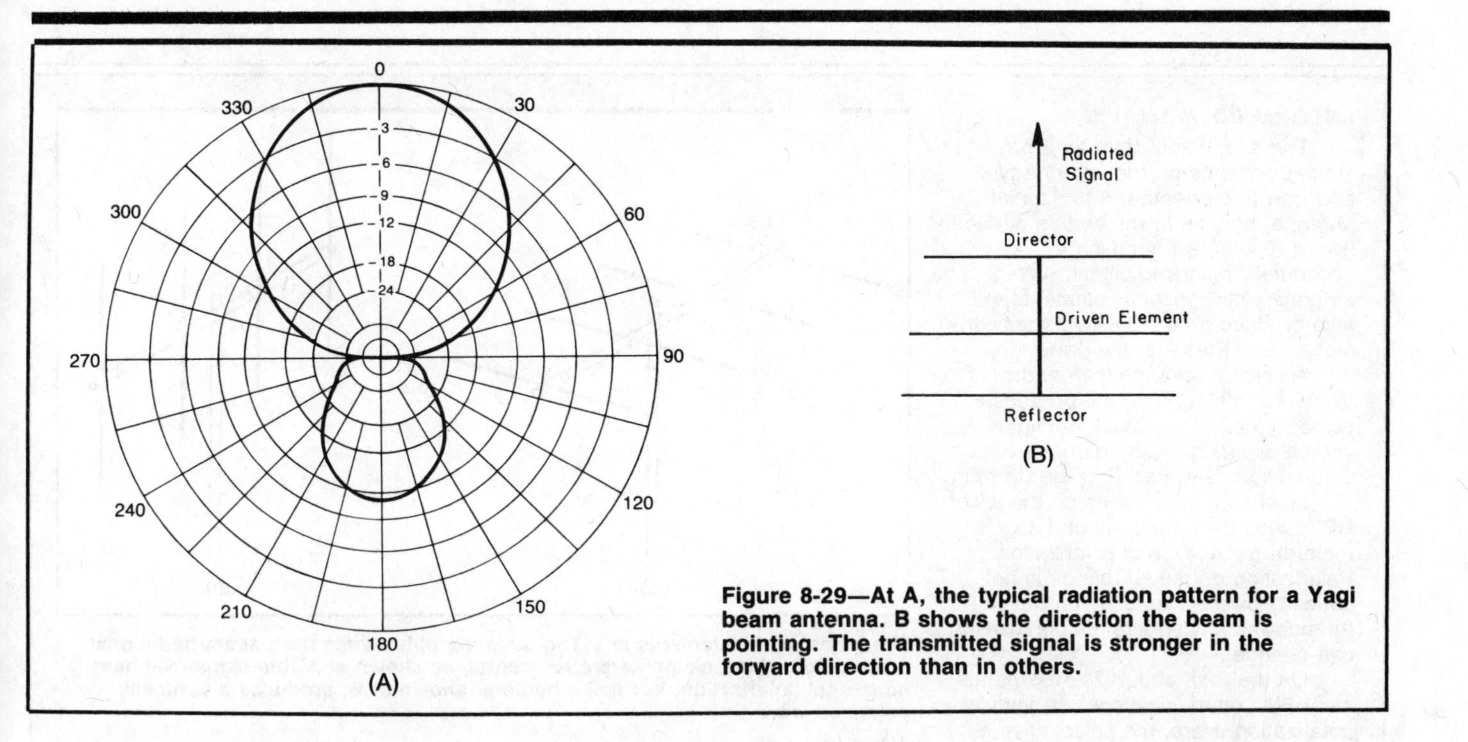

Figure 8-29—At A, the typical radiation pattern for a Yagi beam antenna. B shows the direction the beam is pointing. The transmitted signal is stronger in the forward direction than in others.

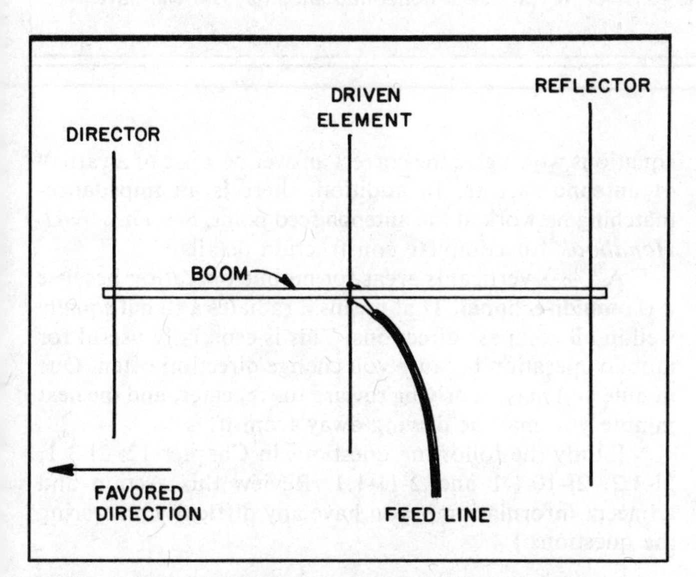

Figure 8-30—A three-element Yagi antenna has a director, a driven element and a reflector. A boom supports the elements.

CONVERTING CB ANTENNAS

It is usually a simple matter to convert a Citizen's Band (CB) antenna to work on the Novice 10-meter band. This is because the Citizens Band at 27 MHz is close in frequency to the Novice 10-meter amateur band. The formula for the correct length is quite simple—468 divided by the frequency in megahertz (Equation 8-2). This answer gives the antenna length in feet. For example, 468 divided by 27 MHz equals 17.3 feet or 17 feet, 4 inches. The most common antenna type used by CBers, however, is a ¼-λ vertical. Its length would be one-half that, or 8 feet, 8 inches long (Equation 8-3).

To have such antennas resonant (or tuned) on 10 meters, you merely shorten them to the required length. Using Equation 8-2, ½ λ for 28.1 MHz works out to be 16 feet, 8 inches. The length for a ¼-λ vertical, using Equation 8-3, is 8 feet 4 inches. This means removing 8 inches from a dipole, or 4 inches from a full-sized vertical (not loaded with a coil). Before shortening any CB antenna, try the antenna on 10 meters. If your transmitter loads and tunes up, it may be a waste of time to prune the antenna. It may work well just as it is!

ANTENNA POLARIZATION

For VHF and UHF base station operation, most amateurs use a vertical or a beam. For VHF and UHF FM and repeater operation, almost everyone uses vertical **polarization**. For SSB and CW VHF/UHF work, virtually everyone uses a beam with horizontal polarization. If you plan to do both VHF weak-signal and FM repeater work, you'll probably need separate antennas for each. This is because signal strength suffers greatly if your antenna has different polarization than the station you're trying to work.

A beam antenna is a bit unwieldy for VHF and UHF FM mobile operation. Most hams use a vertical of some kind. Mobile antennas are available in several varieties. Most

ANTENNA POLARIZATION

The signal sent from an amateur station depends on the antenna type and how it is oriented. A horizontal antenna, parallel to the earth's surface (like a dipole), will produce a *horizontally polarized* signal. A Yagi antenna with horizontal elements will also produce a horizontally polarized signal. See Part A in the drawing.

A vertical antenna (perpendicular to the earth's surface) will produce a *vertically polarized* signal. A Yagi with vertical elements also produces vertical polarization. See Part B in the drawing.

Most communications on the lower HF Novice bands (80, 40 and 15 meters) use horizontal polarization. Polarization on the HF bands is not critical, however. As a signal travels through the ionosphere, its polarization can change.

On the 220- and 1270-MHz bands, most FM communications use vertical polarization. Here, the polarization *is* important. The signals retain their polarization from transmitter to receiver. If you use a horizontal antenna, you will have difficulty working through a repeater with a vertical antenna.

(A) (B)

The plane of the elements in a Yagi antenna determines the transmitted signal polarization. If the elements are horizontal, as shown at A, the signal will have horizontal polarization. Vertical mounting, shown at B, produces a vertically polarized wave.

mount on the automobile roof or trunk lid, but some even mount on a glass window.

Most vertical antennas used at lower frequencies are ¼-λ long. For VHF and UHF, antennas are physically short enough that longer verticals may be used. A popular mobile antenna is a **5/8-λ vertical**, often called a "five-eighths whip." This antenna is popular because it offers more gain than a regular ¼-λ vertical. Simply stated, "gain" means a concentration of transmitter power. A 5/8-λ vertical concentrates the power toward the horizon. Naturally, this is the most useful direction, unless you want to talk to airplanes or satellites. At 220 MHz, a 5/8-λ whip is only 28½ inches long.

Don't use any of the equations in this chapter to calculate how long a 5/8-λ whip is at 220 MHz. The equations won't give the correct answer because of a variety of antenna factors. In addition, there is an impedance-matching network at the antenna feed point. See *The ARRL Handbook* for complete construction details.

A 5/8-λ vertical is great for mobile operation because it is **omnidirectional**. That means it radiates a signal equally well in all compass directions. This is especially useful for mobile operation because you change direction often. One minute you may be driving toward the repeater, and the next minute you may be driving away from it.

[Study the following questions in Chapter 12: 2I-3.1, 2I-3.2, 2I-10-1.1 and 2-11-1.1. Review this section and adjacent information if you have any difficulty answering the questions.]

INSTALLATION SAFETY

No matter what antenna type you choose to build, you should remember a few key points about safety. If you use a slingshot or bow and arrow to get a line over a tree, use caution. Make sure you keep everyone away from the "downrange" area. Hitting a helper with a rock or a fishing sinker is not a good idea!

If you build a wire antenna, be sure the antenna ends are out of reach of passersby. The ends should be high enough that a person on the ground cannot touch them. Even at Novice power levels, enough voltage exists at the antenna ends when you transmit to cause RF burns. If you have a vertical antenna with its base at ground level, you might consider building a wooden safety fence around it. The fence should be at least 4 feet away from the antenna. Do not use a metal fence, as this will interfere with the proper antenna operation. If you have a tower, you should either fence it or put shields directly on the tower to prevent unauthorized climbing.

Antenna work often requires that someone climb up on a tower, into a tree, or onto the roof. Never work alone! Work slowly, thinking out each move before you make it. The person on the ladder, tower, tree or rooftop should

HOW TO LIVE LONG ENOUGH TO UPGRADE

Keep antenna safety in mind when you're setting up your station. Antennas for the ham bands are often large, requiring care and attention to detail when installed. Here are two points to keep in mind when putting up your antenna.

1) Be sure your antenna materials and supports are strong—strong enough to withstand heavy winds without breaking.

2) Keep away from power lines!

If your antenna falls, it could damage your house, garage or property. If it falls into a power line, your house might end up like the unlucky CBer's shown in the photo. A windstorm knocked over his groundplane antenna, sending it into a 34,500-V power line. The resulting fire damaged his house extensively.

Safety pays! Accidents *are* avoidable, if you use good sense.

always wear a safety belt, and keep it securely anchored. Before each use, inspect the belt carefully for damage such as cuts or worn areas. The belt will make it much easier to work on the antenna and will also prevent an accidental fall.

Never try to climb a tower carrying tools or antenna components in your hands. Carry what you can with a tool belt, including a long rope leading back to the ground. Then use the rope to pull other needed objects up to your workplace after securing your safety belt! It is helpful (and safe) to tie strings or lightweight ropes to all tools. You can save much time in retrieving dropped tools if you tie them to the tower. This also reduces the chances of injuring a helper on the ground.

Helpers on the ground should never stand directly under the work being done. All ground helpers should wear hard hats for protection. Even a small tool can make quite a dent if it falls from 50 or 60 feet. A ground helper should always observe the tower work carefully. Have you ever wondered why electric utility crews seem to have someone on the ground "doing nothing"? Now you know that for safety's sake, a ground observer with no other duties is free to notice potential hazards. That person could save a life by shouting a warning.

As mentioned earlier, be especially certain that your antenna is not close to any power wires. That is the only way you can be sure it won't come in contact with them!

When using a hand-held transceiver that runs more than a few watts, be careful! Always keep the antenna away from your head and away from others standing nearby. Some antennas are safer than others. A short, helically wound flexible antenna is not the safest kind. It concentrates its radiation in a small area near your head. A longer antenna, such as a ½-λ whip, would be safer for a hand-held. A ½-λ antenna concentrates its radiation at its center, which will be farther from your head.

[Now study the questions in Chapter 12 with numbers that begin 2D-6 and 2I-5. Review this section if you have any problems.]

KEY WORDS

Antenna switch—A device used to switch an amateur station between several antennas.

Coordinated Universal Time (UTC)—A system of time referenced to time at the prime meridian, which passes through Greenwich, England.

Dummy load (dummy antenna)—A resistor that provides a transmitter with a proper load. The resistor gets rid of the transmitter output power without radiating a signal.

Electronic keyer—A device used to generate Morse code dots and dashes electronically. One input generates dots, the other, dashes. Character speed is usually adjusted by turning a control knob. Speeds range from 5 or 10 words per minute up to 60 or more.

Feed line—The wires or cable used to connect your transmitter and receiver to an antenna.

Field-effect transistor volt-ohm-milliammeter (FET VOM)—A multiple-range meter used to measure voltage, current and resistance. The meter circuit uses an FET amplifier to provide a high input impedance. This leads to more accurate readings than can be obtained with a VOM. This is the solid-state equivalent of a VTVM.

Ground connection—A connection made to the earth for electrical safety.

QSL card—A postcard sent to another radio amateur to confirm a contact.

Safety Interlock—A switch that turns off ac power to a piece of equipment when someone removes the top cover.

Standing-wave-ratio (SWR) meter—A device used for measuring SWR. SWR is a relative measure of the impedance match between an antenna, feed line and transmitter.

Transmit-receive (TR) switch—A device used for switching between transmit and receive operation. This includes changing the antenna between the transmitter and receiver, and making any other changes necessary to go between transmitting and receiving.

Vacuum-tube voltmeter (VTVM)—A multiple-range meter used to measure voltage, current and resistance. The meter circuit includes a vacuum-tube amplifier to provide a high input impedance. This leads to more accurate readings than can be obtained with a VOM.

Volt-ohm-milliammeter (VOM)—A multiple-range meter used to measure voltage, current and resistance. This is the least expensive (and least accurate) type of meter.

Chapter 9

Putting it All Together

After you select your equipment and put up your antenna, you have to connect everything so it becomes a radio station and not just separate pieces of equipment. This is a project that requires careful thought and planning. You may never be completely satisfied with your station. Most hams are always looking for new layout ideas and ways to improve the operating ease of their stations. It is not difficult to come up with an efficient and eye-pleasing arrangement. Simply think about what you want ahead of time.

STATION LOCATION

Your first consideration should be station location. Hams put their equipment in many places. Some use the basement or attic. Others choose the den, kitchen, closet or a spare bedroom. Some hams with limited space build their station into a small closet. A fold-out shelf and folding chair form the operating position! Where you put your station depends on the room you have available and on your personal tastes. There are, however, several requirements to keep in mind while searching for the ideal location. Figure 9-1 shows several amateur stations, with a variety of equipment arrangements and locations.

One often-overlooked requirement for a good station location is adequate electrical service. Eventually you will have several pieces of station equipment, as well as accessories. All of them will require power to operate. Be sure that at least one, and preferably several, electrical outlets are located near your future operating position. Be sure the outlets provide the proper voltage and current for your rig. Most modern radios require only a few amperes. You may run into problems, however, if your shack is on the same circuit as the air conditioner or washing machine. The total current drawn at any one time must not exceed rated limits. Someday you'll probably upgrade and may want to purchase a linear amplifier. If so, you should have a 240-volt line in the shack, or the capability to get one.

Another must for your station is a good **ground connection**. A good ground not only reduces the possibility of electrical shock but also improves the performance of your station. By connecting *all* of your equipment to ground, you will help to avoid stray RF current in the shack. Stray

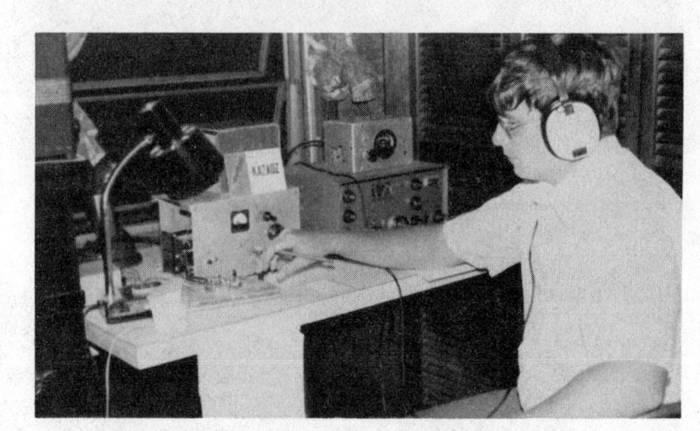

Figure 9-1—There are many ways to arrange your station equipment. The most important considerations are safe operation and a layout that pleases you.

RF can cause equipment to malfunction. A good ground can also help reduce the possibility of interference. The wire connecting your station to an earth ground should be as short as possible. Basement and first-floor locations generally make it easier to provide a good ground connection for your station. But this doesn't rule out the use of a second- or third-floor room. It doesn't mean you can't have a station if you live on the top floor of a high-rise apartment, either.

Your radio station will require an antenna, and a **feed line** of some sort to connect the antenna to the radio. So you will need a convenient means of getting the feed line into the shack. There are many ways of doing this, but one of the easiest and most effective requires only a window. Many hams simply replace the glass pane in a convenient window with a clear acrylic panel. The panel can be drilled to accept as many feed lines as needed. Special threaded "feedthrough" connectors make the job easy. If you decide to relocate your station, the glass pane can be replaced. This will restore the window to its original condition.

A simpler, more temporary approach would be passing the feed lines through the open window and closing it gently. But be careful, because crushed coaxial cable will give you nothing but trouble. Be sure to secure the window with a block of wood or some other "locking" device. Your new equipment seen through an open window may tempt an unwanted visitor!

Another important requirement for your station is comfort. The space should be large enough that you can spread out as needed. Operating from a telephone booth is no fun! Because you will probably be spending some time in your shack (an understatement!), be sure it will be warm in the winter and cool in the summer. It should also be as dry as possible. High humidity can cause such equipment problems as high-voltage arcing and switch-contact failure.

If your station is in an area often used by other family members, be sure they know what they shouldn't touch. You should have some means of ensuring that no "unauthorized" person can use your equipment. One way to do this is to install a key-operated on-off switch in the equipment's power line. When the switch is turned off and you have the key in your pocket, you will be sure that no one can misuse your station.

You will eventually want to operate late at night to snag the "rare ones" on 80 and 40 meters. Although Morse code is music to your ears, it may not endear you to a sleeping family. Putting your station in a bedroom shared with others may not be the best idea. You can keep the "music" to yourself, however, by using a good pair of headphones when operating your station.

WHAT IS A GOOD GROUND?

Most amateurs connect their equipment to a ground rod driven into the ground as close to the shack as possible. For a few dollars you can purchase a 1/2- or 5/8-inch-diameter, 8-foot-long ground rod at any electrical supply store. (Eight feet is the shortest practical length for your station ground rod.) Drive this copper-clad steel rod into the ground outside of your house, as close to your station location as possible. (Don't settle for the short, thin "ground rods" sold by some discount electronics stores. The copper cladding on the outside of most of these steel rods is very thin. They begin to rust almost immediately when they are put in the ground.)

Run a heavy copper wire (number 10 or larger wire) from your shack and attach it to the rod with a clamp. You can purchase the clamp when you buy the ground rod. Heavy copper strap or flashing (sold at hardware or roofing-supply stores) is even better. The braid from a piece of RG-8 coaxial cable also makes a good ground cable. Figure 9-2 shows one method of grounding each piece of equipment in your station. It's important to keep the cable between your station and the earth ground as short as possible.

Many hams ground their station equipment by connecting the ground wire to a cold-water pipe. Caution is in order here. If you live in an apartment or have your shack in an attic, be careful. The cold-water pipe near your transmitter may follow such a long and winding path to the earth that it may not act as a ground at all! It may, in fact, act as an antenna, radiating RF energy, which you don't want it to do.

Beware too, of the nonmetallic cold-water pipes being used more and more! PVC and other plastic pipes are effective insulators. There may be a piece of copper water pipe running close by your station. But if there is a piece of PVC pipe connected between that spot and where the water line enters your house, you will not have a ground connection!

[Study the questions in Chapter 12 that begin with numbers 2D-1 and 2D-3. If you have questions after studying these questions, review the material in this section.]

ARRANGING YOUR EQUIPMENT

Before you set everything up and hook up the cables, think about where you want each piece of equipment. While there is no one best layout for a ham station, some general rules do apply. Of course, a limiting factor is the location you've chosen for your station. For example, if you're going to put the station in the basement, you'll probably have lots of space. If, however, you'll be using a corner of the bedroom or den, you'll want to confine the equipment to a small area.

Generally, the piece of gear that requires the most adjustment is the receiver. It should be located in such a way that you can conveniently reach its controls. In choosing the location, keep in mind which hand you're going to use to make the adjustments. It doesn't make much sense to put the receiver on the left side of the desk or table if you're going to adjust it with your right hand. Once you've found the ideal location for the receiver, you can position the rest of your equipment around it.

In arranging your station, the placement of the telegraph key is also very important. It must be located in a position that is easy to reach. It should also provide the necessary support for your arm. When you choose the key location, try to position it away from anything that could injure you. You should avoid sharp corners and edges, rough surfaces and electrical wires. Place it in a location that is easy to reach with the hand you will use to operate it. Don't put the key on the right hand side of your desk if you use your left hand to send code!

The transmitter is probably the second most often adjusted piece of gear. It too should be located in a position

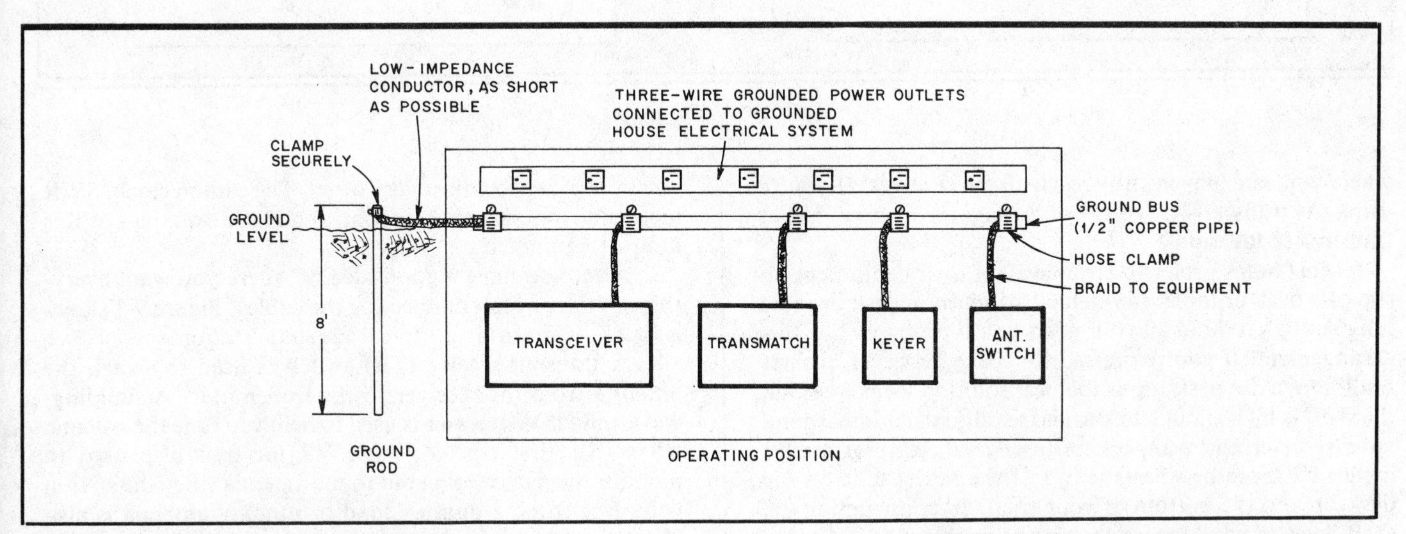

Figure 9-2—An effective station ground bonds the chassis of all your equipment together with low-impedance conductors and ties into a good earth ground.

EARLY RADIO GEAR

Today's compact and efficient radio gear is a far cry from the crude apparatus used at the turn of the century. Early radios often had a range of only a few miles. Today's amateur who "home-brews" really only assembles components. Early ham experimenters often had to create their own parts from scratch.

There were no transistors, not even tubes, to generate a signal. Little printed material existed to explain how. Communication was by a form of man-made static. The simplest transmitter started with a spark coil (a Model T Ford ignition coil was often used). A key in series with a battery energized the coil, electrifying a spark gap across the high-voltage winding. You connected an antenna wire to one side of the gap and a good ground to the other side, as shown in the diagram. Without the ground, you wouldn't be heard at the end of your block. Even with it, you would be fortunate to be heard five miles away. And because there was no real tuning, the output was as broad as the proverbial barn door. It probably occupied a space as large as that between 1500 and 4000 kHz.

More advanced sets used a large coil or "helix" of copper strap or tubing in the antenna lead. This inductor "peaked" the energy at one particular wavelength (measured in meters, not kilocycles or kilohertz). The aerial was often a series of parallel wires shaped like a hammock. These were as high in the air as feasible, and dimensioned to fit the selected wave. Yet the output was still gross, perhaps wider than any present medium-frequency band.

More progressive (and wealthier) amateurs used a spiked, motor-driven metal (but insulated) wheel or "chopper" inserted between spark gap terminals. The speed of rotation determined the interruption rate and imparted a distinctive tone to the output energy. This assisted the receiving operator in discerning the desired signal from other, interfering ones. There was no law, and no licensing. Amateurs had as much right to any part of the spectrum as anyone else, including the US Navy. Each operator had to do the best he could to work through heavy interference.

The transmitter used by Hiram Percy Maxim, 1AW, at his home in Hartford, Connecticut, was fairly advanced for its day. The photo shows a coil or "oscillation transformer" on the left. The large box housed a four-spiked wheel to form a "rotary" spark gap. The meter, a luxury, measured antenna current. Yet his range, too, was limited. When Maxim wanted to get some information by wireless to a colleague in Springfield, Massachusetts, he had to ask another amateur halfway between to relay the message. This was the seed of the idea to form the American Radio Relay League, accomplished in 1914.

Many of the useful advances in two-way wireless and radio (as distinguished from broadcasting) over the years have been in the form of narrowing the emitted signal bandwidth. That way, the finite spectrum can accommodate an ever greater number of communication channels. With today's equipment and techniques, thousands of QSOs, interference-free, can be effected in the space once occupied by a single spark transmitter.

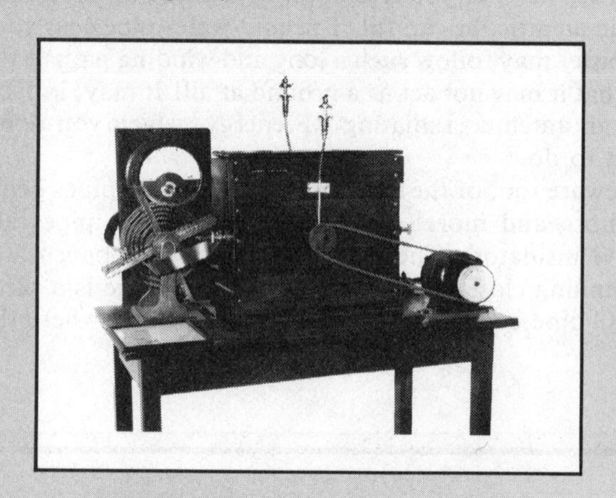

where you can conveniently reach all the controls. If you're using a transceiver, you won't have to worry about transmitter location.

Most hams prefer to arrange all of their equipment on top of a desk or table. If you have room for a desk or table long enough to hold all your gear, you'll probably like this arrangement. If you're pressed for space, however, a shelf built above the desk top is the best solution. Make certain the shelf is high enough to permit free air circulation around your receiver and transmitter. In general, leave at least 3 inches of space between the top of the tallest unit on the desktop and the bottom of your shelf. In constructing the shelf, keep in mind the weight of the equipment you're going to place on it. Try to restrict the shelf space to units that you don't have to adjust very often. The station clock, SWR meter and antenna rotator control box are examples of this kind of equipment.

After you have a good idea of where you want everything, you can start connecting the cables. Figure 9-3 shows a block diagram of a typical amateur station.

A **transmit-receive (TR) switch** is used to switch the antenna from the receiver to the transmitter. A **standing-wave-ratio (SWR) meter** is used to help you tune the antenna when it is first erected. An SWR meter is also used to monitor the power going out to the antenna when the station is in use. Often a **dummy load** or **dummy antenna** is also connected. Many hams use an **antenna switch** to select the dummy load or one of several antennas, as shown in Figure

9-4. The next section covers these accessories in more detail.

If you're using a separate transmitter and receiver, instead of a transceiver with a built-in TR switch, you'll need a TR switch. You can purchase a TR switch or relay, or you can build one from a relay or knife switch. You'll also have to make up two short coaxial cables to connect your receiver and transmitter to the switch.

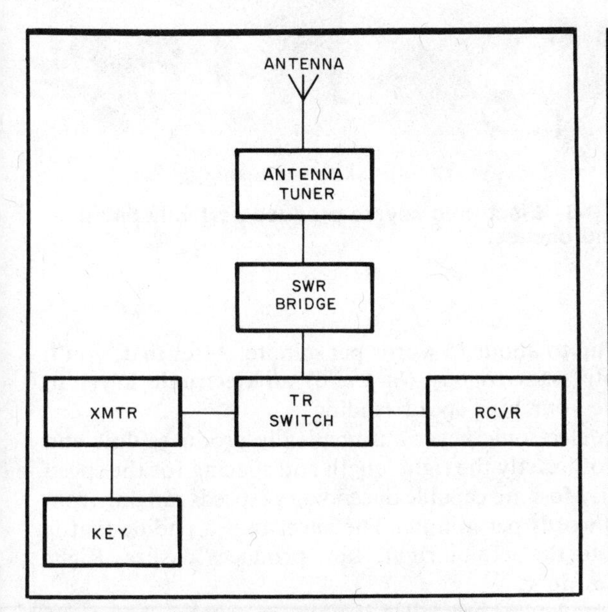

Figure 9-3—This block diagram shows the equipment connections in a typical amateur station.

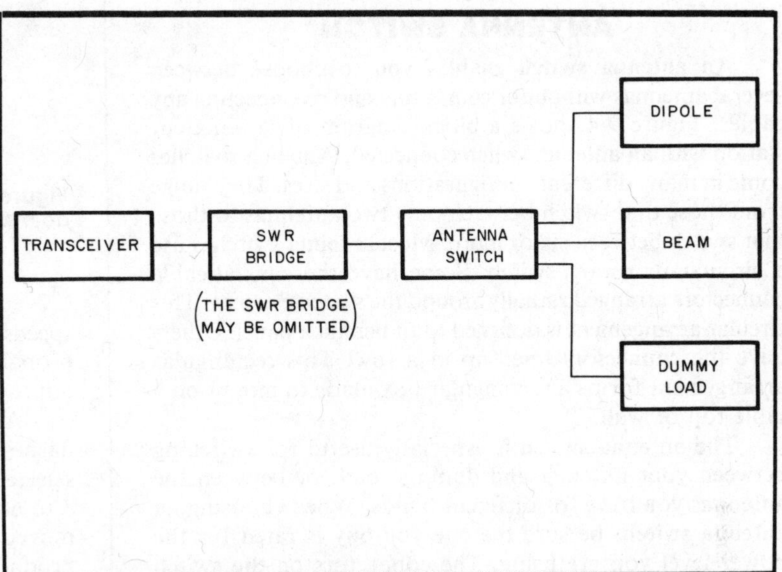

Figure 9-4—A block diagram of an amateur station showing the placement of an antenna switch.

ACCESSORIES

Until now, we have talked about the bare necessities—those items necessary for setting up a basic Amateur Radio station. Few hams are content with the bare necessities, however. Part of the fun in Amateur Radio is adding accessories that make operating more convenient and enjoyable.

DUMMY LOAD

A **dummy load**, sometimes called a **dummy antenna**, is nothing more than a large resistor. This resistor replaces your antenna when you want to operate your transmitter without radiating a signal. The dummy load safely converts the RF energy coming out of your transmitter into heat. The dummy load also dissipates the heat into the air (or into the coolant, depending on the type of dummy load). It does all this while presenting your transmitter with a constant 50-ohm load. This accessory is useful for testing transmitters, or when making preliminary tuning adjustments after changing bands. When looking at dummy loads, be sure the one you choose is rated for the power level you'll be using.

Relatively inexpensive, a dummy load is one of the most useful accessories you can own. Every conscientious amateur should own one. Figure 9-5 shows a homemade dummy

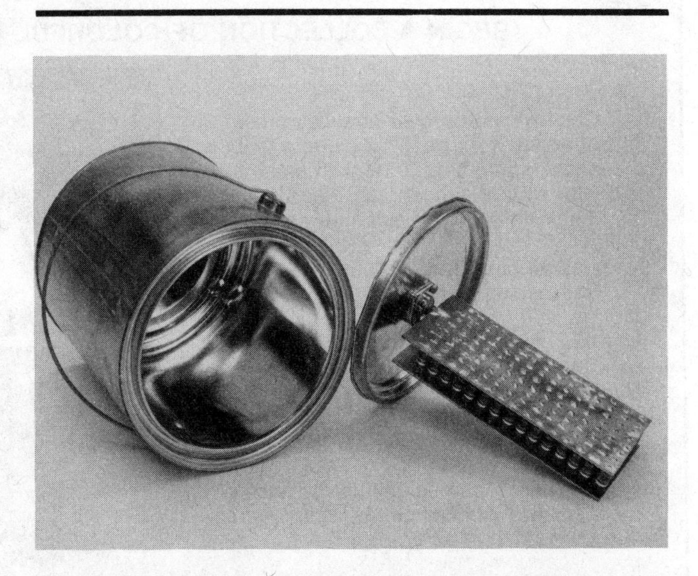

Figure 9-5—A dummy load is really just a resistance that provides your transmitter with a proper load when you are tuning up.

load. You can purchase a kit of parts, including the gallon paint can, from many sources. Besides acting as a shield to keep RF energy from being radiated by the dummy load, the container can be filled with transformer oil. This allows the dummy load to dissipate greater RF power. Commercial units are also available.

ANTENNA SWITCH

An **antenna switch** enables you to choose between several antennas without disconnecting and reconnecting any cables. Figure 9-4 shows a block diagram of an amateur station with an antenna switch connected. Antenna switches come in many different configurations and sizes. They range from those that switch between only two antennas to those that switch between six or more. Most antenna switches use a circular or rotary switch. Some have the coaxial-cable connectors arranged radially around the switch element. This circular arrangement is designed to fit behind a panel. Others have the connectors lined up in a row. This rectangular arrangement forms a rectangular box made to mount on a table top or wall.

The antenna switch is especially useful for switching between your antenna and dummy load, or between the antennas you have for different bands. When choosing an antenna switch, be sure the one you buy is rated for the power level you are using. The connectors on the switch should match the ones on your antenna feed lines and equipment. If they don't, your interconnecting cables will have to have different connector types on each end.

KEYERS

As you operate on the air and gain experience, your code speed will increase. A straight hand key works well at speeds up to about 15 words per minute. After that, you'll probably want to buy (or build) an **electronic keyer** to improve your high-speed sending.

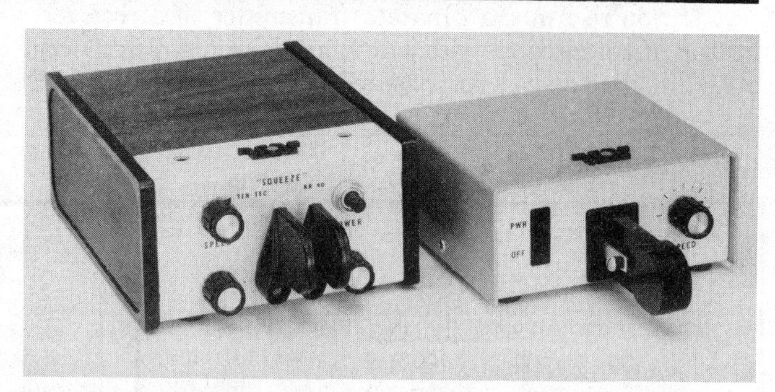

Figure 9-6—Electronic keyers produce perfectly timed dots and dashes.

An electronic keyer automatically produces dots and dashes of exactly the right length and spacing for the speed selected. Most are capable of sending at speeds ranging from 5 to 60 words per minute. The keyer uses a paddle that is moved to the left or right. Left produces dashes. Right produces dots.

Some keyers are called iambic, or squeeze, keyers. If both paddles are squeezed simultaneously, the keyer will produce a string of alternating dots and dashes. Iambic keyers make it easier to send code at higher speeds because less hand movement is required. The iambic technique can be tricky to learn, however. Figure 9-6 shows two electronic keyers with built-in paddles.

BEGIN A COLLECTION OF COLORFUL QSLS AND AWARD CERTIFICATES

Once hams discover how equipment and operator work best, they begin to look around for something to do with those skills and with all the incoming QSL cards that they begin to collect. Each time one ham makes contact with another, they exchange cards confirming that QSO, as it's called.

Hams use these QSL cards as proof of contact, and to earn specific certificates and awards. There are hundreds of awards to work toward. These include the popular Worked All States award and the Worked All Continents certificate. One of the most prestigious awards is the Five-Band DX Century Club, or Five-Band DXCC. Many hams display these colorful cards and awards like wallpaper on the walls of their shack.

Whether you use a straight key or an electronic keyer, you should always place a great emphasis on your "fist." That's the way your CW sounds on the air. Practice until you are proficient in sending. Take care to space your words and letters properly. Your form will greatly affect how people you meet on the air think of you. They certainly won't think much of you if your timing is off and they have a difficult time copying your signal.

QSL CARDS

QSL cards are sent by hams after a contact takes place. They provide written confirmation that the stations have contacted each other via Amateur Radio. Each card should contain information about the contact. Included is information such as date, time, frequency, mode, signal report and the call sign of the station contacted. QSL cards are said to be the final courtesy of a QSO. You will be asked for your card quite often. There are a number of sources for QSL cards. They range from local printers to the many QSL printers that advertise in the ham magazines.

Your QSL card can be an expression of your personality. It can be a way of informing other hams about your part of the country. For example, a ham living in Philadelphia may use the Liberty Bell in some way. A Wyoming ham could use Old Faithful. Often, hams are interested in other hobbies. One ham who is interested in soaring has a picture of his sailplane on the front of his QSL. Some radio clubs even have QSLs available for members, with the club logo worked into the design.

STATION CLOCK

The FCC rules no longer require Amateur Radio operators to keep a station logbook. However, many hams still maintain a complete log. Although it may seem a bother, you'll find it worth the trouble. This is true especially if you intend to exchange QSL cards. If you keep a log, you'll need a clock to keep track of the time.

Although you can use any clock that's handy, a 24-hour clock is more convenient. It eliminates the question of whether a contact was made in the AM or in the PM. In 24-hour time, 0300 is 3 AM, while 2200 is 10 PM. There can be only one time we call 16:15. But what about 4:15? Does it mean AM or PM? Another reason is that most hams interested in communicating with other hams outside of the US keep track of the hour in terms of **Coordinated Universal Time,** or **UTC**. UTC is a 24-hour system, and requires a 24-hour clock. So you might as well invest in a good one. A conversion table that shows the UTC equivalent to your

UTC EXPLAINED

Ever hear of Greenwich Mean Time? How about Coordinated Universal Time? Do you know if it is light or dark at 0400 hours? If you answered no to any of these questions, you'd better read on!

Keeping track of time can be pretty confusing when you are talking to other hams around the world. Europe, for example, is anywhere from 4 to 11 hours ahead of us here in North America. Over the years, the time at Greenwich, England, has been universally recognized as *the* standard time in all international affairs, including ham radio. (We measure longitude on the surface of the earth in degrees east or west of the Prime Meridian, which runs approximately through Greenwich, and which is halfway around the world from the International Date Line.) This means that wherever you are, you and the station you contact will be able to reference a common date and time. Mass confusion would occur if everyone used their own local time.

Coordinated Universal Time (abbreviated UTC) is the name for what used to be called Greenwich Mean Time.

Twenty-four hour time lets you avoid the equally confusing question about AM and PM. If you hear someone say he made a contact at 0400 hours UTC, you will know immediately that this was 4 hours past midnight, UTC, since the new day always starts just after midnight. Likewise, a contact made at 1500 hours UTC was 15 hours past midnight, or 3 PM (15 − 12 = 3).

Maybe you have begun to figure it out: Each day starts at midnight, 0000 hours. Noon is 1200 hours, and the afternoon hours merely go on from there. You can think of it as adding 12 hours to the normal PM time—3 PM is 1500 hours, 9:30 PM is 2130 hours, and so on. However you learn it, be sure to use the time everyone else does—UTC.

The photo shows a specially made clock, with an hour hand that goes around only once every day, instead of twice a day like a normal clock. Clocks with a digital readout that show time in a 24-hour format are quite popular as a station accessory.

local time is easy to make. The sidebar below shows an easy-to-make dial-type device that serves this purpose nicely.

TEST EQUIPMENT

One of the most useful of all ham accessories is a multimeter. Multimeters are used to measure voltage, current and resistance. The most common multimeter is the **volt-ohm-milliammeter**, or **VOM**. These handy meters come in all sizes and price ranges. The biggest advantage of a VOM is its portability. You can take a VOM just about anywhere, even to the top of a 100-foot tower.

In selecting a VOM, consider the number of ranges it has, the ease with which you can change ranges, the readability of the meter and the ohms-per-volt rating. Most VOMs provide about 20,000 ohms/volt of sensitivity. This means that the internal resistance of the meter is equal to the range setting in volts multiplied by 20,000. The greater the ohms/volt rating, the more accurate and, therefore, useful the VOM. Try not to settle for less than 20,000 ohms/volt unless your budget absolutely prohibits the expense. If you settle for less, expect more error in the readings. Whatever VOM you purchase, make certain it's designed to operate in an RF field. Some VOMs are very sensitive to RF. These give wildly inaccurate readings in the presence of an RF field.

Another multimeter used by hams is the **vacuum tube voltmeter**, or **VTVM**. This type of meter uses a vacuum-tube amplifier in its circuitry. This provides a much higher input impedance than an ordinary VOM. A VTVM usually requires an ac power source to operate the vacuum tubes. Because of this, it isn't very portable. However, it does provide a sensitivity in the order of 11,000,000 ohms/volt. As a result, it is much more accurate than the standard VOM.

A recent addition to the multimeter family is the battery-powered **field-effect transistor VOM (FET VOM)**. This VOM uses a solid-state device known as a field-effect transistor. The FET functions in the same manner as the vacuum tube in a VTVM. Most FET VOMs have a sensitivity of at least 1,000,000 ohms/volt. Some have a sensitivity as great as that provided by a VTVM. The FET VOM offers the portability of the VOM combined with the sensitivity of a VTVM.

THE N7BH WORLD TIME FINDER

I eliminated the chance for error in local-to-UTC time and date conversions by keeping a two-function clock set to UTC. This does not help to determine the date and time in other time zones around the world, however. To solve this problem, I devised a circular slide rule that will provide the conversion when set to local time. Notice that for every 15° of longitude around the earth, the time changes one hour. Thus there are 24 Time Zones, each 1 hour different from the next, around the world.

This World Time Finder converts time and date between all time zones. The outer scale is in hours using the 24-hour format. At midnight, the date becomes either "tomorrow" or "yesterday" according to the direction of time conversion indicated by the plus-minus arrows.

The movable second scale indicates the zone identification letter. This letter is used to identify the location of time being used. For example, 2100T would indicate 9 PM Mountain Standard Time or Pacific Daylight Time. Note that adapting daylight time has the same effect as shifting one time zone to the east, and thus one hour closer to UTC. If you pass the International Date Line in doing a time conversion, you again gain or lose a day, depending on which direction you are going.

You can make your own World Time Finder by cutting the two patterns from the appendix at the back of this book. Put each scale between two pieces of clear plastic, or laminate them between sheets of clear Con-Tact® paper (an adhesive-backed plastic material sold in many department stores). Punch a hole in the center of each scale, and use a 1/8-inch rivet with washers for the center pin.

The World Time Finder is operated by aligning the time zone letter for your area with the current hour. In the example shown, at zone U (PST) it is 2000 hours (8 PM). Zone U is 120° from the Prime Meridian, or 8 hours earlier than UTC. If the zone U date is July 1, then the time and date at UTC is 0400 (4 AM) on July 2. In Japan (zone 1), which is 135° from the Prime Meridian and 9 hours later

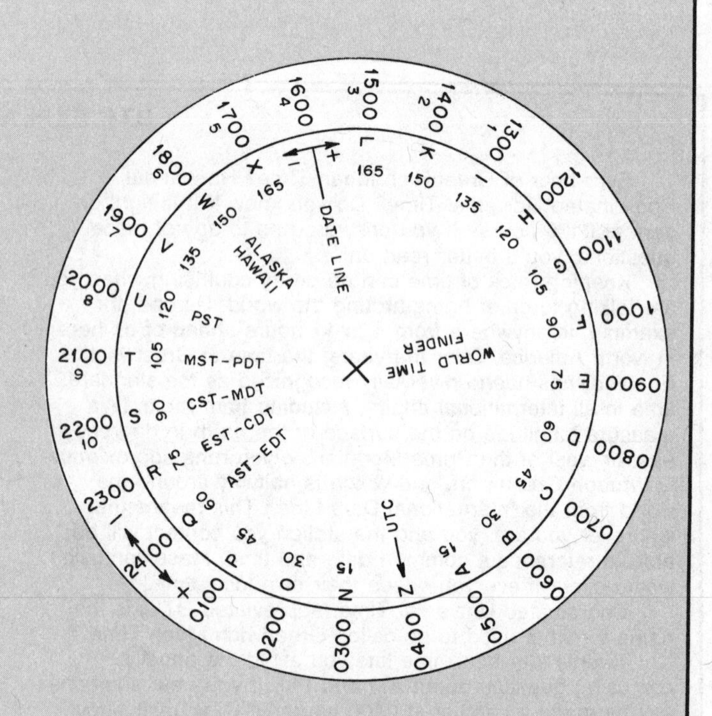

than UTC, it would be 1300 (1 PM) on July 2.

Note that this time conversion can be done by going in either a clockwise or counterclockwise direction around the chart. Converting time by going in a counterclockwise direction from zone U to UTC, you must pass midnight on the outer scale, thus gaining a day. If you perform the conversion by going in a clockwise direction around the scale, you must pass the International Date Line on the center scale, again gaining a day. The gain or loss of date is established easily by the + or − sign next to the arrows.

Measuring Voltage and Current

Using a multimeter is very simple. To get ready to make a measurement, simply plug the black test lead into the negative jack on the meter and the red test lead into the positive jack. Next, set the range switch to the proper scale. Some multimeters have separate jacks for different readings. Always check the instruction sheet that came with the meter before attempting to use it.

Once you have attached the test leads and set the range switch, you're all set to make your measurement. Both voltage and resistance readings are taken by touching the test lead tips to the points between which you want to measure. To measure current, cut or unsolder the conductor through which the current is flowing. Then, connect a test lead to each side of the break. We do this because the circuit current must actually flow through the meter. Figure 9-7 illustrates how to connect a VOM to a circuit to measure both voltage and current.

When you measure current it all flows through the meter itself. Because of that, the internal resistance of the meter is very low when it is set for current measurements. If you accidentally set the meter for current measurement and connect it across a voltage, you'll almost certainly damage the meter. The low internal resistance of the meter will permit high current to flow—higher than the meter is designed to handle. Make certain the range switch is set to voltage when you're measuring a voltage.

Another point to keep in mind when measuring a voltage is that there is, in fact, a voltage present. If you're careless, you might touch something and get an electrical shock. The wire and handles provided on the test leads that came with the meter will protect you from the shock. But if you hold the probes very close to the metal tips you might touch them accidentally. Generally, the best procedure is to connect the appropriate test lead to the chassis. Then, put the hand that would have held that lead in your pocket. That way you've only got one hand near the voltage. It's a lot easier to keep track of one hand then it is to keep track of two!

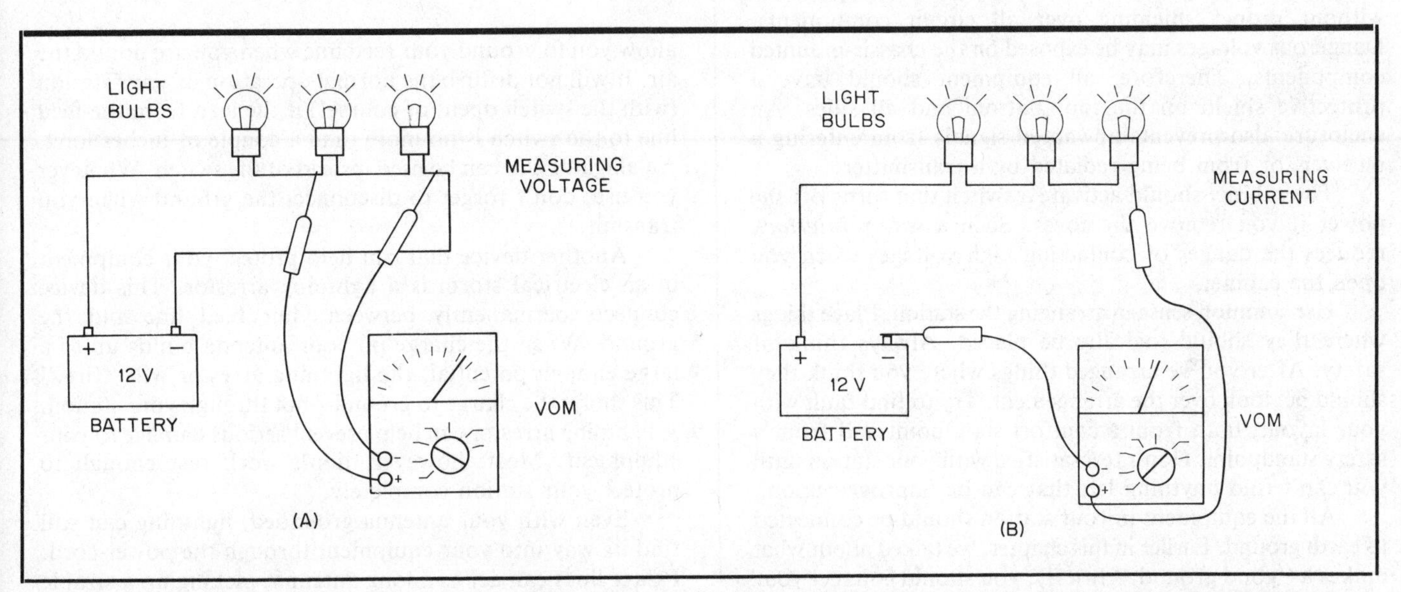

Figure 9-7—When measuring voltage with a VOM, simply connect the test leads across the circuit or component you want to check, as shown at A. To measure current, disconnect a wire at some point, and connect the meter leads at that point, as shown at B.

SAFETY

It is important to construct your station so it is safe for you and for any visitors to the station. Wiring should be neat and out of the way. Arrange your equipment and all wiring in a safe manner. Make sure there is no way you can tangle your feet in loose wires. Don't leave any voltage (regardless how low) exposed.

A main power switch is a convenient item. It allows you to turn all of the station equipment on or off at once. This saves needless wear on the equipment power switches. More than merely convenient, however, the main power switch is an important safety item. Make sure every member of your family knows how to turn off the power to your workbench and/or operating position. If you ever receive an electrical shock and cannot free yourself, the main disconnect switch will help your rescuer come to your aid quickly.

You must take special safety precautions if young children can come into your shack. If you can lock your station into its own room, this is an ideal solution. Few radio amateurs can afford this luxury. There are other, less expensive ways for securing your equipment from unauthorized operation. For example, you could build your station into a closet or cabinet that can be locked. If the equipment must be in a nonsecure area, install a key-operated power switch and *keep track of the keys!* Even a simple toggle switch, if well-hidden, is effective in keeping your station secure.

The same principles apply if you set up your station for public display. FCC rules require that only a licensed control operator may put the station on the air. Therefore you must make sure that unauthorized persons will not be able to transmit. Sometimes that may require temporarily removing a tube, relay, or control cable while the equipment is left unattended.

Whether you use commercially built equipment or homemade gear, you should *never* operate the equipment without proper shielding over all circuit components. Dangerous voltages may be exposed on the chassis-mounted components. Therefore, all equipment should have a protective shield on the top, bottom and all sides. An enclosure also prevents unwanted signals from entering a receiver, or from being radiated by a transmitter.

The cabinet should activate a switch that turns off the power if you remove the cover. Such a *safety interlock* reduces the danger of contacting high voltages when you open the cabinet.

Use common sense in arranging the station. Place things where they should logically be placed. Always think of safety. After you've arranged things where you think they should be, look over the arrangement. Try to find fault with your layout, both from a comfort standpoint and from a safety standpoint. Don't feel satisfied with your station until you can't find anything left that can be improved upon.

All the equipment in your station should be connected to earth ground. Earlier in this chapter, we talked about what makes a "good ground." Briefly, you should connect your equipment to the earth by as short a cable as possible. If you live in an apartment, you may have to use a cold water pipe rather than an outside ground rod. However you ground your equipment, don't neglect this important connection.

When your station is not in use, you should ground the antenna and rotator cables. This will help protect your equipment from nearby lightning strikes. The lightning hazard from an antenna is often exaggerated. Ordinary amateur antennas are no more likely to be hit by a direct stroke than any other object of the same height in the neighborhood. An ungrounded antenna can pick up large electrical charges from storms in the area, however. This can damage your equipment (receiver front ends are particularly susceptible) if you don't take precautions to prevent it.

One simple step you can take is to install a grounding switch, as shown in Figure 9-8. A small knife switch will allow you to ground your feed line when you are not on the air. It will not disturb the normal operation of your station (with the switch open, of course!) if the lead from the feed line to the switch is no more than a couple of inches long. An alligator clip can be used instead of the switch. Whatever you use, don't forget to disconnect the ground when you transmit.

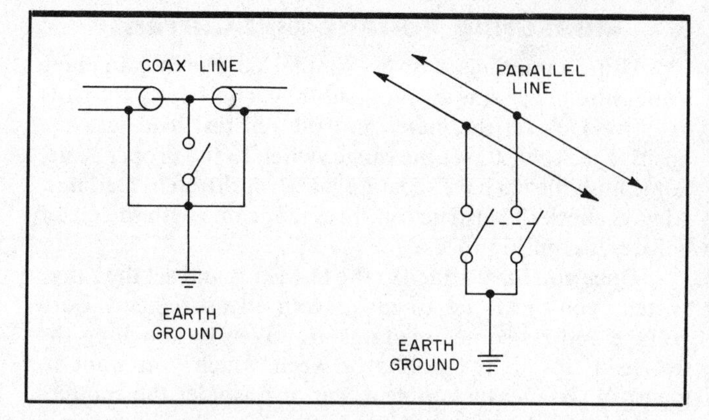

Figure 9-8—A heavy-duty knife switch can be used to connect the wires in your antenna feed line to ground. A clip lead to a ground wire can be used instead of the switch. This will prevent a large static build-up on your antenna. It will also prevent equipment damage caused by voltage on your antenna produced by a nearby lightning strike. There is no sure way to prevent a direct hit by lightning, however.

Another device that can help protect your equipment in an electrical storm is a lightning arrestor. This device connects permanently between your feed line and the ground. When the charge on your antenna builds up to a large enough potential, the lightning arrestor will "fire." This shorts the charge to ground—not through your station. A lightning arrestor can help prevent serious damage to your equipment. Most, however, don't work fast enough to protect your station completely.

Even with your antenna grounded, lightning can still find its way into your equipment through the power cord. Power lines can act as a long antenna, picking up a sizable charge during a storm. It's a good idea to disconnect your station from the power lines when you're not using it. Do so with the main power switch in your radio room, or by unplugging your gear from the wall socket.

[Now turn to Chapter 12 and study questions 2D-2.1 through 2D-2.4, and 2D-5.1 through 2D-5.2. Review this section if you have any difficulty with these questions.]

RF SAFETY

We all know basic safety precautions to follow in the ham shack or when erecting antennas. We know to pull the plug before taking the covers off a transceiver. We know to watch for power lines when hoisting an antenna into the air. Another important safety concern, one we often overlook, is radio-frequency (RF) protection. This involves minimizing human exposure to strong RF fields. These potentially dangerous fields occur near or around antennas.

Biological effects of RF exposure have been studied for

more than 20 years. Body tissues that are subjected to large amounts of RF energy may suffer heat damage. It is possible to receive an *RF burn* from touching an antenna that is being used for transmitting. You don't have to come into direct contact with an antenna to damage body tissues, however. Just being present in a strong RF field can cause problems. Taken to extremes, we could compare the effects of RF exposure to the way a microwave oven cooks food. A typical microwave oven uses a 500-W RF source operating at 2450 MHz. Of course, the microwave oven is designed to concentrate its RF power for heating food and is not directly comparable to Amateur Radio operations. There is no cause for alarm in most amateur installations. But be aware that exposure to strong RF fields—which we cannot see, smell, hear or touch—can cause tissue heating and subsequent damage.

The amount of RF energy that the body absorbs depends on the radio frequency. The body absorbs RF energy most efficiently in the VHF range (30 to 300 MHz). Absorption is greatest if the antenna orientation is parallel to the body (vertically polarized). You should be especially aware of RF safety when operating in the 222- to 224-MHz region.

Most amateur operation is with relatively low RF power, and the transmitter is not operating continuously. Hams spend more time listening than transmitting. Actual transmissions—like keyed CW and SSB—are inherently intermittent. Hams do use modes such as RTTY and FM where the RF carrier is present continuously at full power. Such continuous operation requires more attention to RF safety.

RF SAFETY GUIDELINES

Take the time to study and follow these general guidelines to minimize your exposure to RF fields. Most of these guidelines are just common sense and good amateur practice.

- Confine RF radiation to the antenna, where it belongs. Provide a good earth ground for your equipment. Poor-quality feed line and improperly installed connectors can be a source of unwanted radiation. Use only good-quality coaxial cable. Be sure that the connectors are of good quality and are properly installed.
- Don't operate RF power amplifiers or transmitters with the covers or shielding removed. This practice helps you avoid electric shock hazards as well as RF safety hazards. This is especially important for VHF and UHF equipment. When reassembling transmitting equipment, replace all the screws that hold the RF compartment shielding in place. Tighten all the screws securely before applying power to the equipment.
- In high-power operation in the HF and VHF region, keep the antenna away from people. Humans should not be allowed within 10 to 15 feet of vertical antennas. This is especially important with higher power, high-duty-cycle operation (such as FM or RTTY). Amateur antennas that are mounted on towers and masts, away from people, pose no exposure problem.
- Always install your antennas where people and animals cannot touch them.
- When using mobile equipment with 10-W RF power output or more, do not transmit if anyone is standing within 2 feet of the antenna.
- The best location for a VHF/UHF mobile antenna—from an RF safety standpoint—is in the middle of the automobile roof. This position best protects the car's occupants.
- When using a hand-held transceiver with RF power output of several watts or more, maintain at least 1 to 2 inches separation between the antenna and your forehead.
- Never touch an antenna that has RF power applied. Be sure RF power is off and stays off before working on or adjusting an antenna. Also, make sure any nearby antennas are deactivated. Never have someone else transmit into the antenna and monitor the SWR while you are making adjustments. When matching an antenna, the correct procedure is to turn the transmitter off and make the adjustment. Then, back away to a safe distance before turning the transmitter on again to check your work.
- During transmissions, never point a high-gain UHF antenna (such as a parabolic dish) toward people or animals.
- Never look into the open end of a UHF waveguide feed line that is carrying RF power. Never point the open end of a UHF waveguide that is carrying RF power toward people or animals. Make sure that all waveguide connections are tightly secured.

[Study questions in Chapter 12 that begin with 2D-4 and 2H-6. Review this section if you have difficulty answering any of these questions.]

AND THAT'S IT!

By now, you should have a good idea of what you want to look for when you shop for your radio equipment. You know how to erect an antenna, and how to arrange your equipment into a first-class radio station. We've covered everything you need to know to install a safe and effective Amateur Radio station.

KEY WORDS

AMTOR—Amateur Teleprinting Over Radio. AMTOR provides error-detecting capabilities. See Automatic Repeat Request and Forward Error Correction.

ASCII—American National Standard Code for Information Interchange. This is a seven-bit digital code used in computer and radioteleprinter applications.

Audio-frequency shift keying (AFSK)—A method of transmitting radioteletype information. Two switched audio tones are fed into the microphone input. AFSK RTTY is most often used on VHF.

Automatic Repeat Request (ARQ)—An AMTOR communication mode. In ARQ, also called Mode A, the two stations constantly confirm each other's transmissions. If information is lost, it is repeated until the receiving station confirms correct reception.

Autopatch—A device that allows repeater users to make telephone calls through a repeater.

A1A emission—The FCC emission designator used to describe Morse code telegraphy (CW) by on/off keying of a radio-frequency signal.

Baud—The unit used to describe the transmission speed of a digital signal. For a single-channel signal, one baud is equal to one digital bit per second.

Baudot—A five-bit digital code used in teleprinter applications.

Bleeder resistor—A large resistor connected across the output of a power supply. The bleeder discharges the filter capacitors when the supply is turned off.

Calling frequencies—Frequencies set aside for establishing contact. Once two stations are in contact, they should move their QSO to an unoccupied frequency.

Communications terminal—A computer-controlled device that demodulates RTTY and CW for display by a computer or ASCII terminal. The communications terminal also accepts information from a computer or terminal and modulates a transmitted signal.

Computer-Based Message System (CBMS)—A system in which a computer is used to store messages for later retrieval. Also called a RTTY mailbox.

Connected—The condition in which two packet-radio stations are sending information to each other. Each is acknowledging when the data has been received correctly.

Contests—On-the-air operating events. Different contests have different objectives: contacting as many other amateurs as possible in a given amount of time, contacting amateurs in as many different countries as possible or contacting an amateur in each county in one particular state, to name only a few.

CQ—The general call when requesting a conversation with anyone.

Digipeater—A packet-radio station used to retransmit signals that are specifically addressed to be retransmitted by that station.

Digital communications—The term used to describe Amateur Radio transmissions that are designed to be received and printed automatically. The term also describes transmissions used for the direct transfer of information from one computer to another.

Direct waves—Radio waves that travel directly from a transmitting antenna to a receiving antenna. Also called "line-of-sight" communications.

DX—Distance, foreign countries.

DX Century Club (DXCC)—A prestigious award given to amateurs who can prove contact with amateurs in at least 100 DXCC countries.

Field Day—An annual event in which amateurs set up stations in outdoor locations. Emergency power is also encouraged.

Fills—Repeats of parts of a previous transmission—usually requested because of interference.

Forward Error Correction (FEC)—A mode of AMTOR communication. In FEC mode, also called Mode B, each character is sent twice. The receiving station checks the mark/space ratio of the received characters. If an error is detected, the receiving station prints a space to show that an incorrect character was received.

Frequency-shift keying (FSK)—A method of transmitting radioteletype information by switching an RF carrier between two separate frequencies. FSK RTTY is most often used on HF.

F1B emission—The FCC emission designator used to describe frequency-shift keyed (FSK) digital communications.

F2B emission—The FCC emission designator used to describe audio-frequency shift keyed (AFSK) digital communications.

F3E emission—The FCC emission designator used to describe FM voice communications.

Ground waves—Radio waves that travel along the surface of the earth.

Input frequency—A repeater's receiving frequency.

Ionosphere—A region of charged particles high above the earth. The ionosphere bends radio waves as they travel through it, returning them to earth.

J3E emission—The FCC emission designator used to describe single-sideband suppressed-carrier voice communications.

Line of sight—The term used to describe VHF and UHF propagation in a straight line directly from one station to another.

Maximum usable frequency (MUF)—The greatest frequency at which radio signals will return to a particular location from the ionosphere. The MUF may vary for radio signals sent to different destinations.

Modem—Short for modulator/demodulator. A modem modulates a radio signal to transmit data and demodulates a received signal to recover transmitted data.

Monitor mode—One type of packet-radio receiving mode. In monitor mode, everything transmitted on a packet frequency is displayed by the monitoring TNC. This occurs whether the transmissions are addressed to the monitoring station or not.

Nets—Groups of amateurs who meet on the air to pass traffic or communicate about a specific subject. One station (called the *net control station*) usually directs the net.

Network—A term used to describe several packet stations linked together to transmit data over long distances.

Output frequency—A repeater's transmitting frequency.

Packet Bulletin-Board System (PBBS)—A computer system used to store packet-radio messages for later retrieval by other amateurs.

Packet radio—A system of digital communication whereby information is broken into short bursts. The bursts ("packets") also contain addressing and error-detection information.

Procedural signal (prosign)—One or two letters sent as a single character. Amateurs use prosigns in CW QSOs as a short way to indicate the operator's intention. Some examples are "K" for "Go Ahead," or "$\overline{\text{AR}}$" for "End of Message." (The bar over the letters indicates that we send the prosign as one character.)

Propagation—The study of how radio waves travel from one place to another.

Q signals—Three-letter symbols beginning with "Q." Q signals are used in amateur CW work to save time and for better communication.

QSO—A conversation between two radio amateurs.

Radioteletype (RTTY)—Radio signals sent from one teleprinter machine to another machine. Anything that one operator types on his teleprinter will be printed on the other machine. A type of digital communications.

Ragchew—A lengthy conversation (or QSO) between two radio amateurs.

Repeater—An amateur station that receives a signal and retransmits it for greater range.

RST—A system of numbers used for signal reports: R is readability, S is strength and T is tone.

Secondary station identifier (SSID)—A number added to a packet-radio station's call sign so that one amateur call sign can be used for several packet stations.

Selective-call identifier—A four-character AMTOR station identifier.

Simplex operation—A term normally used in relation to VHF and UHF operation. Simplex means you are receiving and transmitting on the same frequency.

Skip—Radio waves that are bent back to earth by the ionosphere. Skip is also called *sky-wave propagation*.

Skip zone—An area past the maximum range of ground waves and before the range of waves returned from the ionosphere. An area where radio communications between stations is not possible on a certain frequency.

Sky waves—Radio waves that travel through the ionosphere and back to earth. Sky-wave propagation is sometimes called skip.

Speech processor—A device that increases the average power of a sideband signal, making the voice easier to understand under weak signal conditions.

Splatter—The term used to describe a very wide-bandwidth signal. Splatter is usually caused by an improperly adjusted sideband transmitter.

Sunspots—Dark spots on the surface of the sun. When there are few sunspots, long-distance radio propagation is poor on the higher-frequency bands.

Terminal node controller (TNC)—A TNC accepts information from a computer or terminal and converts the information into packets by including address and error-checking information. The TNC also receives packet signals and extracts transmitted information for display by a computer.

Traffic—Messages passed from one amateur to another in a relay system; the amateur version of a telegram.

Tropospheric enhancement—A weather-related phenomenon. Tropo can produce unusually long propagation on the VHF and UHF bands.

Zero beat—When two operators in a QSO are transmitting on the same frequency.

Chapter 10

Over the Airwaves. . .Painlessly

Getting on the air, tuning in and "working" other Amateur Radio operators is the goal of your efforts to obtain an amateur license. Once your ticket arrives, you'll want to indulge yourself in this fabulous communications medium. You'll be joining more than a million radio enthusiasts around the world. Instead of watching TV, you can tune in to a large global network of radio hobbyists. These people actively program their own entertainment.

The operating world is rich and varied. You can relax and enjoy a conversation (a **ragchew**) with another amateur. You can send and receive messages to anywhere in North America via the National Traffic System (handle **traffic**). You can work large numbers of stations in a very short time in competition with other hams **(contests)**. You can try to contact as many different countries as possible (chase **DX**). These are just a few of the activities that Amateur Radio operators enjoy. The more operating time you log, the more quickly your operating proficiency will improve. With the CW practice you can get on the air, you'll soon be ready for that General class exam.

Some prospective amateurs view Morse code as a roadblock. Others think of it as an obsolete torture they must suffer only to obtain phone privileges. Don't fall for that: It is totally false. CW is an art form. It's a combination of skills that *can* be mastered. It's like playing the piano or speaking French—only easier! It's a skill that sets you apart from the average person. More than that, it's a communication mode that will get through when others fail.

In this chapter, we'll give you some tips for operating with your new Novice license. We'll discuss how radio waves travel from one place to another. Hams engage in a wide variety of operating activities. We'll show you some things you might want to try. In the last part of the chapter you'll learn about some of the more specialized operating modes and bands.

SAFETY FIRST

There is probably less danger operating a radio station than driving a car. Yet, as you know, automobile accidents do occur. There is no reason for you to involve yourself in a ham-related accident. That possibility always exists, however, if you're not thinking safety. The following safety rules will make your ham experience more enjoyable. Read them, understand them, and practice them.

1) Kill all power circuits completely before touching anything. That includes circuits behind the panel or inside the chassis or the enclosure.

2) Never allow anyone else to switch the power on and off for you while you're working on equipment.

3) Don't troubleshoot a transmitter when you're tired.

4) Never adjust internal components by hand. Use special care when checking energized circuits.

5) Avoid bodily contact with grounded metal (racks, radiators) or damp floors when working on the transmitter.

6) Never wear headphones while working on gear.

7) Follow the rule of keeping one hand in your pocket.

8) Instruct members of your household how to turn the power off and how to apply artificial respiration. Instruction sheets on the latest approved method of resuscitation can be obtained at your local Red Cross office.

9) If you must climb a tower to adjust an antenna, use a safety harness. Never work alone.

10) If you must climb into a tree or work on a roof, remember that you're not standing on the ground. That first step down can be a long one.

11) Develop your own safety technique. Take time to be careful. Death is permanent.

In addition, you should take the following precautions.

Antenna installations: Follow the advice given in Chapter 8 for installing antennas. Review the safety precautions given there. Follow them whenever you do any antenna work.

Power supplies: All power supplies should be enclosed. Be sure accidental bodily contact with power circuits is impossible. (The *ARRL Handbook for the Radio Amateur* is a good source of information concerning power-supply construction.)

All power-supply ground terminals should connect to the chassis. (This means the negative lead of a positive supply and the positive lead of a negative-voltage supply.) The chassis should connect to the electrical system ground with a three-wire line cord. If your shack does not have grounded outlets, use a water pipe or other good ground connection. See Chapter 9.

Every power supply should use a conservatively rated **bleeder resistor.** A bleeder resistor connects across the power supply output. It discharges the filter capacitors when you turn the supply off. Some amateur equipment uses shielded wire for external power cables. Grounded shielding provides protection if the cable insulation fails.

Panel controls and metering: Every control shaft extending through the panel should be at ground potential. The frames of key or metering jacks should be fastened to the grounded panel. Meters should be recessed to avoid danger of contact with the adjusting screw.

Audio equipment: The following rule should be followed for speech equipment. Use a microphone-cable with a shield conductor connected to the microphone stand and enclosure. The other end of the cable shield must connect to chassis ground.

Lightning protection: Merely "grounding everything" is not adequate protection from lightning strikes or static buildup. Lightning has devious methods of going into the wrong places. Protect your radio by disconnecting it completely from power lines and antennas. Also, ground all disconnected antenna lines. "Pulling the big switch" is a phrase used by many hams. You can run all power to your radio equipment through one main switch. This is a good way to disconnect your equipment from the power lines. Finally, pay special attention to rotator control boxes. A lightning strike can come into the box via the control cable. It can go through the control box, out the power cord, through the outlet and up the power cord of your radio. This can happen even with the power switch off! Please handle every circuit with care.

Those lowly 120-V circuits cause more electrocutions than any other voltage. Remember these four safety rules. Know your equipment. Never work alone. Use common sense. Don't take chances or safety shortcuts. A foolhardy ham is one with a short career!

AMATEUR RADIO FREQUENCIES AND PROPAGATION

Amateurs enjoy the privilege of using many different frequency bands. Some of these are scattered throughout the high-frequency (HF) spectrum. Some of these frequencies work better during the day. Some work better at night. Some frequencies are good for long-distance communications. Others provide good short-range communications. Amateurs often want to work a certain part of the country or world. They need to know which frequency to use and when to be there. Experience combined with lots of careful listening is a good way to gain this knowledge.

Amateurs have allocations in certain frequency bands in the radio spectrum. Novices have privileges on six of these bands. Table 10-1 summarizes the HF privileges granted to radio amateurs. We often refer to amateur bands by approximate wavelength rather than frequency. In an average day, you can reach any part of the world using one of the Novice

Table 10-1
High-Frequency Ham Bands

HF Bands For Full Amateur Privileges (MHz)	*Novice Privileges (MHz)*	*Band (Meters)*
1.8 - 2.0	None	160
3.5 - 4.0	3.70- 3.75	80
7.0 - 7.3	7.10- 7.15	40
10.10 -10.15	None	30
14.00 -14.35	None	20
18.068-18.168	None	17
21.00 -21.45	21.1 -21.2	15
24.89 -24.99	None	12
28.00 -29.70	28.1 -28.5	10

bands. The amateur community won new frequency allocations at 10, 18 and 24 MHz during the 1979 World Administrative Radio Conference. These new bands will help to ensure good worldwide communications, despite changing ionospheric conditions.

Radio waves travel to their destination in three ways. They may travel directly from one point to another. They may travel along the ground. They may be refracted, or bent, back to earth by the **ionosphere**. The ionosphere is a layer of charged particles in the earth's atmosphere. The study of how radio waves travel from one point to another is the science of **propagation**. Radio-wave propagation is a complicated topic. You will need to know only a few fundamentals to enjoy your new hobby.

DIRECT WAVES

Direct waves travel in a straight line from one antenna to another. We think of direct waves mostly at VHF and higher frequencies. You have some experience with these. TV and FM broadcasts are usually received as direct waves. Direct waves travel from your 220-MHz hand-held transceiver to an FM repeater. From there, direct waves travel to other fixed, portable and mobile stations. This kind of radio-wave propagation is also called **line-of-sight** propagation.

GROUND WAVE

In **ground-wave** propagation, the radio waves travel along the surface of the earth. The waves are able to travel over hills. They follow the curvature of the earth for some distance. Daytime standard AM broadcasting uses ground-wave propagation. Ground wave works best at lower frequencies. You might have an 80-meter QSO with a station a few miles away. The station may be located on the other side of a hill. For that contact, you will be using ground-wave propagation.

Ground-wave propagation on the ham bands means relatively short-range communications. Stations at the top of the AM broadcast band do not carry far during the day. Stations near the low end of the dial can be heard at much greater distances. Amateur Radio frequencies are higher than the standard broadcast band. As a result, the ground-wave range is shorter.

SKIP

There is a layer of the atmosphere, about 25 to 200 miles above the earth. There, the thin air can be electrically charged (ionized) by radiation from the sun. When ionized, this layer, the ionosphere, can refract (or bend) radio waves. If the wave is bent enough, it will come back down to the Earth. If it is not bent enough, it will just travel off into space. Communications of as much as 2500 miles are possible with one **skip** off the ionosphere. Worldwide communications using several skips (or hops) are possible at times.

Two factors determine skip propagation possibilities. They are the frequency in use and the level of ionization in the ionosphere. The higher the frequency of the radio wave, the less it is bent by the ionosphere. At any one time there is a specific frequency above which radio waves will penetrate the ionosphere and not return to earth. The highest-frequency waves that will be bent enough when they leave one location to return to another particular location on the earth are at the **maximum usable frequency (MUF)**. At frequencies higher than the MUF, the signal may not bend enough in the ionosphere to return to the desired location. Radio waves that travel beyond the horizon by refraction in the ionosphere are called **sky waves**. This is why we sometimes refer to skip propagation as sky-wave propagation.

Ionization of the ionosphere results from the sun's radiation striking the upper atmosphere. Ionization is greatest during the day and during the summer. The amount of radiation coming from the sun varies. Factors include the time of day, season of the year and others. This radiation is closely related to visible **sunspots** (grayish-black blotches on the sun's surface). See Figure 10-1. Sunspots vary over

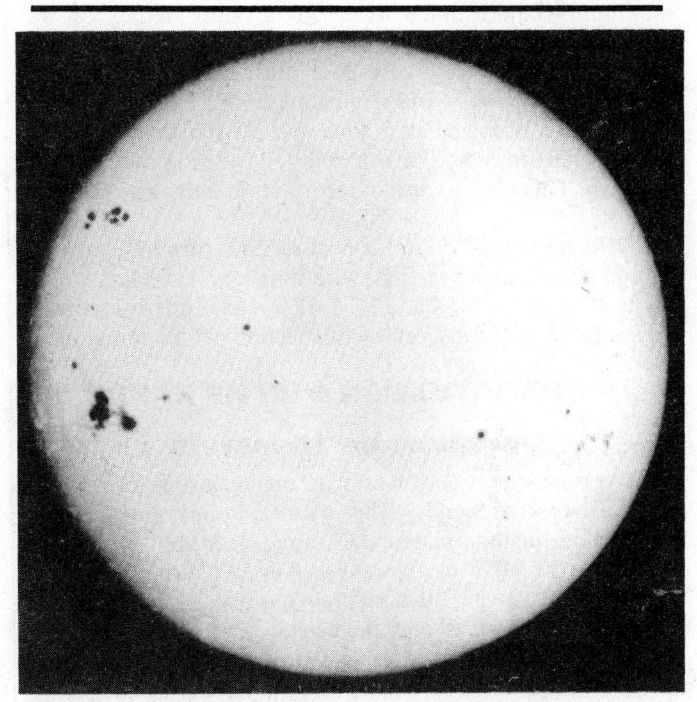

Figure 10-1—Cool sunspots allow hot propagation on earth!

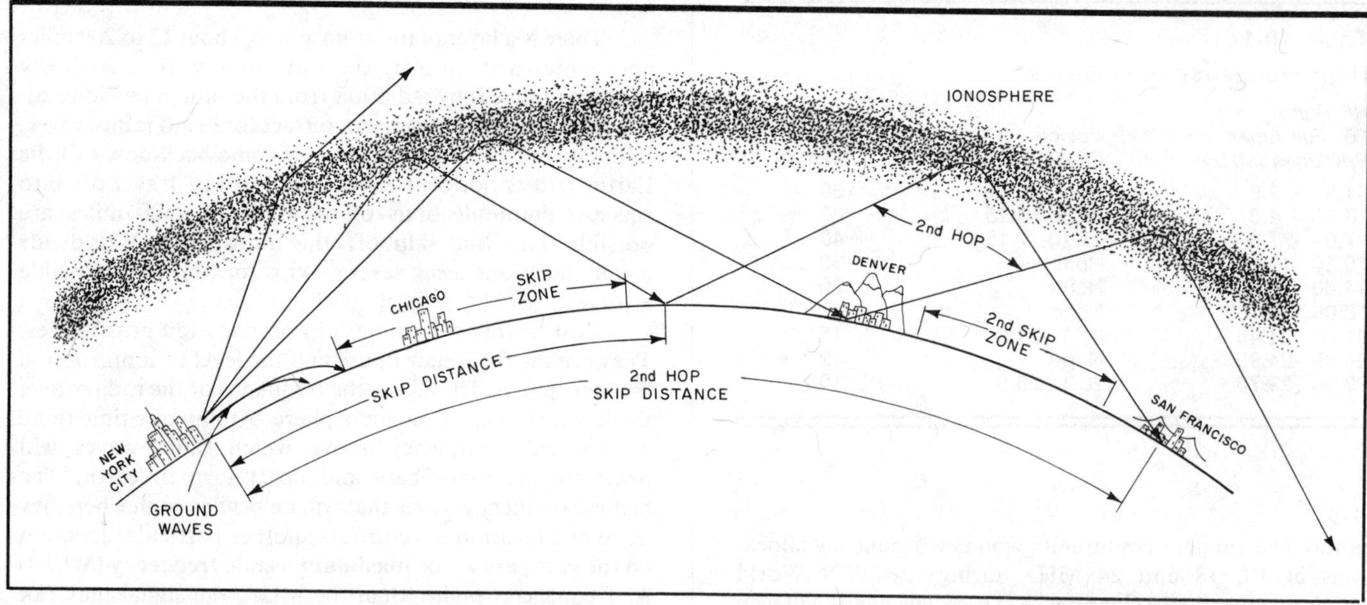

Figure 10-2—Illustration of how skip zone, skip distance and ground waves are related. See text for details.

an 11-year cycle. When the sunspots are at a maximum, solar radiation is greatest. As a result, ionization of the ionosphere is maximum. That means that the MUF tends also to be higher.

By contrast, when sunspots are low, radiation, and consequently the MUF, is lower. That is why most hams prefer times of greater sunspot activity.

Skip propagation not only has a maximum range limit, it also has a minimum range limit. That minimum is often greater than the ground-wave range. There is an area between the maximum ground-wave distance and the minimum skip distance where your radio signals on a particular frequency will not reach. This "dead" area is called the **skip zone**. Figure 10-2 illustrates the difference between ground-wave and skip propagation. The figure also shows the concept of skip zone and skip distance. Some radio signals may not be bent enough to bring them back to Earth. There is a wealth of information illustrated in this figure.

[Before you go on to the next section, turn to Chapter 12 and study those questions with numbers that begin with 2C-1, 2C-2, 2C-3, 2C-4 and 2C-5. If you have difficulty with any of these questions, review this section before going on.]

PROPAGATION ON OUR HF BANDS

Operating on 10 meters

At times the best of bands, 10 meters also can seem to be the worst of bands. This band contains the highest frequencies in the amateur HF range. It is subject to the whims of the MUF and the sunspot cycle. During years of high sunspot activity, 10 meters remains open all day. Strong signals arrive from all over the world. In years of low solar activity, just the opposite occurs. You may not hear signals, except local ones, for days at a time. Even during the peak of the sunspot cycle, the MUF may not reach 28 MHz. There are many days when 10 should be "open," but isn't.

Gil, F2SI, is very active on 10-meter SSB.

When operating on 10 meters, remember that the Novice CW subband is a full 200 kHz wide. The voice subband is also 200 kHz wide. Some signals may be weak, but perfectly readable. Tune the entire Novice subband carefully.

During years of high solar activity, you can tune across the band during the daylight hours and find many signals. If there aren't any signals in the Novice portion, listen in the rest of the band. That will help you to determine if any signals can be heard there. A good place to listen is around 28.5 MHz. If the band is open, you'll likely hear someone.

During the sunspot minimum, you'll have to work a little harder to see if 10 is open. Several automated beacons operate between 28.2 and 28.3 MHz. They transmit continuous signals to test propagation. If you can hear one of these beacons, chances are the band is open. You'll find these beacons listed occasionally in *QST* and other ham publications.

Even if 10 seems dead, the band may still be open to some part of the world. Try calling CQ. It may just be a case of everyone listening and no one calling. Because it takes a few moments to tune the full 10-meter subband, make your CQ longer than normal.

Ten meters offers many Novices their first taste of DX. That is, the chance to communicate with hams in foreign lands. Familiarize yourself with some of the common DX call signs. That way, you won't be surprised the first time you hear one. Besides US call signs like W6ISQ, N2ATD, or KD8LL, there are many others in use worldwide. In a typical week, you might hear JA7YAA in Japan, EA2IA in Spain or CT2QN in the Azores. Some countries also use call signs beginning with a number. Don't be surprised if you work 4X6AS in Israel or 9Y4VU in Trinidad. Table 10-2 lists all of the currently assigned international prefixes.

In addition to DX potential, 10 meters also provides excellent short-range communications. Many clubs and local groups meet regularly on 10 meters. Check with a local radio club to find out if there are any **nets** in your area.

Spend some time on 10 meters. It can be one of the most exciting bands to work. You can talk to hams from all over. It can also be one of the most frustrating because of the quirks of propagation. With experience you will gain a better understanding of the band. The time you do spend on 10 will be worthwhile and educational.

Ten meters is also the lowest-frequency Novice band that allows emissions other than CW (emission type **A1A**). From 28.1 to 28.3 MHz, Novices are allowed to use CW and frequency-shift keying (FSK), which is emission type **F1B**. FSK modes include packet, Baudot radioteletype (RTTY), ASCII and AMTOR. From 28.3 to 28.5 MHz, Novices are authorized to use CW (A1A) and single-sideband, suppressed-carrier (SSB) voice (emission type **J3E**). The maximum power level that Novices may use for any mode on 10 meters is 200 watts PEP output.

Operating on 15 meters

The 15-meter band shares several characteristics with the 10-meter band. It is not, however, as dependent on high sunspot activity. The 21-MHz frequency range is still high enough so that propagation conditions change dramatically with sunspot variations. Because of its lower frequency, 15 meters is open for longer periods and more often than 10. Also, the band is more stable in years of low sunspot activity.

DX is plentiful on 15 meters, even in the lean years of the solar cycle. During the day, stations in Europe, Africa and South America may be workable. You may hear signals from the Pacific and Japan in the late afternoon through early evening hours. Fifteen meters provides good communications to areas all over the US.

Many foreign stations operate within the Novice band using phone transmissions. American hams are prohibited from doing this, but many other countries do allow it. If you hear a foreign station on voice, give him a CW call at the end of his QSO. Many are glad to work Novices. This kind of operation won't give you much practice copying CW. It will give you a chance to hear what phone DX stations sound like, though. You might even work a "new country."

Because 15 meters is so popular, you'll most likely hear stations if the band is open. Again, listen at the bottom of the CW and phone portions. This is where stations usually congregate, if you don't hear signals in the Novice band.

Fifteen is one of the best bands around because of its DX possibilities. More stable than 10 meters, it is the most reliable Novice band for DX. It's usable even at the low point of the sunspot cycle. Higher in frequency than 40 and 80 meters, it provides good long-distance contacts.

Operating on 40 meters

The 40-meter band is one of the best US ham bands. It suffers, however, from some of the worst interference you'll ever hear. International treaties allow the world outside of the Americas to use the 40-meter band above 7.1 MHz for high-power shortwave broadcasting. This causes quite a racket in North America.

During the morning and early afternoon hours, 40 meters provides good, reliable communications. Skip distance ranges from about 400 to 1200 miles. Later in the afternoon, however, the MUF drops. Skip distance then increases to several thousand miles. You'll have no trouble detecting the increase in range. You'll find yourself suddenly listening to those foreign broadcast stations. The first time you try 40 when the broadcasters are coming in, you'll be tempted to switch to 80 meters. Don't. If you listen around carefully, you'll hear quite a few hams still using 40. If you're patient and have a good antenna, you should be able to work some interesting DX on 40 meters.

Even though it has interference from the broadcasters, 40 meters is one of the most popular Novice bands. It is open to some part of the country 24 hours a day. During the morning and afternoon hours, you'll be able to chat with other hams to your heart's content. In the late afternoon, evening and at night, you'll be able to work more distant locations. Once you learn to live with the broadcast interference, you'll find yourself using 40 meters for many of your contacts.

Operating on 80 meters

At 3.5 MHz, 80 meters is the lowest in frequency of all the Novice bands. It is very popular among Novices for stateside contacts. A half-wavelength dipole antenna (see Chapter 8) for 80 is rather large. Shape, however, is not critical. Most Novices manage to put up one that works fairly well.

During the daylight hours, 80 meters provides reliable communications out to about 350 miles. Although 80 is not as heavily populated as 40 during the day, you will probably find someone eager to talk with you.

Eighty comes into its own in the evening and nighttime hours. At that time, 10 and 15 meters shut down for the night. Forty meters opens up to foreign broadcasters. Many hams switch to 80 meters. The band provides strong, reliable local propagation. It has none of the uncertainty of 10 and 15, nor the interference of 40.

The long-range capabilities of the 80-meter band are subject to seasonal changes. During the summer months, nighttime communications may extend to 1000 miles.

Table 10-2
Allocation of International Call Signs

Call Sign Series	*Allocated to*
AAA-ALZ	United States of America
AMA-AOZ	Spain
APA-ASZ	Pakistan
ATA-AWZ	India
AXA-AXZ	Australia
AYA-AZZ	Argentina
A2A-A2Z	Botswana
A3A-A3Z	Tonga
A4A-A4Z	Oman
A5A-A5Z	Bhutan
A6A-A6Z	United Arab Emirates
A7A-A7Z	Qatar
A8A-A8Z	Liberia
A9A-A9Z	Bahrain
BAA-BZZ	China
CAA-CEZ	Chile
CFA-CKZ	Canada
CLA-CMZ	Cuba
CNA-CNZ	Morocco
COA-COZ	Cuba
CPA-CPZ	Bolivia
CQA-CUZ	Portugal
CVA-CXZ	Uruguay
CYA-CZZ	Canada
C2A-C2Z	Nauru
C3A-C3Z	Andorra
C4A-C4Z	Cyprus
C5A-C5Z	Gambia
C6A-C6Z	Bahamas
C7A-C7Z*	World Meteorological Organization
C8A-C9Z	Mozambique
DAA-DRZ	Federal Republic of Germany
DSA-DTZ	Republic of Korea
DUA-DZZ	Philippines
D2A-D3Z	Angola
D4A-D4Z	Cape Verde
D5A-D5Z	Liberia
D6A-D6Z	Comoros
D7A-D9Z	Republic of Korea
EAA-EHZ	Spain
EIA-EJZ	Ireland
EKA-EKZ	Union of Soviet Socialist Republics
ELA-ELZ	Liberia
EMA-EOZ	Union of Soviet Socialist Republics
EPA-EQZ	Iran
ERA-ESZ	Union of Soviet Socialist Republics
ETA-ETZ	Ethiopia
EUA-EWZ	Byelorussian Soviet Socialist Republic
EXA-EZZ	Union of Soviet Socialist Republics
FAA-FZZ	France
GAA-GZZ	United Kingdom of Great Britain and Northern Ireland
HAA-HAZ	Hungary
HBA-HBZ	Switzerland
HCA-HDZ	Ecuador
HEA-HEZ	Switzerland
HFA-HFZ	Poland
HGA-HGZ	Hungary
HHA-HHZ	Haiti
HIA-HIZ	Dominican Republic
HJA-HKZ	Colombia
HLA-HLZ	Republic of Korea
HMA-HMZ	Democratic People's Republic of Korea
HNA-HNZ	Iraq
HOA-HPZ	Panama
HQA-HRZ	Honduras
HSA-HSZ	Thailand
HTA-HTZ	Nicaragua
HUA-HUZ	El Salvador
HVA-HVZ	Vatican City
HWA-HYZ	France
HZA-HZZ	Saudi Arabia
H2A-H2Z	Cyprus
H3A-H3Z	Panama
H4A-H4Z	Solomon Islands
H6A-H7Z	Nicaragua
H8A-H9Z	Panama
IAA-IZZ	Italy
JAA-JSZ	Japan
JTA-JVZ	Mongolian People's Republic
JWA-JXZ	Norway
JYA-JYZ	Jordan
JZA-JZZ	Indonesia
J2A-J2Z	Djibouti
J3A-J3Z	Grenada
J4A-J4Z	Greece
J5A-J5Z	Guinea-Bissau
J6A-J6Z	Saint Lucia
J7A-J7Z	Dominica
J8A-J8Z	St. Vincent and the Grenadines
KAA-KZZ	United States of America
LAA-LNZ	Norway
LOA-LWZ	Argentina
LXA-LXZ	Luxembourg
LYA-LYZ	Union of Soviet Socialist Republics
LZA-LZZ	Bulgaria
L2A-L9Z	Argentina
MAA-MZZ	United Kingdom of Great Britain and Northern Ireland
NAA-NZZ	United States of America
OAA-OCZ	Peru
ODA-ODZ	Lebanon
OEA-OEZ	Austria
OFA-OJZ	Finland
OKA-OMZ	Czechoslovakia
ONA-OTZ	Belgium
OUA-OZZ	Denmark
PAA-PIZ	Netherlands
PJA-PJZ	Netherlands Antilles
PKA-POZ	Indonesia
PPA-PYZ	Brazil
PZA-PZZ	Suriname
P2A-P2Z	Papua New Guinea
P3A-P3Z	Cyprus
P4A-P4Z	Aruba
P5A-P9Z	Democratic People's Republic of Korea
QAA-QZZ	(Service abbreviations)
RAA-RZZ	Union of Soviet Socialist Republics
SAA-SMZ	Sweden
SNA-SRZ	Poland
SSA-SSM	Egypt
SSN-STZ	Sudan
SUA-SUZ	Egypt
SVA-SZZ	Greece
S2A-S3Z	Bangladesh
S6A-S6Z	Singapore
S7A-S7Z	Seychelles
S9A-S9Z	Sao Tome and Principe
TAA-TCZ	Turkey
TDA-TDZ	Guatemala
TEA-TEZ	Costa Rica
TFA-TFZ	Iceland
TGA-TGZ	Guatemala
THA-THZ	France
TIA-TIZ	Costa Rica
TJA-TJZ	Cameroon
TKA-TKZ	France
TLA-TLZ	Central African Republic
TMA-TMZ	France
TNA-TNZ	Congo
TOA-TQZ	France
TRA-TRZ	Gabon
TSA-TSZ	Tunisia
TTA-TTZ	Chad
TUA-TUZ	Ivory Coast
TVA-TXZ	France

Call Sign Series	Allocated to
TYA-TYZ	Benin
TZA-TZZ	Mali
T2A-T2Z	Tuvalu
T3A-T3Z	Kiribati
T4A-T4Z	Cuba
T5A-T5Z	Somalia
T6A-T6Z	Afghanistan
T7A-T7Z	San Marino
UAA-UQZ	Union of Soviet Socialist Republics
URA-UTZ	Ukrainian Soviet Socialist Republic
UUA-UZZ	Union of Soviet Socialist Republics
VAA-VGZ	Canada
VHA-VNZ	Australia
VOA-VOZ	Canada
VPA-VSZ	United Kingdom of Great Britain and Northern Ireland
VTA-VWZ	India
VXA-VYZ	Canada
VZA-VZZ	Australia
V2A-V2Z	Antigua and Barbuda
V3A-V3Z	Belize
V4A-V4Z	St. Christopher and Nevis
V8A-V8Z	Brunei
WAA-WZZ	United States of America
XAA-XIZ	Mexico
XJA-XOZ	Canada
XPA-XPZ	Denmark
XQA-XRZ	Chile
XSA-XSZ	China
XTA-XTZ	Burkina Faso
XUA-XUZ	Kampuchea
XVA-XVZ	Viet Nam
XWA-XWZ	Laos
XXA-XXZ	Portugal
XYA-XZZ	Burma
YAA-YAZ	Afghanistan
YBA-YHZ	Indonesia
YIA-YIZ	Iraq
YJA-YJZ	New Hebrides
YKA-YKZ	Syria
YLA-YLZ	Union of Soviet Socialist Republics
YMA-YMZ	Turkey
YNA-YNZ	Nicaragua
YOA-YRZ	Romania
YSA-YSZ	El Salvador
YTA-YUZ	Yugoslavia
YVA-YYZ	Venezuela
YZA-YZZ	Yugoslavia
Y2A-Y9Z	German Democratic Republic
ZAA-ZAZ	Albania
ZBA-ZJZ	United Kingdom of Great Britain and Northern Ireland
ZKA-ZMZ	New Zealand
ZNA-ZOZ	United Kingdom of Great Britain and Northern Ireland
ZPA-ZPZ	Paraguay
ZQA-ZQZ	United Kingdom of Great Britain and Northern Ireland
ZRA-ZUZ	South Africa
ZVA-ZZZ	Brazil
Z2A-Z2Z	Zimbabwe
2AA-2ZZ	United Kingdom of Great Britain and Northern Ireland
3AA-3AZ	Monaco
3BA-3AZ	Mauritius
3CA-3CZ	Equatorial Guinea
3DA-3DM	Swaziland
3DN-3DZ	Fiji
3EA-3FZ	Panama
3GA-3GZ	Chile
3HA-3UZ	China
3VA-3VZ	Tunisia
3WA-3WZ	Viet Nam
3XA-3XZ	Guinea

Call Sign Series	Allocated to
3YA-3YZ	Norway
3ZA-3ZZ	Poland
4AA-4CA	Mexico
4DA-4IZ	Philippines
4JA-4LZ	Union of Soviet Socialist Republics
4MA-4MZ	Venezuela
4NA-4OZ	Yugoslavia
4PA-4SZ	Sri Lanka
4TA-4TZ	Peru
4UA-4UZ*	United Nations Organization
4VA-4VZ	Haiti
4WA-4WZ	Yemen Arab Republic
4XA-4XZ	Israel
4YA-4YZ*	International Civil Aviation Organization
4ZA-4ZZ	Israel
5AA-5AZ	Libya
5BA-5BZ	Cyprus
5CA-5GZ	Morocco
5HA-5IZ	Tanzania
5JA-5KZ	Colombia
5LA-5MZ	Liberia
5NA-5OZ	Nigeria
5PA-5QZ	Denmark
5RA-5SZ	Madagascar
5TA-5TZ	Mauritania
5UA-5UZ	Niger
5VA-5VZ	Togo
5WA-5WZ	Western Samoa
5XA-5XZ	Uganda
5YA-5ZZ	Kenya
6AA-6BZ	Egypt
6CA-6CZ	Syria
6DA-6JZ	Mexico
6KA-6NZ	Republic of Korea
6OA-6OZ	Somalia
6PA-6SZ	Pakistan
6TA-6UZ	Sudan
6VA-6WZ	Senegal
6XA-6XZ	Madagascar
6YA-6YZ	Jamaica
6ZA-6ZZ	Liberia
7AA-7IZ	Indonesia
7JA-7NZ	Japan
7OA-7OZ	Yemen
7PA-7PZ	Lesotho
7QA-7QZ	Malawi
7RA-7RZ	Algeria
7SA-7SZ	Sweden
7TA-7YZ	Algeria
7ZA-7ZZ	Saudi Arabia
8AA-8IZ	Indonesia
8JA-8NZ	Japan
8OA-8OZ	Botswana
8PA-8PZ	Barbados
8QA-8QZ	Maldives
8RA-8RZ	Guyana
8SA-8SZ	Sweden
8TA-8YZ	India
8ZA-8ZZ	Saudi Arabia
9BA-9DZ	Iran
9EA-9FZ	Ethiopia
9GA-9GZ	Ghana
9HA-9HZ	Malta
9IA-9JZ	Zambia
9KA-9KZ	Kuwait
9LA-9LZ	Sierra Leone
9MA-9MZ	Malaysia
9NA-9NZ	Nepal
9OA-9TZ	Zaire
9UA-9UZ	Burundi
9VA-9VZ	Singapore
9WA-9WZ	Malaysia
9XA-9XZ	Rwanda
9YA-9ZZ	Trinidad and Tobago

Note

The series of call signs with an asterisk indicate the international organization to which they are allocated.

Atmospheric static levels, however, are often high. Contacts can be difficult to establish and maintain. The winter months are a different story. Static levels are usually very low. Cross-country and DX contacts are normal occurrences.

Like 40 meters, the 80-meter band is a 24-hour-a-day happening. It provides excellent local communications almost all the time. It offers long-distance possibilities at night during the winter months. As on 15 meters, you may hear foreign and Canadian amateurs operating phone in the Novice band. Feel free to call them using CW, when they are finished with their present QSO.

Eighty is a band often used to make and maintain schedules with friends. After operating on the band for a while, you will come to know many of your fellow 80-meter Novices. They will come to know you too. You'll hear many of the same stations each time you operate. Propagation is quite constant. Plan to spend some of your evening operating time on 80 meters making new friends. You'll likely meet them face-to-face at local conventions or hamfests.

PROPAGATION ON OUR VHF AND UHF BANDS

Normally, you'll be contacting nearby stations—within 100 miles or so—on 1.25 meters and 23 cm. The usual propagation mode on these bands is **line of sight**. The signal travels directly between stations, so if you're using a directional antenna, you normally point it toward the station that you are trying to contact. VHF and UHF signals, however, are easily reflected by buildings, hills or other large objects. Some of your signal reaches the other station by a direct path. Some of it reaches the station by a reflected path. When such reflections occur, it is possible to contact other stations by pointing your antenna toward the reflecting object, rather than directly at the station you're trying to contact.

Reflections can be a significant problem during mobile FM operation. This happens because the propagation path is constantly changing. The direct and reflected waves may cancel each other and then reinforce each other. This causes a rapid fluttering sound called "picket fencing."

Sometimes, especially during the spring, summer and fall months, it is possible to make VHF and UHF contacts over long distances—up to 1000 miles or more. This occurs during certain weather conditions that cause **tropospheric enhancement** and tropospheric ducting. When such "tropo" openings occur, the VHF and UHF bands are filled with excited operators eager to work DX. The troposphere is the layer of the atmosphere just below the stratoshpere. It extends upward approximately 7 to 10 miles. In this region clouds form and temperature decreases rapidly with altitude. You can find out more about VHF and UHF propagation from *The ARRL Handbook*.

Operating on 1.25 Meters

Well into the VHF portion of the radio spectrum, 1.25 meters (220 MHz) feels entirely different than the HF Novice bands. Although many operating modes are used here, FM repeaters (discussed later) dominate 220 MHz.

Novices are allowed to operate in the segment from 222.1 to 223.91 MHz. They may use any and all emission modes authorized for that band. Popular modes on 1.25 meters include frequency-modulated (FM) voice (emission type **F3E**), SSB voice, CW, packet radio, and Baudot and ASCII RTTY. Unlike the HF bands, FCC rules and regulations do not set aside portions of the 1.25-meter band for specific modes. Any authorized mode may be used on any frequency within the band. To avoid confusion and interference problems, amateurs have agreed to a voluntary band plan. The plan allocates segments of the band for different uses. For example, there is a section set aside around 222.1 MHz for "weak signal" CW and SSB operation. Another segment is set aside for FM voice operation, and so on. For complete information on the 1.25-meter band plan, see the latest edition of *The ARRL Repeater Directory*. The maximum power level that Novices may use for any mode on 1.25 meters is 25 watts PEP output.

Operating on 23 Centimeters

Twenty-three centimeters (1270 MHz) is the highest-frequency Novice band. It lies in the UHF portion of the radio spectrum, and is the least-populated of all Novice frequencies. Novices are allowed to operate in the segment from 1270 to 1295 MHz. They may use any and all emission modes authorized for that band. Popular modes on 23 cm include FM voice, SSB, CW and amateur television (ATV). The 23-cm band is similar to 1.25 meters. There are no FCC-mandated segments for the different modes. Rather, frequency usage is determined by a voluntary band plan. The 23-cm band plan calls for FM voice and ATV operation from 1270 to 1295 MHz. For complete information on the 23-cm band plan, see the latest edition of *The ARRL Repeater Directory*. The maximum power level that Novices may use for any mode on 23 cm is 5 watts PEP output.

[Now turn to Chapter 12 and study questions 2B-1-1.2, 2C-6.1 and 2C-6.2. You should also study questions 2H-1-1.1, 2H-1-1.2, 2H-1-3.1 and 2H-1-4.1. Review this section if you have any difficulty with these questions.]

OPERATING SKILL

Poor operating procedure is ham radio's version of original sin. It is a curse that will not go away. Usually, a new ham picks bad habits up early in his or her amateur career. Other operators (who don't know better) can reinforce poor operating practices. Hams call these poor operators *lids*. No one wants to be a lid or to have a QSO with a lid. Calling someone a lid is the ultimate insult. (Don't do it on the air. Be sure your operating practices are such that no one *wants* to call *you* a lid.)

It is actually easier to be a good operator than to fall prey to sloppy habits. You can transmit a message without needless repetition and without unnecessary identification. You don't have to spell out each and every word. Good operating makes hamming more fun for everyone.

Clearly, the initial glamour of Amateur Radio is the opportunity to talk to people you can't see. It doesn't matter if they're across town or in another country. Novices have plenty of chances to get in on DX action on the 10- and 15-meter bands. The 40- and 80-meter bands (7 and 3.7 MHz) are useful primarily for domestic contacts.

Chapter 8 presents information on antennas for all types of locations. You must determine the frequency bands that you want to try. The success of your antenna installation will be quite evident once you actually get on the air. Either they'll hear you or they won't. A beam antenna is not mandatory for working DX. At times of high sunspot

Table 10-3
Q Signals

Given below are a number of Q signals whose meanings most often need to be expressed with brevity and clarity in amateur work. (Q abbreviations take the form of questions only when they are sent followed by a question mark.)

Signal	Meaning
QRG	Will you tell me my exact frequency (or that of . . .)? Your exact frequency (or that of . . .) is . . . kHz.
QRH	Does my frequency vary? Your frequency varies.
QRI	How is the tone of my transmission? The tone of your transmission is . . . (1. Good; 2. Variable; 3. Bad).
QRK	What is the intelligibility of my signals (or those of . . .)? The intelligibility of your signals (or those of . . .) is . . . (1. Bad; 2. Poor; 3. Fair; 4. Good; 5. Excellent).
QRL	Are you busy? I am busy (or I am busy with . . .). Please do not interfere.
QRM	Is my transmission being interfered with? Your transmission is being interfered with . . . (1. Nil; 2. Slightly; 3. Moderately; 4. Severely; 5. Extremely.)
QRN	Are you troubled by static? I am troubled by static . . . (1-5 as under QRM).
QRO	Shall I increase power? Increase power.
QRP	Shall I decrease power? Decrease power.
QRQ	Shall I send faster? Send faster (. . . WPM).
QRS	Shall I send more slowly? Send more slowly (. . . WPM).
QRT	Shall I stop sending? Stop sending.
QRU	Have you anything for me? I have nothing for you.
QRV	Are you ready? I am ready.
QRW	Shall I inform . . . that you are calling him on . . . kHz? Please inform . . . that I am calling him on . . . kHz.
QRX	When will you call me again? I will call you again at . . . hours (on . . . kHz).
QRY	What is my turn? Your turn is numbered . . .
QRZ	Who is calling me? You are being called by . . . (on . . . kHz).
QSA	What is the strength of my signals (or those of . . .)? The strength of your signals (or those of . . .) is . . . (1. Scarcely perceptible; 2. Weak; 3. Fairly good; 4. Good; 5. Very good).
QSB	Are my signals fading? Your signals are fading.
QSD	Is my keying defective? Your keying is defective.
QSG	Shall I send . . . messages at a time? Send . . . messages at a time.
QSK	Can you hear me between your signals and if so can I break in on your transmission? I can hear you between signals; break in on my transmission.
QSL	Can you acknowledge receipt? I am acknowledging receipt.
QSM	Shall I repeat the last message which I sent you, or some previous message? Repeat the last message which you sent me [or message(s) number(s) . . .].
QSN	Did you hear me (or . . .) on . . . kHz? I did hear you (or . . .) on . . . kHz.
QSO	Can you communicate with . . . direct or by relay? I can communicate with . . . direct (or by relay through . . .).
QSP	Will you relay to . . .? I will relay to . . .
QST	General call preceding a message addressed to all amateurs and ARRL members. This is in effect "CQ ARRL."
QSU	Shall I send or reply on this frequency (or on . . . kHz)? Send or reply on this frequency (or on . . . kHz).
QSV	Shall I send a series of Vs on this frequency (or . . . kHz)? Send a series of Vs on this frequency (or . . . kHz).
QSW	Will you send on this frequency (or on . . . kHz)? I am going to send on this frequency (or on . . . kHz).
QSX	Will you listen to . . . on . . . kHz? I am listening to . . . on . . . kHz.
QSY	Shall I change to transmission on another frequency? Change to transmission on another frequency (or on . . . kHz).
QSZ	Shall I send each word or group more than once? Send each word or group twice (or . . . times).
QTA	Shall I cancel message number . . .? Cancel message number . . .
QTB	Do you agree with my counting of words? I do not agree with your counting of words. I will repeat the first letter or digit of each word or group.
QTC	How many messages have you to send? I have . . . messages for you (or for . . .).
QTH	What is your location? My location is . . .
QTR	What is the correct time? The time is . . .

activity, many Novices report outstanding results using dipoles or converted CB antennas. Your goal is to put the best possible signal on the air. Of course, this will vary depending on your location and your budget. Don't worry about competing with the ham down the street who has a huge antenna atop a 70-foot tower.

In Amateur Radio, it's important to enjoy and take pride in your own accomplishment. You've heard it said "It's what you do with what you've got that really counts." That is especially true in Amateur Radio. To make up for any real or imagined lack of equipment, concentrate on improving your own operating ability.

Good operating skills can be like adding 10 dB to your signal. (That's like increasing your power by 10 times!) More detailed operating information can be found in *The ARRL Operating Manual*. This book is available from your local radio dealer or directly from ARRL.

A good rule for Amateur Radio contacts **(QSOs)** is to talk just like you would during a face-to-face conversation. When you meet someone for the first time, you introduce yourself once. You don't repeat your name like a broken record. An exception would be if the other person is hard of hearing or in a subway tunnel. Even then you would only repeat if the person couldn't hear you. When you are on the air, don't assume that you must continually repeat information. The other operator will tell you if she needs repeats (**"fills"**). Problem situations include static (QRN), interference (QRM) or signal fading (QSB). Table 10-3 lists many common **Q signals**.

The following example may help you understand the situation. Imagine that you are meeting someone for the first time. You strike up a conversation by saying "Hi, my name is John, John, Name is John. I live in Newington, Connecticut. Newington. Newington. Connecticut. Connecticut." What would you expect the other person to do? Well don't be surprised if he or she doesn't want to talk to you anymore!

The following sections on operating procedures are important. Follow the guidelines and you'll get off to a good start.

CW OPERATING PROCEDURES

Although Novices are allowed phone privileges, CW is the traditional mode for most hams' first contact. Even if you're in a rush to key the mike, don't pass up the information in this section. There are many similarities between CW and voice operation. Non-CW modes such as SSB, FM and specialized digital modes are thoroughly covered later in the chapter.

Calling CQ

There is no point in wasting words, and that includes trying to make a radio contact. Hams establish a contact when one station calls **CQ** and another replies. CQ means "Hey you hams out there in radioland, I'm here and I want to talk to someone." You can usually tell a good ham from a less-experienced operator by the length of the CQ call. A good operator will send short calls separated by concentrated listening periods. The not-so-good ham will send CQ forever (or longer!). By that time, the patience of the audience will have worn thin. Generally, a 3 × 3 call is more than sufficient. Here is an example:

CQ CQ CQ DE WA6YBT WA6YBT WA6YBT K

Perhaps the best way to get started is to listen for someone else's CQ. The Novice bands are alive with signals. There is usually no shortage of stations. You may choose to call CQ yourself, however, and that's okay. Always listen first before you transmit, even if the frequency appears clear. To start, send QRL? ("Is this frequency busy?") If you hear a C ("yes") in reply, then try another frequency. It is not uncommon to be able to hear only one of the stations in a QSO. A frequency may *seem* clear even though there is a contact in progress. It's the worst of bad manners to jump on a frequency that's already busy. Admittedly, signals on the Novice bands, especially on 40 and 80 meters, are often crowded. Even so, you should exercise as much courtesy as possible.

If, when tuning for a clear frequency on which to call CQ, you hear a distress call for help—immediately stand by and copy the message. Do not transmit or in any way interfere with the emergency transmission. Too many stations rushing to offer assistance can often confuse the distressed operator and hinder any response efforts.

Whatever you do, don't send faster than you can reliably copy. That will surely invite disaster! It's best to send at the speed you wish to receive. That is especially true when you call CQ. If you answer a CQ, you should answer at a speed no faster than that of the sending station. Don't be ashamed to send PSE QRS (please send more slowly).

As you gain experience, you will develop the skill of sorting out the signal you want from those of other stations. There is limited space available in the ham bands. It makes good sense to make sure your QSO takes up as little space as possible. When two stations are in communication, their transmitters should be on exactly the same frequency. We call this procedure "operating **zero beat**." This means there is no difference between the operating frequencies of the transmitters involved in the contact. Both transmissions will have the same pitch when you hear them in a receiver. When you tune in the other station, adjust your receiver for the strongest signal and the "proper" tone for your radio. This will usually be between 500 and 1000 Hz.

Zero-beat operating is a good idea. It helps the other operator know where to listen for your signal. This is easy with a transceiver. The transmitter will automatically be close to zero beat when you tune in the received signal to a comfortable pitch. With a separate receiver and transmitter, you must be sure your transmitter is zero beat with that of the other station. This procedure limits the frequency space (bandwidth) required for your radio conversation. It also improves operating convenience and avoids needless interference.

For amateurs to understand each other, we must standardize our communications. You'll find, for instance, that most hams use abbreviations on CW. Why? It's faster to send a couple of letters than it is to spell out a word. But there's no point in using an abbreviation if no one else understands you. Over the years, amateurs have developed a set of standard abbreviations (Table 10-4). If you use these abbreviations, you'll find that everyone will understand you,

Table 10-4

Some Abbreviations for CW Work

Although abbreviations help to cut down unnecessary transmission, it's best not to abbreviate unnecessarily when working an operator of unknown experience.

Abbreviation	Meaning
AA	All after
AB	All before
ABT	About
ADR	Address
AGN	Again
ANT	Antenna
BCI	Broadcast interference
BCL	Broadcast listener
BK	Break; break me; break in
BN	All between; been
BUG	Semi-automatic key
B4	Before
C	Yes
CFM	Confirm; I confirm
CK	Check
CL	I am closing my station; call
CLD-CLG	Called; calling
CQ	Calling any station
CUD	Could
CUL	See you later
CW	Continuous wave (i.e., radiotelegraphy)
DE	From, this is
DLD-DLVD	Delivered
DR	Dear
DX	Distance, foreign countries
ES	And, &
FB	Fine business, excellent
FM	Frequency modulation
GA	Go ahead (or resume sending)
GB	Good-by
GBA	Give better address
GE	Good evening
GG	Going
GM	Good morning
GN	Good night
GND	Ground
GUD	Good
HI	The telegraphic laugh; high
HR	Here, hear
HV	Have
HW	How
LID	A poor operator
MA, MILS	Milliamperes
MSG	Message; prefix to radiogram
N	No
NCS	Net control station
ND	Nothing doing
NIL	Nothing; I have nothing for you
NM	No more
NR	Number
NW	Now; I resume transmission
OB	Old boy
OC	Old chap
OM	Old man
OP-OPR	Operator
OT	Old-timer; old top
PBL	Preamble
PSE	Please
PWR	Power
PX	Press
R	Received as transmitted; are
RCD	Received
RCVR (RX)	Receiver
REF	Refer to; referring to; reference
RFI	Radio frequency interference
RIG	Station equipment
RPT	Repeat; I repeat
RTTY	Radioteletype
RX	Receiver
SASE	Self-addressed, stamped envelope
SED	Said
SIG	Signature; signal
SINE	Operator's personal initials or nickname
SKED	Schedule
SRI	Sorry
SSB	Single sideband
SVC	Service; prefix to service message
T	Zero
TFC	Traffic
TMW	Tomorrow
TNX-TKS	Thanks
TT	That
TU	Thank you
TVI	Television interference
TX	Transmitter
TXT	Text
UR-URS	Your; You're; yours
VFO	Variable-frequency oscillator
VY	Very
WA	Word after
WB	Word before
WD-WDS	Word; words
WKD-WKG	Worked; working
WL	Well; will
WUD	Would
WX	Weather
XCVR	Transceiver
XMTR (TX)	Transmitter
XTAL	Crystal
XYL (YF)	Wife
YL	Young lady
73	Best regards
88	Love and kisses

and you'll understand them. In addition to these standards, we use a set of **procedural signals**, or **prosigns**, to help control a contact. The "Correct CW Procedures" sidebar explains these prosigns in more detail.

Answering a CQ

What about our friend WA6YBT? In our last example he was calling CQ. What happened? Another ham, K7LYA in Tucson, Arizona, heard him and answered the CQ:

WA6YBT WA6YBT DE K7LYA K7LYA $\overline{\text{AR}}$

Use the prosign $\overline{\text{AR}}$ (the letters A and R run together with no separating space) in an initial call to a specific station before officially establishing contact. When calling CQ, use the prosign K, because you are inviting replies. WA6YBT is considered a good operator. He comes back to K7LYA in a to-the-point manner (one you should copy).

K7LYA DE WA6YBT R GE UR RST 599 HARRISBURG PA $\overline{\text{BT}}$ NAME ERICK HW BK

In this transmission, **RST** refers to the standard readability, strength and tone system of signal-quality reporting. You'll exchange an RST signal report in nearly every Amateur Radio QSO. Don't spend a lot of time worrying about what signal report to give to a station you're in contact with. The scales are rather broad, and are simply an indication of how you are receiving the other station. As you gain experience with the descriptions given in Table 10-5, you'll be more comfortable estimating the proper signal report.

A report of RST 368 would be interpreted as "Your signal is readable with considerable difficulty, good strength, with a slight trace of modulation." The tone report is a useful indication of transmitter performance. When the RST system was developed, the tone of amateur transmitters varied widely. Today, a tone report of less than 9 is cause to ask a few other amateurs for their opinion of the transmitted signal. Consistently poor tone reports may mean that you have problems in your transmitter.

The basic information is only transmitted once. The other station will request repeats if necessary. Notice too that $\overline{\text{BT}}$ (B and T run together) is used to separate portions of the text. This character is really the double dash (=), and is usually written as a long dash or hyphen on your copy paper. HW implies "how copy?" BK signifies that

CORRECT CW PROCEDURES

Establishing a contact—The best way to do this, especially at first, is to listen. When you hear someone calling CQ, answer them. If you hear a CQ, wait until the ham indicates she is listening, then call her. For example: KA1KOW KA1KOW DE WL7AGA WL7AGA $\overline{AR}$ ($\overline{AR}$ is equivalent to over).

In answer to your call, the called station will reply WL7AGA DE KA1KOW R . . . That R (roger) means that she has received your call correctly. That's all it means—received. It does not mean (a) correct, (b) I agree, (c) I will comply, or anything else. It is not sent unless everything from the previous transmission was received correctly. Perhaps KA1KOW heard someone calling her but didn't quite catch the call because of interference (QRM) or static (QRN). In this case, she might come back with QRZ? DE KA1KOW $\overline{AR}$ ("Who is calling me?").

Calling CQ—CQ means "I wish to contact any amateur station." Avoid calling CQ endlessly. It clutters up the air and drives off potential new friends. The typical CQ goes like this: CQ CQ CQ DE KA1DYZ KA1DYZ KA1DYZ K. The letter K is an invitation for any station to go ahead.

The QSO—During a contact, it is only necessary to identify your station once every 10 minutes. Keep the contact on a friendly and cordial level. Remember, the conversation is not private. Many others, including nonamateurs, may be listening. Both on CW and phone, it is possible to be informal, friendly and conversational. This is what makes the Amateur Radio QSO enjoyable. During the contact, when you stand by, use K ("go") at the end of your transmission. If you don't want someone else to join the QSO, use $\overline{KN}$. That says you want only the contacted station to come back to you. Most of the time K is sufficient (and shorter!).

Ending the QSO—When it's time to end the contact, don't keep talking. Briefly express your pleasure at having worked the other operator. Then, sign out: $\overline{SK}$ WL7AGA DE KA1KOW. If you are leaving the air, add CL to the end, right after your call sign.

These ending signals establish Amateur Radio as a cordial and fraternal hobby. At the same time, they foster orderliness and organization. These signals have no legal standing, however. FCC regulations say little about our internal procedures. The procedures we ourselves adopt are even more important. They show we're not just hobbyists, but that we are an established communications service. We take pride in our distinctive procedures—tailored to our special needs.

VOICE EQUIVALENTS TO CODE PROCEDURE

Voice	*Code*	*Meaning*
over	$\overline{AR}$	after call to a specific station
end of message	$\overline{AR}$	end of message
wait, stand by	$\overline{AS}$	please stand by
roger	R	all received correctly
go	K	any station transmit
go only	$\overline{KN}$	addressed station only
clear	$\overline{SK}$	end of contact
closing station	CL	going off the air
break or back to you	BK	the receiving station's turn to transmit

Table 10-5

The RST System

READABILITY

1—Unreadable.
2—Barely readable, occasional words distinguishable.
3—Readable with considerable difficulty.
4—Readable with practically no difficulty.
5—Perfectly readable.

SIGNAL STRENGTH

1—Faint signals barely perceptible.
2—Very weak signals.
3—Weak signals.
4—Fair signals.
5—Fairly good signals.
6—Good signals.
7—Moderately strong signals.
8—Strong signals.
9—Extremely strong signals.

TONE

1—Sixty-cycle ac or less, very rough and broad.
2—Very rough ac, very harsh and broad.
3—Rough ac tone, rectified but not filtered.
4—Rough note, some trace of filtering.
5—Filtered rectified ac but strongly ripple-modulated.
6—Filtered tone, definite trace of ripple modulation.
7—Near pure tone, trace of ripple modulation.
8—Near perfect tone, slight trace of modulation.
9—Perfect tone, no trace of ripple or modulation of any kind.

The "tone" report refers only to the purity of the signal. It has no connection with its stability or freedom from clicks or chirps. Most of the signals you hear will be a T-9. Other tone reports occur mainly if the power supply filter capacitors are not doing a thorough job. If so, some trace of ac ripple finds its way onto the transmitted signal. If the signal has the characteristic steadiness of crystal control, add X to the report (e.g., RST 469X). If it has a chirp or "tail" (either on "make" or "break") add C (e.g., 469C). If it has clicks or noticeable other keying transients, add K (e.g., 469K). Of course a signal could have both chirps and clicks, in which case both C and K could be used (e.g., RST 469CK).

MY FIRST ON-THE-AIR CONTACT

I slowly reach for the key, my other hand on the TR switch. All I have to do is throw the TR switch, which connects the antenna to the transmitter, and push down on the key. Then I'll be on the air and able to talk to the world. But I can't. My hand doesn't move, except to tremble slightly. It's not the rig, not the key. It's me! I can't find the nerve to do it. I can't send a signal to the world.

That's how it was for me when I tried to make my first contact. It had been weeks since I passed the Novice exam. Each day that passed without the arrival of my new ticket seemed endless. I knew that if that day would just end, the next would surely see the arrival of my license. Days, weeks passed. Then a month had gone by. The station was there waiting for me. All the equipment was in place and ready for operation. If only the license would arrive.

Finally, after what seemed to be several years, it did arrive. Can you believe it? It arrived on my birthday. What a present! The letter carrier had hardly made it to the sidewalk before I was at the rig. I turned on the transmitter and receiver. Nothing! What could be wrong? All the wires connected? Oh no! The power cords were just hanging there. I'd forgotten to plug them into the wall socket. The few seconds it took for the receiver to warm up seemed like hours. Then I heard the sound of radio. I was ready to enter the world of ham radio. Or so I thought.

After several minutes of total failure at my attempt to send CQ, I gave up. I needed help. But where to get it? Of course, the instructor of my Novice training class. He could, and would help me.

The next day I got in touch with him and explained my problem. Would he help me? Yes, he would be glad to stand behind me during that first contact. But, he was busy with something else. I'd have to wait a day or so. Another day without a contact? It might as well have been another year.

Although it didn't seem possible, the next day rolled around, and with it, my instructor. A few minutes of preparation and I was ready. I would give it my best, to sink or swim. Tuning the receiver to the frequency of my crystal-controlled transmitter, I found another station calling CQ. His sending was slow and steady. It had to be for me to copy it. "Go ahead, give him a try," advised my instructor. "What have you got to lose?"

. . . CQ CQ CQ CQ. His call seemed endless. Finally, he signed

. . . CQ DE W3AOH. I flipped the TR switch and reached confidently for the key. W3AOH DE WN2VDN $\overline{\text{AR}}$.

"That was a pretty short call you gave him. You should have sent your call sign more than once. He may have missed it." But he didn't. There it was, slow and steady. WN2VDN DE W3AOH $\overline{\text{BT}}$ TNX CALL $\overline{\text{BT}}$ UR . . . "He had answered my call . . . my call. I'd done it! I made a contact!"

"Pay attention and copy the code!" my instructor yelled at me. "He's still sending." Dahdididit dahdidah dah dahdahdah dididah dahdahdah . . . dahdididit . . . What on earth did that mean? I'd forgotten the code. Those strange sounds coming out of the speaker didn't make any sense at all. What was I going to do?

Fortunately, my instructor was copying it all. "He's just signed it over to you. Your turn now."

I flipped the TR switch again, and hit the key. This time there was no fear. The key was part of my hand, part of my mind. W3AOH DE WN2VDN . . . Until, that is, after I sent my call. I didn't know his name, his location, anything. So I sent the only thing I could think of. "Thanks for coming back to me OM. You're my first contact. I'm a bit nervous." Behind me, the instructor is saying, "You should have sent him your name and location, and his signal report." Too late, I was already signing it over to W3AOH.

Slow and steady came the reply. "Welcome aboard. Hope you enjoy ham radio as much as I do . . ." This time I had no trouble copying him. After the contact was completed, I breathed a very large sigh of relief. But, I had done it. I had actually "talked" to another person via Morse code. I was a bona fide ham.

My instructor asked if he should hang around a little longer. But I didn't hear him. I had already begun looking around for another CQ to answer. I didn't need his help any longer; I was a ham.

I turned on the transmitter and receiver. Nothing!

WA6YBT is turning it over (back) to K7LYA for his basic info. WA6YBT does not sign both calls all over again. FCC rules require identification only at the end of a QSO, and once every 10 minutes.

Conversation is a two-way phenomenon. There is no reason that an Amateur Radio QSO can't be a back-and-forth process. No one wants to listen to a long, unnecessary monologue. Propagation conditions might change, hampering the QSO. WA6YBT and K7LYA will relate better if each party contributes equally. Sometimes there is interference or marginal copy because of weak signals. It may help to transmit your call sign when you are turning it back to the other station.

CW Operating Summary

In summary, here are the points to keep in mind:

1) Listen before transmitting. Send QRL? ("Is this frequency busy?") before transmitting. Listen again! It's worth repeating: Listen!

2) Send short CQs and listen between each.

3) Send no faster than you can reliably copy.

4) Use standard abbreviations whenever possible—become familiar with them.

5) Use prosigns and Q-signals properly.

6) Identify at the end of a QSO (the entire contact, not each turnover) and every 10 minutes.

7) Use R only if you've received 100 percent of what the other station sent.

8) Be courteous.

9) Zero beat the other station's frequency before calling.

TUNING UP

What is often the most exciting, most memorable and perhaps most terrifying moment in your entire ham experience? It is your first on-the-air contact with another station! Before you have that experience, you'll want to know how to operate your station equipment.

The best source of specific information about your equipment is the instruction manual. Before you even turn on your radio, you should study the instructions carefully so you'll be familiar with each control. Without turning on the equipment, try adjusting the controls. Nothing will happen, but you'll learn the location and feel of each important control.

After studying the manual and finding the important controls, you'll be ready to tune up. You must have your FCC license in hand before you can transmit or even tune up on the air!

If your transmitter requires tuning when you change bands, connect your transmitter output to a *dummy load* or *dummy antenna* (see Chapter 8) while you tune. Never tune up your transmitter on the air because you could interfere with other hams.

Once you have tuned up the transmitter according to the instruction manual, disconnect the dummy load. Then connect the antenna (an antenna switch makes this easy). Now, you're ready to operate! If you use a Transmatch, you may have to transmit a brief low-power signal to adjust the circuit.

[Before you go on to the next section, turn to Chapter 12 and study questions 2B-1-1.1, 2B-1-1.3 and 2B-1-2.1. You should also study questions 2B-2-1.1 through 2B-2-6.5. Review this section if you have any difficulty with these questions.]

VOICE OPERATING PROCEDURES

Among the most exciting Novice privileges is voice operation on 10 meters, 1.25 meters and 23 cm. To make the most of your voice privileges, you'll need to learn a new set of operating procedures. The operating procedures described here apply to voice operation on 10-meter SSB and on the "weak-signal" SSB frequencies around 222.1 MHz. Some, but not all, of the procedures apply to FM voice operation as well. While SSB voice operation and FM voice operation share some operating practices, there are a number of important differences. Techniques unique to FM and repeaters are covered later in this chapter.

Operating techniques and procedures vary from band to band. If you're accustomed to 80-meter CW, you may be uncertain about how to make a 10-meter SSB contact. A great way to become familiar with new techniques is to spend time listening to operators who are already using the band. Be discriminating, though. Take a few moments to understand the techniques used by the proficient operators. Proficient operators are the ones who are the most understandable (least confusing) and who sound the best. Don't simply mimic whatever you hear. This is especially true if you're going on the air for the first time.

Whatever band or mode you are undertaking, there are three fundamental things to remember. These fundamentals apply for any type of voice operating you might try. The first is that courtesy costs very little. It is often rewarded by bringing out the best in others. Second, the aim of each radio contact should be 100% effective communication. A good operator is never satisfied with anything less. Third, our "private" conversation with another station is actually open to the public. Many amateurs are uncomfortable discussing controversial subjects over the air. Also, never give any confidential information on the air. You never know who may be listening.

Keep it Plain and Simple

Correct voice operation is more challenging than it may appear. Even though it does not require the use of code or special abbreviations, the proper procedure is very important. Voice operators *say* what they want to have understood. CW operators have to spell it out or abbreviate. The speed of transmission on voice is generally between 150 and 200 words per minute. Readability and understandability are critical to good communications.

It is important to speak clearly and not too quickly. This is an excellent practice to follow. Whether you're working a DX operator who may not fully understand our language, or talking to your friend down the street, speak slowly and clearly. That way, you'll have fewer requests to repeat information.

Avoid using CW abbreviations and prosigns such as "HI" and "K" for voice communications. Also, the Q-code (QRX, QRV and so forth) is a CW, not voice, procedure. The use of QSL, QSO and QRZ has become accepted practice on voice, however. Abbreviations are used on CW to say more in less time. On voice, you have plenty of time

Table 10-6

Voice Equivalents to Code Procedure

Voice	*Code*	*Meaning*
over	$\overline{AR}$	after call to a specific station
end of message	$\overline{AR}$	end of message
wait, stand by	$\overline{AS}$	please stand by
roger	R	all received correctly
go	K	any station transmit
go only	$\overline{KN}$	addressed station only
clear	$\overline{SK}$	end of contact
closing station	CL	going off the air
break or back to you	BK	the receiving station's turn to transmit

to say what you mean. On CW, for example, it's convenient to send "K" at the end of a transmission. On voice, it takes less than a second to say "go ahead." Table 10-6 shows the voice equivalents for common CW prosigns.

Use plain language and keep jargon to a minimum. In particular, avoid the use of "we" when you mean "I" and "handle" or "personal" when you mean "name." Also, don't say "that's a roger" when you mean "that's correct." Taken individually, any of these sayings is almost harmless. Combined in a conversation, however, they give a false-sounding "radioese" that is actually less effective than plain language.

Phonetic Alphabet

Sometimes you will have difficulty getting your call sign or other information across to the operator on the other end. This might be because conditions are poor, there is nearby interference, or the operator on the other end is not proficient in English. There is a standard International Telecommunication Union (ITU) *phonetic alphabet* to assist you. The standard phonetic alphabet is shown in Table 10-7. It is generally used when signing your call or passing information that must be spelled out. For example, KA5CHW would sign his call Kilo Alfa Five Charlie Hotel Whiskey. W9RE would sign Whiskey Nine Romeo Echo, and YU1EXY would say Yankee Uniform One Echo X-ray Yankee.

The ITU phonetic alphabet is generally understood by hams in all countries. If you want people to understand you, stick to the standard phonetics. For example, KA1MJP should sign Kilo Alfa One Mike Juliett Papa, not something like Kilowatt America One Mexican Jalapeno Pepper or King Adam One Mary John Peter. Use of nonstandard phonetics is often more confusing than using no phonetics at all. Save the "cute" phonetics for talking with friends who already know your call sign.

Initiating a Contact

The procedures described in this section generally apply to SSB operation on 10 meters. They also apply to 222-MHz SSB operation, but FM and repeater operation on 1.25 meters and 70 cm is somewhat different. See the FM and repeaters section later in this chapter.

There are two ways to initiate an SSB voice contact: call CQ (a general call to any station) or answer a CQ. At first, you may want to tune around and find another station to answer. If activity on a band seems low, a CQ call may be worthwhile.

Before calling CQ, it is important to find a frequency that appears unoccupied by any other station. This may not be easy, particularly during crowded band conditions. Listen carefully—perhaps a weak DX station is on frequency. If you're using a beam antenna, rotate it to make sure the frequency is clear. If, after a reasonable time, the frequency seems clear, ask if the frequency is in use, then sign your call. "Is the frequency in use? This is KA1IFB." If, as far as you can determine, no one responds, you are ready to make your call.

As in CW operation, keep voice CQ calls short. Long calls are considered poor operating technique. You may interfere with stations already on frequency. Often, stations will not hear your initial frequency check. Also, stations intending to reply to the call may become impatient and move to another frequency. No one wants to listen to an endless CQ. Call CQ three times, followed by "this is,"

Table 10-7

International Telecommunication Union Phonetics

A—Alfa (**AL** FAH)
B—Bravo (**BRAH** VOH)
C—Charlie (**CHAR** LEE or **SHAR** LEE)
D—Delta (**DELL** TAH)
E—Echo (**ECK** OH)
F—Foxtrot (**FOKS** TROT)
G—Golf (GOLF)
H—Hotel (HOH **TELL**)
I—India (**IN** DEE AH)
J—Juliett (**JEW** LEE ETT)
K—Kilo (**KEY** LOH)
L—Lima (**LEE** MAH)
M—Mike (MIKE)
N—November (NO **VEM** BER)
O—Oscar (**OSS** CAH)
P—Papa (PAH **PAH**)
Q—Quebec (KEH **BECK**)
R—Romeo (**ROW** ME OH)
S—Sierra (SEE **AIR** RAH)
T—Tango (**TANG** GO)
U—Uniform (**YOU** NEE FORM or **OO** NEE FORM)
V—Victor (**VIK** TAH)
W—Whiskey (**WISS** KEY)
X—X-RAY (**ECKS** RAY)
Y—Yankee (**YANG** KEY)
Z—Zulu (**ZOO** LOO)

Note: The **Boldfaced** syllables are emphasized. The pronunciations shown in this table were designed for speakers from all international languages. The pronunciations given for "Oscar" and "Victor" may seem awkward to English-speaking people in the US.

followed by your call sign three times and then listen. If no one comes back, try again. If two or three calls produce no answer, it may be that interference is present on frequency, or that the particular band is not open. At that point, change frequency and try again. If you still get no answer, try looking around and answer someone else's CQ.

An example of a good CQ call is: "CQ CQ Calling CQ. This is KA1MJP, Kilo Alfa One Mike Juliett Papa, Kilo Alfa One Mike Juliett Papa, calling CQ and standing by." There is no need to say what band is being used, and certainly no need to add "tuning for any possible calls, dah-di-dah!" or "K someone please!" and the like.

When replying to a CQ, say both call signs clearly. It's not necessary to sign the other station's call phonetically. You should, however, always sign yours with standard phonetics. It is good practice to keep calls short. Say the call sign of the station you are calling once or twice only. Follow with your call sign repeated several times. For example, "W1AW, W1AW, this is WA3VIL, Whiskey Alfa Three Victor India Lima, WA3VIL, over." Depending on conditions, you may need to sign your call phonetically several times. Repeat this calling procedure as required until you receive a reply or until the station you are calling has come back to someone else. (During FM repeater operation, signals are generally clear. It is not necessary to sign your call with phonetics unless the other station is having difficulty understanding you.)

Listening is very important. If you're using PTT (push-to-talk), be sure to let up on the transmit button between calls so you can hear what is going on. With VOX (voice operated switch), you key the transmitter simply by talking into the microphone. VOX operation is helpful because, when properly adjusted, it enables you to listen between words. Remember: It is extremely poor practice to make a long call without listening. Also, don't continue to call after the station you are trying to contact replies to someone else.

Conducting the QSO

Once you've established contact, it is no longer necessary to use the phonetic alphabet or sign the other station's call. FCC regulations stipulate that you need only sign your call every ten minutes and at the conclusion of the contact. (The exception is when handling international third-party traffic. Then, you must sign both calls.) This allows you to enjoy a normal two-way conversation without the need for continual identification. Use "over" or "go ahead" at the end of a transmission to indicate that it's the other station's turn to transmit. (During FM repeater operation, it is obvious when you or the other station stops transmitting because the repeater carrier drops. In this case it may not be necessary to say "over.")

Signal reports on SSB are two-digit numbers using the RS portion of the RST system. No tone report is required. The maximum signal report would be "59." That is, readability 5, strength 9. On FM repeaters, RS reports are not appropriate. FM signal reports are generally given in terms of quieting. "Full quieting" means that no noise is present with the signal.

Voice contacts are often similar in content to CW QSOs. Aside from signal strength, it is customary to exchange name, location and information on equipment being used. Once these routine details are out of the way, you can discuss virtually any appropriate topic.

Working DX

During the years around the sunspot maximum, 10-meter worldwide communication on a daily basis is commonplace. Ten meters is an outstanding DX band when conditions are right. A particular advantage of 10 meters for DX work is that effective beam-type antennas tend to be small and light, making for relatively easy installation.

CU2AN from the Azores is typical of the many DX contacts awaiting you on 10-meter SSB.

There are a few things to keep in mind when you contact amateurs from outside the United States. While many overseas amateurs have an exceptional command of English (which is especially remarkable since very few US amateurs understand foreign languages) they may not be intimately familiar with many of our local sayings. Because of the language differences, some DX stations are more comfortable with the "bare-bones" type contact, and you should be sensitive to their preferences. A further point is that during unsettled band conditions it may be necessary to keep the contact short in case fading or interference occurs. Take these factors into account when expanding on a basic contact. Also, during a band opening on 10 meters or on VHF, it is crucial to keep contacts brief. This allows many stations to work whatever DX is coming through.

When the time comes to end the contact, end it. Thank the other operator (once) for the pleasure of the contact and say goodbye: "This is KA9MAN, clear." This is all that is required. Unless the other amateur is a good friend, there is no need to start sending best wishes to everyone in the household including the family dog! Nor is this the time to start digging up extra comments on the contact which will require a "final final" from the other station (there may be other stations waiting to call in).

Additional Recommendations

• Listen with care. It is natural to answer the loudest station that calls, but you should answer the best signal. Not all amateurs can run high power, but there is no reason every amateur cannot have a signal of the highest quality. Do not reward the operator who has cranked up the transmitter gain to the point of being unintelligible, especially if a station with a nice sounding signal is also calling.

• Use VOX or PTT. If you use VOX, don't defeat its purpose by saying "aah" to keep the transmitter on the air. If you use PTT, let go of the mike button every so often to make sure you are not "doubling" with the other station. A QSO should be an interactive conversation. Don't do all the talking.

• Don't rush. The speed of voice transmission (with perfect accuracy) depends almost entirely on the skill of the two operators concerned. Use a rate of speech that allows perfect understanding. The operator on the other end should have time to record important details of the contact. If you go too fast, you'll end up repeating a lot of information.

[Before going on to the next section, turn to Chapter 12 and study the questions with numbers that begin 2B-3-1 and 2B-3-2. Review this section if you have difficulty with any of these questions.]

VHF AND UHF PROCEDURES

"Weak-signal" VHF/UHF operating practices are not very different from those on HF. There are, however, some differences that you should know about. Most of these differences evolved because of the nature of VHF and UHF propagation. This is just a brief overview of VHF and UHF procedures. See *The ARRL Operating Manual* for more information.

Grid Squares

One of the first things you'll notice when you tune the weak-signal frequencies is that most QSOs include an exchange of grid squares. Grid squares are just a way of dividing up the surface of the earth into 1° latitude × 2° longitude rectangles. The grid square designator consists of two letters and two numbers. It's a shorthand way of describing your general location. For example, W1AW in Newington, CT is in grid square FN31.

You can determine your grid square identifier from the *QST* article, "VHF/UHF Century Awards," published in January 1983, pp 49-51. (Reprints are available from the Awards Branch at ARRL HQ. Please include an SASE with your request.) ARRL also publishes a map (Figure 10-3) that shows the grid squares for the continental United States and

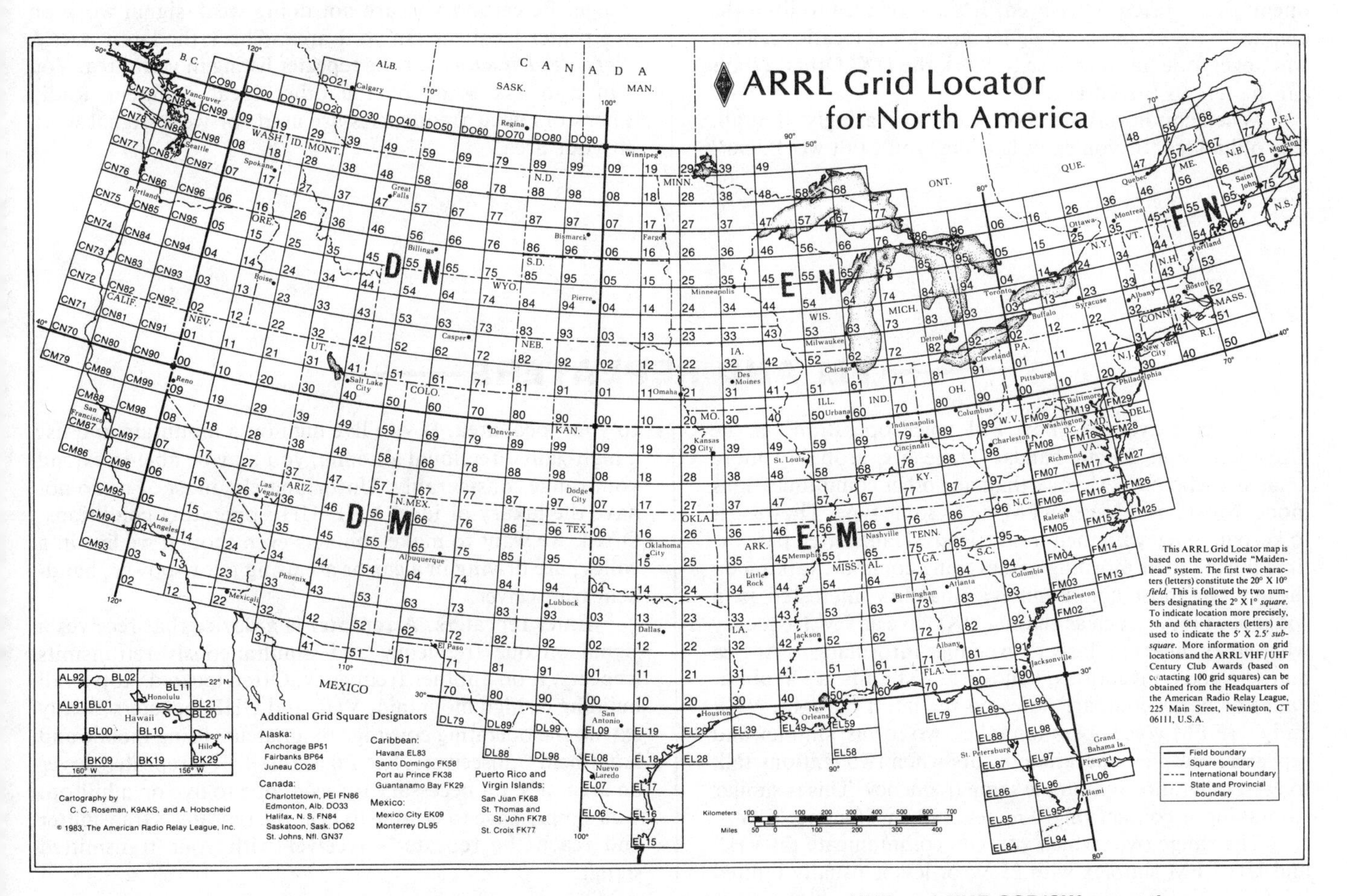

Figure 10-3—This map shows grid squares, used to specify your location for VHF and UHF SSB/CW operation.

most of Canada. This 12- × 18-inch map is available from ARRL HQ for $1.

VHF/UHF Operating

On HF, you never know who you will run into. On any given day, the HF bands are open for cross-country contacts. At least one band will be open for transcontinental QSOs. This makes for a vast number of potential contacts. Most HF QSOs are with operators you've never met before.

It's different on VHF and UHF. Transmission range is normally limited to your local area. You'll tend to hear the same people regularly on 220 and 1270 MHz. A lot of the activity on these bands is relaxed ragchewing. Many groups meet regularly on VHF and UHF to discuss topics of mutual interest. You'll quickly identify one or more groups of "regulars." They'll probably be happy to have you join the group. These regular get-togethers are a wonderful way to meet people. The "regulars" can answer many of your questions about operating your Amateur Radio station.

When there is enhanced propagation, such as during a tropo opening, the VHF and UHF bands sound more like HF. They are teeming with operators eager to work stations not heard during normal conditions. Pileups occur during good VHF/UHF propagation conditions, just like on HF. In this case, the DX is a station from outside your local area, not from another country. Good VHF or UHF band openings are often short-lived. It's a good idea to limit the contact to the exchange of signal reports and locations. That way, everyone has a chance to work the DX. Other details can be left to the QSL card.

When conditions are good, call CQ sparingly, if at all. The old adage "if you can't her 'em, you can't work 'em" is just as true at VHF as at HF. Stations can often be heard calling CQ DX on the same frequency as distant stations that they cannot hear.

One feature of VHF operating that is rarely found on the HF bands is the use of **calling frequencies** to provide a meeting place for operators using the same mode. Once a contact has been set up, a change to another frequency (a working frequency) is arranged. That lets others use the calling frequency.

The ARRL recommends band plans to coordinate various activities on each amateur band. The 1.25-meter band plan specifies 222.1 MHz as a calling frequency. This frequency marks the bottom edge of the Novice subband, however. You should not operate right on 222.1 MHz with a Novice license. You must always be aware of the bandwidth of your transmitted signal. Your transmitted signal will vary from the operating frequency. You must keep your signal within your band limits.

Theoretically, by using the upper sideband for SSB operation, you could operate on a carrier frequency of 222.100 MHz and all of your transmitted signal would be within the Novice subband. In practice, this isn't a good idea, however. You should operate a few kilohertz above the bottom edge of the band, just to be safe. Check the frequencies near the bottom edge of the band for other stations doing weak-signal work.

There are also repeaters operating in this frequency range. Be certain you are not doing weak-signal work on a repeater frequency in your area. Check the latest *ARRL Repeater Directory* for the repeater listing in your area. You can also ask some of the other local Amateur Radio operators what frequencies are used for weak-signal work in your area.

FM AND REPEATERS

Novice privileges allow FM voice operation on the 1.25-meter and 23-cm bands. There are probably more amateurs who use FM voice than any other communications mode. Most hams have an FM rig of some type. They use it to keep in touch with their local friends. Hams often pass the time during their morning and evening commute talking on the air. In most communities, amateurs interested in a specialized topic (such as chasing DX) have an FM frequency where they meet regularly to exchange information. At flea markets and conventions, hand-held FM units are in abundance as hams compare notes on the latest bargain. VHF and UHF FM voice operation takes two forms: simplex and repeater. **Simplex operation** occurs when two stations talk to each other directly, on the same frequency. This is similar to making a contact on 10-meter SSB or 80-meter CW.

The range over which you can communicate on VHF and UHF FM simplex with 25 W or less is usually limited to your local area. If you live high on a mountain and use a high-gain directional antenna, you may be able to extend your range considerably. Unfortunately, most of us do not have the luxury of ideal VHF/UHF operating conditions. Often, we want to make contacts even though we live in a valley, are driving in a car or are using a low-power, hand-held transceiver.

Enter repeaters. A **repeater** is a device that receives a signal on one frequency and simultaneously retransmits (repeats) it on another frequency. Often located atop a tall building or high mountain, VHF and UHF repeaters greatly extend the operating coverage of amateurs using mobile and hand-held transceivers. See Figure 10-4. If a repeater serves an area, it's not necessary for everyone to live on a hilltop. You simply have to be able to hear the repeater's transmitter and reach the repeater's receiver with your transmitted signal.

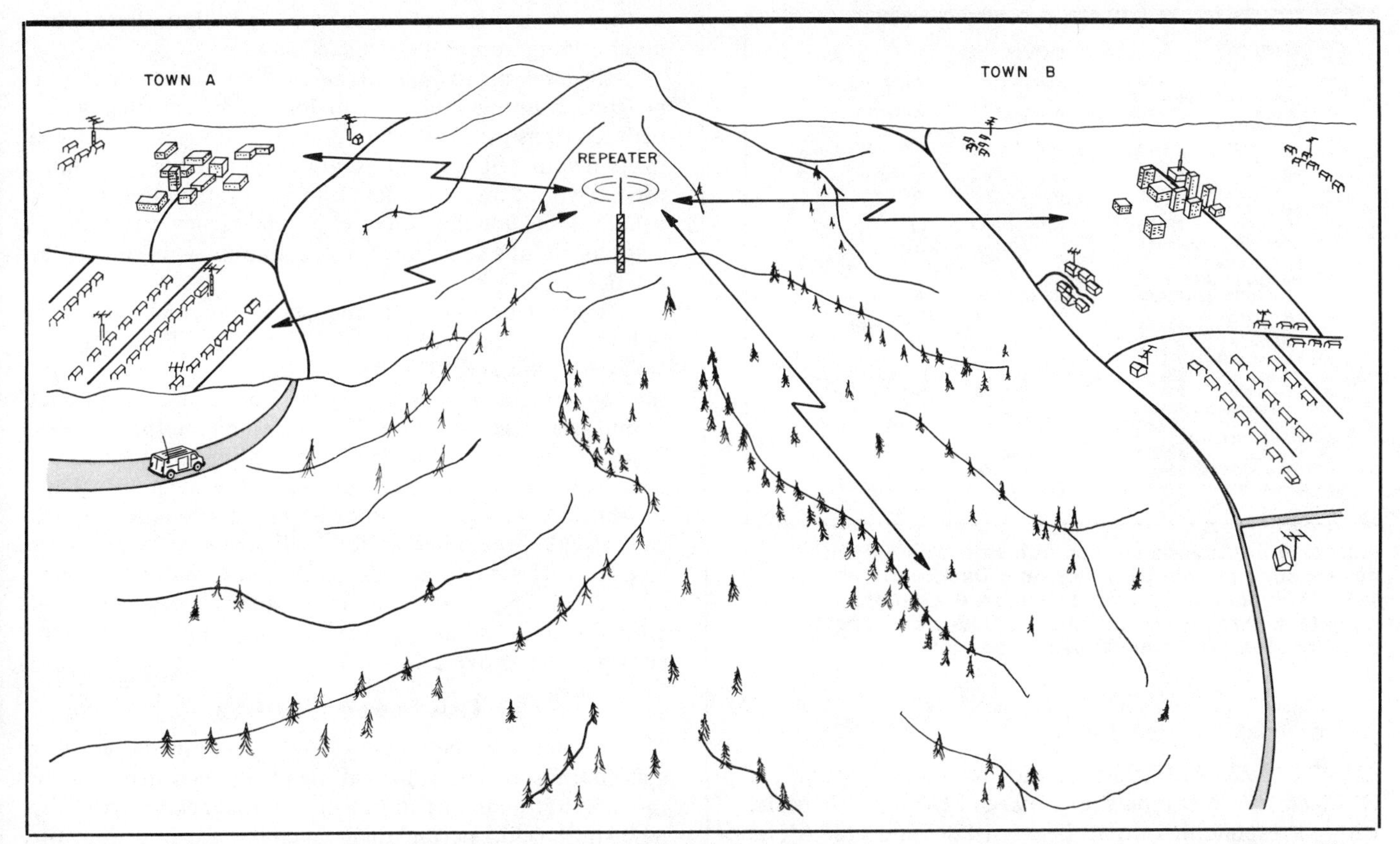

Figure 10-4—Stations in town A can easily communicate with each other, but the hill blocks their communications with town B. The hill-top repeater enables the groups to communicate with each other.

EQUIPMENT CONSIDERATIONS

To use a repeater, you must have a transceiver with the capability of transmitting on the repeater's **input frequency** (the frequency that the repeater listens on) and receiving on the repeater's **output frequency** (the frequency the repeater transmits on). The input and output frequencies are spaced apart ('offset') by a predetermined amount that is different for each band. The offset on 1.25 meters is 1.6 MHz, while on 23 cm it is 12 MHz. For example, a repeater on 1.25 meters might have its input frequency on 222.32 MHz and its output on 223.92 MHz.

Most transceivers designed for FM repeater operation are set up for the right offset. They usually have a switch to change between simplex (transmit and receive on the same frequency) and duplex (transmit and receive on different frequencies). So, if you wanted to use the repeater in the preceding example, you would switch your transceiver to the duplex mode and dial up 223.92 to listen to the repeater. When you transmit, your rig will automatically switch to 222.32 MHz (1.6 MHz away), the repeater input frequency.

When you have the correct frequency dialed in, all you need do is key your microphone button to transmit through ("access") the repeater. Most repeaters are "open"—that is, available for use by anyone in range. Some repeaters, however, have limited access. Their use is restricted to exclusive groups, such as members of a club. Such "closed" repeaters require the transmission of a continuous subaudible tone or a short "burst" of tones for access. There are also some repeaters available for use by everyone that require the use of special codes or subaudible tones to gain access. The reason for requiring access tones for "open" repeaters is to prevent interference from extraneous transmissions that might accidentally key the repeater.

FINDING A REPEATER

Most communities in the United States are served by repeaters. While the majority of repeaters are on 2 meters, there are more than 1400 repeaters in the 1.25-meter band and more than 160 in the 23-cm band. More repeaters are being put into service all the time.

There are several ways to find the local repeater(s). Ask local amateurs or contact the nearest radio club. Each spring, the ARRL publishes *The ARRL Repeater Directory*. This is a comprehensive listing of repeaters throughout the United States, Canada and other parts of the repeater world. Besides finding out about local repeater activity, the *Directory* is handy for finding repeaters during vacations and business trips. See Figure 10-5.

Certain segments of each band are set aside for FM operation. On 1.25 meters, repeater inputs are found between 222.32 and 223.28 MHz. The corresponding outputs are between 223.92 and 224.98 MHz. Frequencies between

KENTUCKY					
ASHLAND					
Ashland	223.94	–	KC4QK	o	WR4ALH RG
CINCINNATI AREA					
Covington	224.90	–	WB9MRB		WB9MRB
LEXINGTON					
Lexington	223.74	–	WB8VMI	o	+KA8MJQ
LOUISVILLE					
Brooks	224.50	–	N4FND	o	N4FND
Louisville	224.82	–	W4MOP	o(ca)e	W4MOP
SOUTHEAST					
Pikeville	224.58	–	WB4F	al	EKRC
Pikeville	224.62	–	N4ETG	l	EKRC
LOUISIANA					
BATON ROUGE					
Baton Rouge	224.06	–	W5RWF	o	W5RWF
Baton Rouge	224.62	–	KD5SL	oe	KD5SL
Gonzales	224.38	–	KA5LUR	o	ASCE ARC
Rosedale	224.78	–	W5OVV	ol	Rsdl RA
HAMMOND/SLIDELL					
Blond	224.66	–	W5VUH	oe	W5VUH
Covington	224.74	–	WB9TBX	oael	Bayou220Gp
Madisonville	224.14	–	W5NJJ	o	N Lake RC
Slidell	223.94	–	WB5LOC	o	WB5LOC
NEW ORLEANS					
Gretna	224.32	–	WA5RKD	o	CARS
New Orleans	224.52	–	W5VAS	oel	TPRG
New Orleans	224.94	–	WB5LOC	o	WB5LOC

Figure 10-5—Anyone who operates on VHF FM should have a copy of *The ARRL Repeater Directory*. The 1989-1990 edition lists more than 1400 220-MHz repeaters. *Directory* information includes the repeater call sign, location, frequency and sponsor.

223.42 and 223.9 MHz are set aside for simplex operation. On 23 cm, repeater inputs run between 1270 and 1276 MHz, with corresponding outputs between 1282 and 1288 MHz. Simplex operation is between 1294 and 1295 MHz.

REPEATER OPERATING PROCEDURES

FM repeaters provide a means to communicate efficiently. To operate effectively, you'll need to know some fundamental techniques that are unique to repeaters. It's worth a few minutes to listen and familiarize yourself with the procedures used by other hams before you make your first FM repeater contact. Accepted procedures can vary slightly from repeater to repeater.

Your First Transmission

Making your first transmission on a repeater is as simple as signing your call. If the repeater is quiet, pick up your microphone, press its switch, and transmit your call sign—"KA1CV" or "KA1CV listening"—to attract someone's attention. After you stop transmitting, you will usually hear a short, unmodulated carrier transmitted by the repeater. This is to let you know that the repeater is working. If someone is listening and is interested in talking to you, he or she will call you after your initial transmission. Some repeaters have specific rules for making yourself heard. In general, however, your call sign is all you need to do the trick.

Don't call CQ to initiate a conversation. It takes a lot longer to complete a long CQ than to simply transmit your call sign. (In some areas, a solitary "CQ" is permissible.) Efficient communication is the goal. You are not on HF, trying to attract the attention of someone who is casually tuning his receiver across the band. In the FM and repeater mode, stations are either monitoring their favorite frequency or they are not. Except for scanner operation, there is not much tuning across the repeater bands.

If you want to join a conversation that is already in progress, transmit your call sign during the break that occurs between transmissions. The station that transmits after you drop in your call sign will usually acknowledge you. Don't use the word "break" to join into a conversation. "Break" usually indicates that there is an emergency and that all stations should stand by for the station with emergency traffic.

If you want to call another station and the repeater is not in use, simply call the other station. For example, "N2DRR, this is KA1MDH." If the repeater is in use, but the conversation that is in progress sounds like it is about to end, wait until it is over before calling another station. If the conversation sounds like it is going to continue for a while, however, transmit your call sign between their transmissions. After you are acknowledged, ask to make a quick call. Usually, the other stations will acquiesce. Make your call short. If your friend responds to your call, try to meet on another repeater or a simplex frequency. Otherwise, ask your friend to stand by on frequency until the present conversation is over.

Courtesy Counts

If you are in the midst of a conversation and another station transmits his or her call sign between transmissions, the next station in line to transmit should acknowledge the new station and permit them to make a call or join the conversation. It is impolite not to acknowledge new stations. Furthermore, it is impolite to acknowledge them but not let them speak. You never know; the calling station may need to use the repeater immediately. He or she may have an emergency to handle, so let him or her make a transmission promptly.

A brief pause before you begin each transmission allows other stations to participate in the conversation. Don't key your microphone as soon as someone else releases theirs. If your exchanges are too quick, you can prevent other stations from getting in.

The "courtesy beepers" found on some repeaters force users to leave a space between transmissions. The beeper sounds a second or two after each transmission to permit new stations to transmit their call signs in the intervening time period. The conversation may continue only after the beeper sounds. If a station is too quick and begins transmitting before the beeper sounds, the repeater may indicate the violation, sometimes by shutting itself down.

Keep each transmission as short as possible. Short transmissions permit more people to use the repeater. All repeaters promote this by having timers that "time-out," (that is, shut down) the repeater whenever someone transmits too long. Learn the length of the repeater's timer and stay well within its limits. Some timers are as short as 15 seconds and some are as long as three minutes. Others automatically vary their length depending on the amount of traffic on frequency. The other purpose of a repeater timer is to prevent extraneous signals from keeping the repeater on the air continuously. That could damage the repeater's transmitter.

Many hams use VHF FM rigs for mobile operation. Donna, N2FFY, particularly enjoys VHF FM during "drive time," the daily commute to and from the office. ***(WA4PFN photo)***

You must transmit your call sign at the end of a contact and at least every 10 minutes during the course of any communication. You do not have to transmit the call sign of the station to whom you are transmitting.

Never transmit without identification. For example, keying your microphone to turn on the repeater without identifying is illegal. If you do not want to engage in conversation, but simply want to check if you are able to access a particular repeater, simply say "KJ4KB testing." Thus, you have accomplished what you wanted to do, legally.

Fixed Stations and Prime Time

During commuter rush hours, mobile stations have preference for use of repeaters. Originally, repeaters were intended to enhance mobile communications. During mobile operating prime time, fixed stations should generally yield to mobile stations. (Some repeaters have a specific policy to this effect.) When you're operating as a fixed station, don't abandon the repeater completely, though. Monitor mobile activity. Your assistance may be needed in an emergency.

SIMPLEX OPERATION

After you have made a contact on a repeater, move the conversation to a simplex frequency if possible. The repeater is not a soapbox. You may like to listen to yourself, but others, who may need to use the repeater, will not be as appreciative.

The function of a repeater is to provide communications between stations that would not normally be able to communicate because of terrain and/or equipment limitations. It logically follows that if stations are able to communicate without the need of a repeater, they should not use a repeater. In other words, when it is possible to use a simplex frequency for communications, use simplex. That allows the repeater to be available for stations that need its facilities. (Besides, communication on a simplex frequency offers a degree of privacy that is impossible to achieve on a repeater, You can usually carry on an extensive conversation on a simplex channel without interruption.)

You will be surprised how well you can communicate with modest equipment on a simplex frequency. There is almost never any need for fixed stations in the same community to use a repeater to communicate with each other. Similarly, amateurs walking around the same ham radio flea market or convention floor should be able to stay in touch on a simplex frequency.

When selecting a simplex frequency, make sure that it is a frequency designated for FM simplex operation. If you select a simplex frequency indiscriminately, you may interfere with stations operating in other modes (and you may not be aware of it). Table 10-8 lists the established simplex frequencies in the 1.25-meter band. Each VHF and UHF band has a frequency designated as the national FM simplex calling frequency, which is the center for most simplex operation. On 1.25 meters, the national simplex calling frequency is 223.5 MHz, while on 23 cm it is 1294.5 MHz.

Table 10-8
1.25 Meter Simplex Frequencies

223.42	223.52	223.62	223.72	223.82
223.44	223.54	223.64	223.74	223.84
223.46	223.56	223.66	223.76	223.86
223.48	223.58	223.68	223.78	223.88
223.50*	223.60	223.70	223.80	223.90

* National simplex frequency

AUTOPATCH

An **autopatch** is a device that allows repeater users to make telephone calls through the repeater. In most repeater autopatch systems, the user simply has to generate the standard telephone company tones to access and dial through the system. Usually, this is accomplished with a telephone tone pad connected to the transceiver. Tone pads are usually available from equipment manufacturers as standard or optional equipment. They are often mounted on the front of a portable transceiver or on the front of a fixed or mobile transceiver's microphone. Whatever equipment is used, the same autopatch operating procedures apply.

There are strict guidelines for autopatch use. The first question you need to ask is "Is the call necessary?" If it is an emergency situation, there is no problem. Any other reason falls into a gray area. As a result, autopatch use, except for emergency situations, is expressly forbidden by some repeater groups.

Never use an autopatch where regular telephone service is available. A common example of poor operating practice can be heard most evenings in any metropolitan area. Someone will call home to announce departure from the office. Why not make that call from work before leaving?

Never use the autopatch for anything that could be considered business communications. The FCC strictly forbids any business communications in Amateur Radio. Don't use the autopatch to call the local radio store to find out if they have a modified trimmer capacitor in stock. Don't call the local pizza palace to order a large mozzarella. You may, however, call a business if the call is related to an emergency situation. Calling an ambulance or a tow truck is okay.

Never use an autopatch to avoid a toll call. Autopatch operation is a privilege granted by the FCC. Abuses of autopatch privileges may lead to their loss.

Now that you have a legitimate reason to use the autopatch, how do you use it? First, you must access (turn on) the autopatch. This is usually done by pressing a designated key on the telephone tone pad. Ask the other hams on a repeater how to learn the access code. Many clubs provide this information only to club members. When you hear a dial tone, you know that you have successfully accessed the autopatch. Now, simply punch in the telephone number you wish to call.

Once a call is established, remember that you are still on the air. Unlike a normal telephone call, only one party at a time may speak. Both you and the person you're talking to should use the word "over" to indicate that you are finished talking and expect a reply. Keep the call short. Many autopatches have timers that terminate the connection after a certain period.

Turning off the autopatch is similar to accessing it. A key or combination of keys must be punched to return the repeater to normal operation. Autopatch users should consult the repeater group sponsoring the autopatch for specific information about access and deaccess codes, as well as timer specifics.

[Before you go on to the next section, turn to Chapter 12 and study the questions with numbers that begin 2B-6-1, 2B-6-2, 2B-6-3, 2B-6-4, and 2B-6-5. Review this section if you have difficulty with any of these questions.]

INTERFERENCE AND SPLATTER

Chapter 11 covers various types of interference, and describes how to solve these problems. Voice modes require careful transmitter adjustment to produce the best signal quality. It's easy to avoid problems if you know what to do.

Most SSB transmitters and transceivers have a microphone gain control. This control may be on the front panel, or it may be inside the rig. Different microphones and different voices have different characteristics. You'll have to adjust the microphone gain control for your particular situation. Instructions for properly setting the mike gain are included in your instruction manual.

Most FM transceivers do not have a front-panel mike-gain control. The microphone gain control, sometimes called the deviation control, is preset. It requires no adjustment unless there is a problem.

Don't continuously adjust the mike gain. Set it according to the manual and leave it alone. Follow the manufacturer's instructions for use of the microphone. Some require close talking, while some need to be turned at an angle to the speaker's mouth. Always try to speak in an even amplitude, and at the same distance from the microphone. If you set the gain properly, you will minimize background room noise on your signal. You'll always hear one or two stations on the band with their mike gain set so high that you hear people talking in the background, music playing, dogs barking and the like. This is not good practice.

If you operate with the microphone gain set too high, you will most likely cause **splatter**—interference to stations on nearby frequencies. This applies to FM voice operation as well as SSB operation. If you adjust the mike gain or deviation control on your FM rig too high, your signal will be too wide and will cause interference to stations on nearby frequencies.

A **speech processor** (sometimes built into contemporary transceivers or available as an accessory) is often a mixed blessing. It can give your audio more "punch" to help you cut through the interference and static. If too much processing is used, however, the audio quality suffers greatly. You'll sometimes hear operators with their speech processor set so high that they are difficult to understand. That's exactly the opposite of what speech processing is supposed to do.

If you set the speech processing level too high, your signal will sound bad. It will also cause splatter and interfere with stations on nearby frequencies. Make tests to determine the maximum speech processing level that can be used effectively. Make a note of the control settings. Be ready to turn your speech processor control down or off if it is not required during a contact.

[It is time to turn to Chapter 12 and study a few more questions. You should be able to answer questions 2H-7.4, 2H-7.5 and 2H-7.6. If you have any difficulty, review this section.]

DIGITAL COMMUNICATIONS

Digital communications is simply a term used to describe amateur communications designed to be received and printed automatically. Digital communications often involve direct transfer of information between computers. You type the information into your computer. The computer then (with the help of accessory equipment) processes the signal and sends it over the air from your transceiver. The station on the other end receives the signal, processes it and prints it out on a computer screen or printer. In this section, you will learn about two very exciting forms of digital communication: **radioteletype (RTTY)** and **packet radio**.

EMISSION MODES

RTTY and packet communications are similar to CW. They both use two states to convey information. Instead of switching a carrier on and off as in CW operation, the carrier is left on continuously and switched between two different frequencies. The transmitter carrier frequency "shifts" between two frequencies, called MARK and SPACE. MARK is the ON state; SPACE is the OFF state. This is called **frequency-shift keying (FSK)**. The FCC emission designator used to describe FSK is **F1B**. On 10-meter RTTY, the MARK and SPACE frequencies are normally 170 Hz apart. This is known as 170-Hz shift RTTY.

On VHF, **audio-frequency shift keying (AFSK**, or emission type **F2B)** is used. AFSK is similar to FSK, except that an FM transmitter is used. Audio tones corresponding to MARK and SPACE are fed into the microphone jack and used to modulate the carrier. The most common VHF shift is 170 Hz, but you may hear 850-Hz shift as well.

[Now turn to Chapter 12 and study questions 2H-1-2.1 and 2H-1-2.2. Review this section if you have any difficulty with these questions.]

SENDING SPEEDS

Just as you can send CW at a variety of speeds, RTTY and packet transmissions are sent at a variety of speeds. The signaling speed depends on the type of transmission and on the frequency band in use. A **baud** is the unit used to describe transmission speeds for digital signals.

For a single-channel transmission, a baud is equivalent to one digital bit of information transmitted per second. A 300-baud signaling rate represents a transmission rate of 300 bits of digital information per second in a single-channel transmission. A digital bit of information has two possible conditions, ON or OFF. We often represent a bit with a 1 for ON or a 0 for OFF. The MARK and SPACE tones of a radioteletype signal also represent the two possible bit conditions.

On HF, we use signaling rates of up to 300 bauds. Sending speeds on VHF are generally faster than on HF. A common VHF sending speed is 1200 bauds. The most important thing to remember about sending speeds is that both stations use the same speed during a contact.

RTTY COMMUNICATIONS

Radioteletype (RTTY) is a popular form of digital communications. Among your new privileges is the opportunity to operate RTTY on 10 meters, and on the 222-MHz and 1270-MHz bands. On 10 meters, you're allowed to operate RTTY from 28.1 to 28.3 MHz. On 222 and 1270 MHz, you are allowed to operate RTTY on all of the frequencies that the FCC authorizes you to operate on. You should, however, follow the band plans as described earlier. On VHF, most RTTY activity is on repeaters. Check with local amateurs to find out where the 222-MHz RTTY activity is in your area.

This section will tell you about three popular types of RTTY: **Baudot, AMTOR** and **ASCII**. Baudot and AMTOR are the most popular RTTY modes on HF. Until recently, high-speed ASCII was most often heard on VHF, but this mode is quickly being replaced by packet radio.

BAUDOT RADIOTELETYPE

Radioteletype communication using the **Baudot** code (also known as the International Telegraph Alphabet number 2, or ITA2) is widely used on the Amateur Radio HF bands in most areas of the world.

The Baudot code represents each character with a string of five bits of digital information. Each character has a different combination of bits. There are only 32 possible Baudot code combinations. This limits the number of possible characters. All text is in upper-case characters. To provide numbers and punctuations, you shift between the letters case (LTRS) and the figures case (FIGS).

There are three common speeds for 10-meter Baudot RTTY communications: 60 WPM (45 bauds), 75 WPM (56 bauds) and 100 WPM (75 bauds). You'll often hear 60, 75 and 100 WPM RTTY referred to as "60 speed," "75 speed," and "100 speed," respectively. Both stations must use the same sending speed to make a RTTY contact.

Setting Up

To set up a RTTY station using Baudot, the first thing you will need is a mechanical teleprinter or a computer-based RTTY terminal. Most radioteletype operators are now using computer-based **communications terminals**. One popular way of gearing up for radioteletype is to use a communications terminal designed specifically for amateur RTTY service. See Figure 10-6. Such units are manufactured by

Figure 10-6—This Tono EXL-5000E is a complete RTTY communications terminal. It sends and receives CW, Baudot, ASCII and AMTOR, and has its own keyboard and display screen.

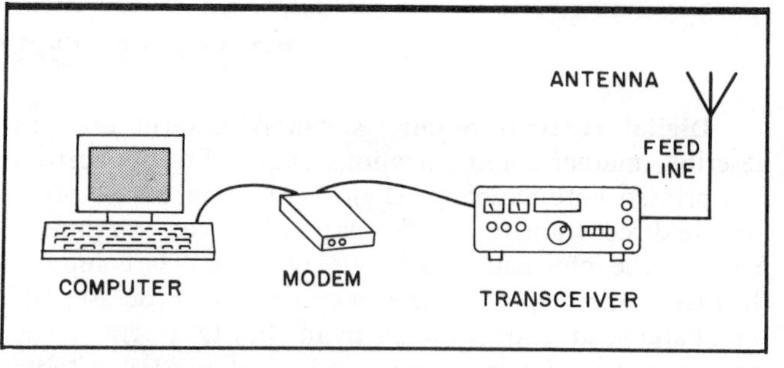

Figure 10-7—A RTTY modem connects between your transceiver and your computer.

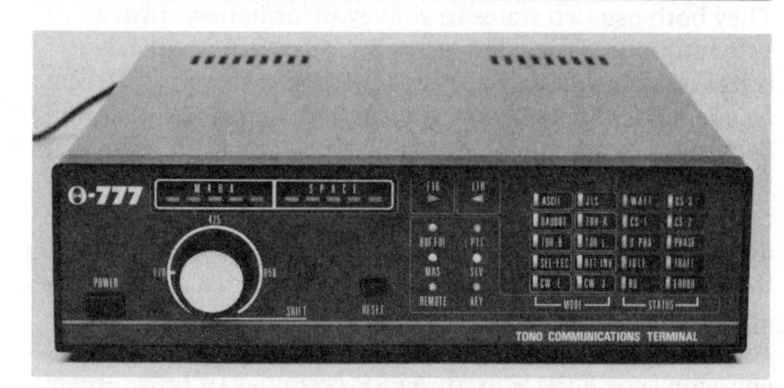

Figure 10-8—A full-feature RTTY modem, the Tono Theta-7777 has an LED bar-graph tuning indicator to help you tune in stations you're trying to contact. It is used between a computer or ASCII terminal and transceiver.

Hal Communications, Microlog and Tono.

Another popular way of getting on RTTY is to use a personal computer, such as those manufactured by Commodore®, Apple® or IBM®, along with appropriate software. Such a system is shown in Figure 10-7. If you use a personal computer, you will also need an external **modem** (a contraction of *mo*dulator-*dem*odulator, also called a "terminal unit" or "TU"). See Figure 10-8. (Dedicated RTTY terminals usually have built-in modems). A modem takes information from your computer and modulates the transmitted radio signal. It also demodulates the received signal. In choosing a modem or a complete RTTY system with a built-in modem, look for versatility. Look for a modem with shift capabilities of 170 and 850 Hz. Check the *QST* Product Review column, and articles and ads in *QST* and other amateur magazines for information about these products.

What about a rig for RTTY? You can use almost any HF transceiver. Ideally, the receiver used for 170-Hz-shift RTTY should have the minimum practical bandwidth, preferably between 270 and 340 Hz. Many CW filters have bandwidths around 500 Hz and are good for 170-Hz RTTY reception. Many receivers (and transceivers) can only use the SSB filter in the SSB or RTTY modes. Some RTTY operators have modified their receivers to use the CW filter when the RTTY mode is selected. A switch can be added so it is possible to select the CW filter with the mode switch set for SSB operation. A transceiver with a frequency display of 10-Hz resolution is also helpful, although not a necessity.

Many modern transceivers include an FSK mode. If yours does not, you can operate RTTY with the transceiver in the SSB mode. It is normal to use the lower-sideband mode for RTTY on HF SSB radio equipment. Some modern transceivers have two positions on the sideband-selection knob, labeled NORMAL and REVERSE. This labeling may cause some uncertainty when you use the rig for RTTY. On 10 meters, normal SSB operation uses upper sideband, so you will have to set the switch to the REVERSE position to select the lower sideband for 10-meter RTTY.

If you are not using the correct sideband, then your signal will be "upside down," and other operators will have to change their normal operating setup in order to copy your signals. Be sure you select the lower sideband and transmit signals that are "right-side up." Consult your radio operator's manual for details.

On VHF, the most common practice is to use AFSK. Since most VHF RTTY work is done on local FM repeaters, almost any FM transceiver will work.

Your transmitter must be able to withstand 100% duty cycle when transmitting conventional Baudot and ASCII radioteletype, as well as AMTOR Mode B. This means that it must be able to operate at full power for extended periods. Some transceivers, designed for SSB and CW operation (which do not require constant transmission at full power), may overheat and possibly fail if subjected to a long RTTY transmission. Many operators reduce to half power during long RTTY transmissions to avoid overheating problems. AMTOR Mode A transmits data in shorter blocks and does not require constant transmission at full power.

Receiver-tuning accuracy is important. Thus, a tuning

RTTY EQUIPMENT

Years ago, all RTTY operation was done with mechanical teleprinters. You may still see some of these machines in old movies—they are big, slow and *noisy*. Today, most amateurs use a computer-based communications terminal for RTTY operation. The *communications terminal* (sometimes called a "terminal unit" or "interface") connects between a transceiver and a computer and decodes the RTTY signals for display by the computer. Most communications terminals will send and receive CW and RTTY (Baudot and ASCII). Some terminals also send and receive AMTOR, and some add packet-radio operation for "all-mode" operation.

When you shop for a communications terminal, there are a few things to keep in mind. Some terminals are computer-specific—they are designed to work with only one type of computer. Some terminals require additional software in the computer, and others require only a simple terminal program. As always, the best way to find out about a particular unit is to ask someone who has one. The next best way is to read the product review columns in *QST* and the other ham magazines. The following list is a representative sample of what is available today. Some older used units may be available at hamfests and flea markets. If you plan to buy a used communications terminal, try to take an experienced RTTY operator along to the hamfest with you. Remember, most hams love to talk about their favorite operating mode—ask around at your local club.

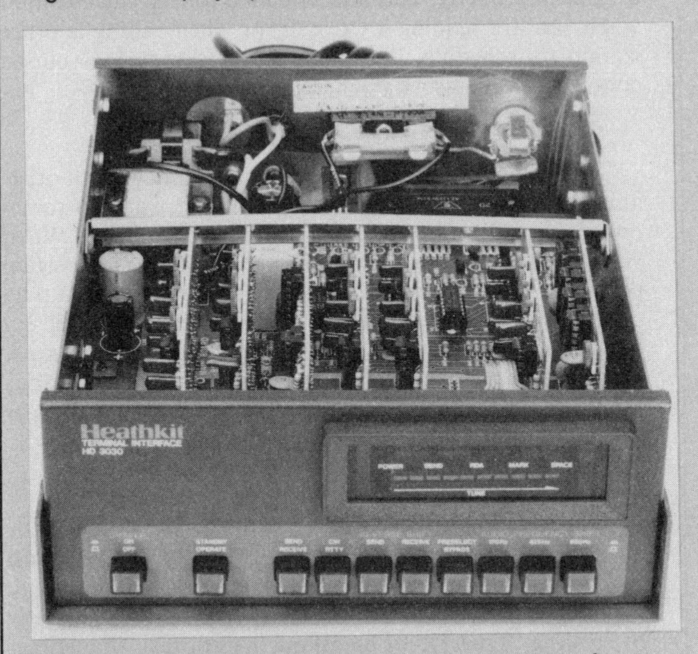

The Heath HD-3030 communications terminal sends and receives CW, Baudot and ASCII RTTY.

The trend in RTTY equipment today is toward the "all-mode" communications terminal. This MFJ-1278 provides CW, Baudot, ASCII, AMTOR, FAX and packet-radio capability. It is used with any of a number of popular personal computers.

RTTY Communications Terminals

Manufacturer	*Model Number*	*Modes*	*Product Review*
AEA	CP-1	CW, Baudot, ASCII	April 1984 *QST*
AEA	PK232	CW, Baudot, ASCII, AMTOR, packet	January 1988 *QST*
AEA	PK-64†	CW, Baudot, ASCII, AMTOR, packet	June 1986 *QST*
Heath	HD3030	CW, Baudot, ASCII	February 1985 *QST*
Kantronics	UTU	CW, Baudot, ASCII, AMTOR	
Kantronics	UTU-XT/P	CW, Baudot, ASCII, AMTOR, packet	
Kantronics	KAM	CW, Baudot, ASCII, AMTOR, packet	June 1989 *QST*
MFJ	MFJ-1224	CW, Baudot, ASCII	
Tono	EXL-5000E‡	CW, Baudot, ASCII, AMTOR	July 1985 *QST*

†For use with the Commodore 64 or 128 only
‡Includes display screen and keyboard. No computer needed.

aid can be a great help in proper receiver adjustment. Most modems have some type of tuning indicator, possibly just flashing LEDs. Some have oscilloscopes that produce patterns such as those shown in Figure 10-9. The MARK signal is displayed as a horizontal line and the SPACE signal as a vertical line. Theoretically this should appear as a "+" sign on the 'scope screen, but in practice it may look more like a pair of crossed bananas! To avoid the high cost of an oscilloscope, some manufacturers are offering LED displays that imitate the 'scope display.

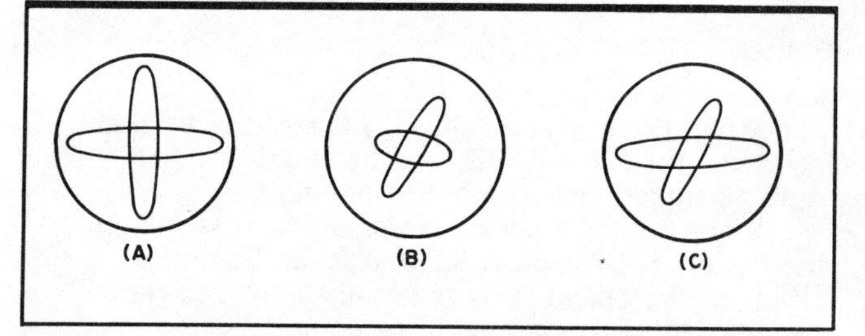

Figure 10-9—Oscilloscope RTTY tuning patterns. A shows the pattern produced by a properly tuned signal. The displays at B and C indicate improperly tuned signals.

Typing the Message

No one can disagree that touch typing (typing without looking at the keys) is the best way to go. But, you can have enjoyable contacts even if you can only type with two fingers! It is a good practice to keep the number of characters per line to a maximum of 69. That line length can be handled by virtually any teleprinter designed for Baudot. Many US teleprinters will print 72 or 74 characters, but some foreign printers take only 69. Computer screens and printers vary widely and may have 22 (Commodore VIC 20™), 32 (Sinclair ZX81), 40 (Apple II® and Commodore 64™), or 80 (IBM PC® and compatibles).

After a mistake, the "oops" signal is XXXXX (some operators may use EEEEEE). There is no need to overdo this, though, because generally the other operator will figure out what you meant to say anyway, and a page filled with XXXXX can be confusing.

Station Identification

When you are transmitting RTTY, simply send your call sign at least once every 10 minutes, just as you would on any other mode. FCC rules permit a station engaged in digital communications using Baudot, AMTOR or ASCII to identify in the digital code used for the communication. On the Novice 222- and 1270-MHz bands, identification may be given using one of the methods just mentioned or by voice. (Radiotelephone operation is permitted on the same frequency being used for the RTTY communication on these bands.) Identifying the station you are communicating with, in addition to your own station, is not legally required, but it will help those monitoring your QSO to determine band conditions.

Test Messages

A message consisting of repetitive RYs is useful for local and on-the-air testing of Baudot RTTY equipment:

RYRYRYRYRYRYRYRYRYRYRYRYRYRY DE W1AW

The use of the RY sequence goes back to the early Teletype® machine days. To set the mechanical machines, it was necessary to send a string of Rs and Ys while the receiving operator adjusted a knob called the "rangefinder." The military and commercial services also used RYRYRYRYRY as a channel-holding signal during idle periods. The letters R and Y both alternate between MARK and SPACE tones in the Baudot code, so it was natural to use these letters for the range-setting operation. (If we represent the SPACE tone with a 0 and the MARK tone with a 1, then R is 01010 and Y is 10101. So you can see that a string of RY characters sends alternating SPACE and MARK tones.)

Use the RY sequence sparingly. RYs have a musical sound that is easily recognizable, and is useful for tuning a signal in properly. This makes it good for rousing attention when you are sending a CQ call, or when calling another station on schedule, but it is unnecessary during routine contacts. Don't use RY when you are working DX.

Making Contact

There are two ways of establishing contact with other amateur stations using radioteletype. You can answer someone else's call, or you can try calling CQ yourself. When you call CQ, you depend on someone else to tune your signal properly, and if you are a newcomer to the mode, that may be an advantage. By listening for other calls, you have a good chance of finding stations that you might not contact otherwise. You will want to use both methods at various times, but whichever one you choose, learn the proper procedure and follow it.

CALLING CQ

Since it is a little easier to explain, we will start by describing the procedure for calling CQ on RTTY. You must first locate a clear frequency on which to make your call. Courtesy should be the rule. Ask if the frequency is busy by sending "QRL?," or by asking in clear text. Sometimes it is possible that there is a QSO on frequency, but you may only be able to hear one side of it. The frequency may sound clear to you while the other station is transmitting. Once you have established that you have a clear frequency, it is time to make the CQ call.

The general CQ call, like RY, has a recognizable, musical sound. Take advantage of this fact, and send CQ in a pattern something like this:

CQ CQ CQ CQ CQ DE KT1N KT1N KT1N K

Transmit "CQ" three to six times, followed by "DE," followed by your call sign three times. Many operators add their name and QTH to the last line of the CQ call. This is a good idea, but keep it simple and short. As on voice, several short calls with pauses to listen are much better than one long call.

CALLING ANOTHER STATION

You have just heard 9H1EL in Malta calling CQ. He is booming in with an S9 signal, and this would be a new country for you. Make sure that he is tuned in properly, wait until he stops calling CQ and give him a call. Keep your calls short. Forget the RYs. Send 9H1EL's call sign only once, and your call sign no more than five times:

9H1EL DE KA3KHZ KA3KHZ KA3KHZ KA3KHZ K

Then stand by and listen. If he doesn't answer, try another short call.

Line Feeds and Other Things

When another station answers your call, gives you a report and then turns it back to you, what should you do? Always send a line feed (L/F) and a carriage return (C/R) first. This will put his computer and/or printer into the "letters" mode, and will eliminate the possibility of part of your message being garbled by numbers and/or punctuation marks.

Next, send his call and your call once each. On your first exchange, send your name and QTH message from the computer memory, if your system has such a message. If you have a type-ahead buffer on your system, you can compose some of your reply before it is actually transmitted. This saves time, especially if you are not a fast typist.

If you give another station a signal report of RST 599, it means you're copying solid. If you receive a similar report, it is unnecessary to repeat everything twice. Give honest reports—if the copy is solid, then the readability report is five. Adapt your operating technique to the reported conditions. If copy is marginal, repeats or extra spaces are in order. But with solid copy, you can zip right along. It's that simple.

Another caution is to not send a string of carriage returns to clear the screen. The other operator may be copying on a printer, and the carriage returns will waste a lot of paper. The last two characters of each transmission should be carriage returns, however.

Signing Off

You have now completed your QSO and are about to sign off. What prosign do you use? Well, the standard prosigns mean the same thing on RTTY as they do on CW, and should be adequate. Table 10-9 summarizes the meaning of the most commonly used prosigns.

Table 10-9
Common Prosigns Used On RTTY

K—Invitation to transmit.
KN—Invitation to the addressed station only to transmit.
SK—Signing off. End of contact.
CL or CLEAR—I am shutting my station down.
SK QRZ—Signing off and listening on this frequency for any other calls. The idea is to indicate which station is remaining on the frequency.

Some operators use various other combinations of prosigns:

SK KN—Signing off, but listening for one last transmission from the other station.
SK SZ—Signing off and listening on this frequency for any other calls.

[It is time to turn to Chapter 12 again, to study some questions. You should be able to answer questions 2B-4-1.1 and 2B-4-2.1. Review this section is you have any difficulty.]

AMTOR OPERATION

Amateur Teleprinting Over Radio (AMTOR) is a more reliable transmission system than Baudot RTTY. AMTOR has the ability to detect errors, so you can be assured of getting perfect copy throughout the QSO. In many respects, AMTOR operation is the same as for Baudot RTTY. AMTOR characters are commonly sent at 100 bauds, and 170-Hz is the normal shift.

AMTOR Modes

There are several modes of AMTOR operation: **Automatic Repeat Request (ARQ)** or Mode A, and **Forward Error Correction (FEC)** or Mode B. Mode B is further subdivided into Collective B-Mode and Selective B-Mode. Amateurs have added Mode L for monitoring AMTOR transmissions.

Stations in contact using AMTOR are classified either as Information Sending Stations (ISS) or Information Receiving Stations (IRS). The stations change identities as information is exchanged during a QSO. The station that originates the communication is also called the master station and the other station is called the slave station. The master/slave relationship does not change during the contact.

Mode A

Mode A is a synchronous system in which the ISS transmits blocks of three characters to the IRS. The ISS sends the block, listens for replies from the IRS, then sends three more characters. The transmitting and listening times are less than a quarter of a second, so your transceiver must be able to change from transmit to receive very quickly. The ISS keeps the three characters in memory until it gets a receipt from the IRS. When the IRS acknowledges, the ISS moves on to the next three characters. If the IRS does not acknowledge, the ISS keeps sending the three characters until they are acknowledged. AMTOR keeps trying until it gets it right!

Mode A is used only for contacts between two stations. It should not be used for calling CQ or in nets. This mode has a characteristic "chirp" sound on the air. It has a 47% duty cycle, allowing transmitters to operate at full power output without worry about overheating. Mode L is used by stations to monitor, or listen to, Mode A contacts.

Collective B-Mode

Collective B-Mode is basically a "broadcast" mode, although most hams don't use that word because FCC rules say that we are not allowed to "broadcast" in the entertainment sense of the word. In this mode, there is only one sending station, called the Collective B Sending station (CBSS) and any number of Collective B Receiving Stations (CBRSs). The CBSS sends each character twice: the first transmission of a character, called DX, is followed by four other characters, and then the repetition, called RX, is sent.

This is the mode that W1AW uses for bulletins and the one that should be used for calling CQ, for nets and for making multi-way contacts. It has a 100% duty cycle, so you may have to reduce your transmitter power output as explained before.

Selective B-Mode

Selective B-Mode is intended for transmissions to a single station or group of stations. It is similar to collective B-Mode. Only stations set up to accept this mode and recognize the specific selective-call identifier are intended to receive Selective B-Mode transmissions.

AMTOR Selective-Calling Identities

AMTOR is designed so that four-letter identities are used as the **selective-call identifier** (similar to a call sign). Amateurs can't simply use their call signs as their AMTOR selective-call identifier because amateur call signs consist of letters and numerals. Most AMTOR stations use the first letter and the last three letters of their call sign for the selective-call identifier. Using this algorithm, the selective-call identifier for W1AW would be WWAW. This scheme doesn't provide unique selective-call identifiers for every call sign (for example, WWAW would also be used for W2AW, W2WAW and others), so be aware that someone may ask you to choose another selective-call identifier if yours is already in use.

Making Contact on AMTOR

If you try to make your first contact using Mode A (ARQ) and there is a problem somewhere in the system, you will not make contact. It takes a two-way contact to make Mode A work at all.

Try your first contact using Collective B-Mode. Use this mode for calling CQ, and include your selective-call identifier. Use of Mode B should ensure that you can get the circuit working one way. In Mode B, you can check to confirm that you are using the correct frequency-shift polarity for both receiving and transmitting.

After making contact using Mode B, ask the other station to go into the monitor (listen) mode and try to receive your Mode A transmission. If there is a problem with Mode A, the most likely cause is that your transceiver is taking too long to change between receive and transmit. If that's the case, check with other AMTOR operators having the same equipment setup. If you are using commercial AMTOR gear, the manufacturer will probably know which transceivers will switch fast enough and which require specific modification.

If the other station (in monitor mode) is copying your Mode-A transmission, ask the station to make a Mode-A call to your selective-call identifier. The other station is acting as the master and your station the slave in this case. Finally, reverse the roles and try being the master station.

Because of the timing cycle specified in the AMTOR standards, it is not practical to contact AMTOR stations on the opposite side of the earth using Mode A. Mode B can be used for communication with stations too distant for Mode-A contacts.

ASCII OPERATION

ASCII (pronounced "askee") stands for **American National Standard Code for Information Interchange**. ASCII operation differs from Baudot RTTY operation in that it has a larger character set than ITA2. This is possible because ACSII characters consist of 7 digital information bits. Besides a number of control characters, ASCII also provides both upper- and lower-case letters and a more complete set of punctuation marks. ASCII is a character set that was designed for computer and data communications uses. For more information about the complete ASCII coded character set, and for the technical details about methods and transmission rates used for the ASCII code, see the latest *ARRL Handbook*.

HF ASCII transmissions are normally sent at 110 or 300 bauds, using 170-Hz shift. Higher data rates are used on higher frequency bands. On 10 meters, the maximum permitted speed is 1200 bauds, while up to 19,600 bauds is permitted on 220 MHz and above, On VHF and UHF, 1200 bauds is the commonly used RTTY sending speed.

COMPUTER-BASED MESSAGE SYSTEMS

No discussion of modern radioteletype operation could be complete without some information about mailbox operation. Since the mid 1970s, a number of **Computer-Based Message Systems (CBMSs)** have appeared on the ham bands. You may hear them called one of these terms: MSO (Message Storage Operation), bulletin board or mailbox.

All of these systems have some features in common. They will automatically respond to calls on their operating frequency if the calling station uses the correct character sequence. You can send messages to be stored in the mailbox and retrieved later by another amateur. You may also request a listing of the messages on file and read any addressed to you, as well as any bulletin messages intended for general consumption. Some systems may allow messages to have password protection to restrict access to the information.

Message handling in this manner is third-party traffic, and the system operator (SYSOP) is required to observe appropriate rules concerning message content. Also the SYSOP is responsible for maintaining control of the transmitter and removing it from the air in the event of malfunction.

PACKET RADIO

By conservative estimate, there are 50,000 amateur **packet radio** stations on the air today. Meanwhile, manufacturers work night and day to keep up with the demand for new equipment as more and more amateurs put packet stations on the air. Why has this relatively new amateur mode of digital communications caught on so fast?

Packet radio is communications for the computer age. In the 1980s, a computer in a ham shack is as common as a 2-meter hand-held transceiver was in the 1970s. As computers appeared in more and more shacks, computer programs were written to allow computers to send and receive CW and RTTY. That was great, but writing computer programs for CW and RTTY today is like building a mechanical horse for transportation at the turn of the century. It could be done, but what a waste of technology! Rather than building a mechanical horse, some inventors used the technology of the day to build the "horseless carriage" (the automobile). And instead of writing computer programs to emulate keyers and teleprinters, some farsighted hams developed a new amateur mode of communications that unleashes the power of the computer. That mode is packet radio.

Being a child of the computer age, packet radio has the computer-age features that you would expect.

- It is data communications; high speed and error-free packet-radio communications lends itself to the transfer of large amounts of data.
- It is fast, much faster than the highest speed CW or RTTY.
- It is error-free; no "hits" or "misses" caused by propagation variations or electrical interference.
- It is spectrum efficient; many stations can share one frequency at the same time.
- It is networking; packet stations can be linked together to send messages over long distances.
- It is message storage; packet-radio bulletin boards (PBBS) provide storage of messages for later retrieval.

So, the many hams with computers in their shacks are naturally attracted to packet radio. Now, they are able to unleash the power of their machines.

HOW DOES PACKET WORK?

Packet radio uses a **terminal node controller (TNC)** as the interface between computer and transceiver. A TNC can be considered an "intelligent" modem. We discussed what a modem is and what it does in the section on RTTY. The TNC accepts information from your computer or ASCII terminal and breaks the data into small pieces called "packets." In addition to the information from your computer, each packet contains addressing, error-checking and control information. The addressing information includes the call sign of the station that sent the packet, and the call sign of the station the packet is being sent to. The address may also include call signs of stations that are being used to relay the packet. The error-checking information allows the receiving station to determine whether the received packet contains any errors. If the received packet contains errors, the receiving station asks for a repeat transmission until the packet is received error-free.

Breaking up the data into small parts allows several users to share the channel. Packets from one user are transmitted in the spaces between packets from other users. The address section allows each user's TNC to separate packets intended for him from packets intended for other users. The addresses also allow packets to be relayed through several stations before they reach their ultimate destination. Having information in the packet that tells the receiving station if the packet has been received correctly assures perfect copy.

WHAT DO I NEED TO GET ON

All you need to set up a VHF packet-radio station is a 220-MHz or 1270-MHz FM transceiver (with an antenna), a computer or ASCII terminal and a terminal node controller. The TNC connects between the radio and the computer, as shown in Figure 10-10. For HF operation (on 10 meters) you will need a 10-meter SSB transceiver in addition to the TNC and computer. The sidebar entitled "Packet Radio Equipment" covers the available equipment in detail.

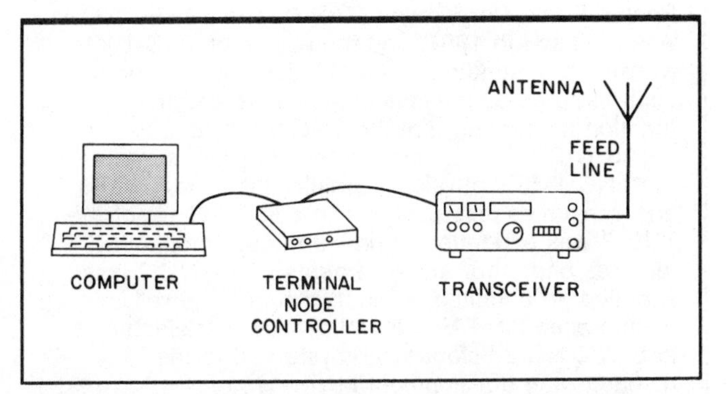

Figure 10-10—A terminal node controller connects between your transceiver and your computer, like a RTTY modem.

Your TNC manual should contain detailed instructions for wiring the TNC, radio and computer together. So many hams are on packet now that someone in the local radio club will probably be able to help you if you have problems, or ask around on the local voice repeater. Many of the hams on VHF FM are also on packet!

WHAT DOES IT LOOK LIKE?

So far we've talked about what equipment you need for packet and how packet works. But what does a packet contact look like? In the following examples, we'll look at the procedures used by most of the TNCs on the market today. Some TNCs may use different command formats—consult your operating manual if you're not sure.

First, you must tell the TNC your call sign. Most TNCs allow you to change your call sign at any time and have a way to remember it when the power is switched off. Before you can enter commands into the TNC, it must be in command mode. When the TNC is in command mode, you will see a prompt:

cmd:

This indicates that the TNC is waiting for input. To tell the TNC her call sign, KA1MJP types:

MYCALL KA1MJP <CR>

<CR> means "carriage return." On some computers this key may be labeled "ENTER" or have an arrow (←).

PACKET RADIO EQUIPMENT

Terminal Node Controllers

There are many terminal node controllers available today, and new models are appearing almost daily. The following list is by no means complete, but it is a sample of what is available new and used. For more details on the TNCs mentioned here (and others), see "The Shopper's Guide to Packet-Radio TNCs" in March 1987 *QST* and *Your Gateway to Packet Radio,* published by ARRL.

TNC 1 or TNC 2?

Most of the TNCs on the market today can trace their roots to two designs by the Tucson Amateur Packet Radio Corporation (TAPR). The TAPR TNC 1 was released in 1981, and thousands of TNC 1 kits were built by amateurs. The TNC 1 does not provide such features as multiple connections and full-function monitoring that the TNC 2 provides.

The TAPR TNC 2 was the next logical step after the TNC 1. It is physically smaller than the TNC 1 and consumes less power. The command set of the TNC 2 was expanded to provide better monitoring features and other enhancements. The TNC 2 was supplied (in a limited production run) as a well-documented kit. TAPR then licensed manufacturers to build TNC 2 "clones" and returned to the research and development business.

TNC 1 Compatibles

AEA PKT-1. The PKT-1 was the first commercial preassembled TNC. It is functionally the same as the original TAPR TNC 1.

Heath HD-4040. The Heath TNC kit is another functional equivalent of the TAPR TNC 1 in a different enclosure. Heath also manufactures the HD-4040-1 status indicator and connect alarm kit and the HCD-4040-2 HF modem filter. The HD-4040 was covered in a Product Review in November 1985 *QST.*

Kantronics Packet Communicator. This discontinued TNC was the first to provide built-in modems for both HF and VHF. The packet communicator is smaller than the TNC 1, but its commands and messages are the same as those used with the TNC 1.

TNC 2 Compatibles

AEA PK-80. This preassembled version of the TNC 2 has beefed-up circuitry to suppress RFI. Its power requirements are relatively high—12 to 15 V dc at 400 mA. Otherwise it is identical to the TNC 2.

GLB Electronics TNC2A. A kit TNC 2 clone. Other than cosmetic differences, the TNC2A and the TAPR TNC 2 are identical.

The AEA-PKT-1 was the first commercial TAPR TNC 1 compatible.

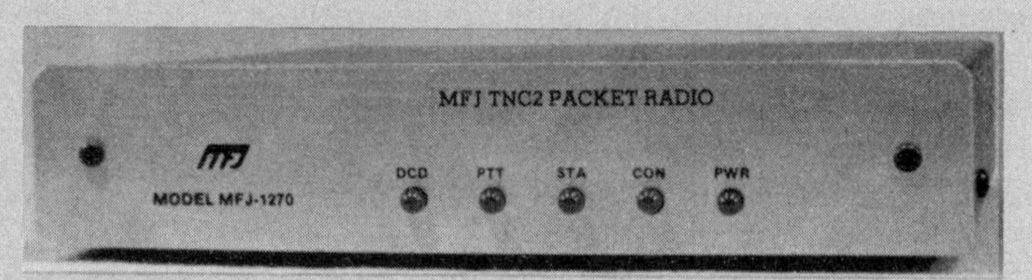

The MFJ 1270 is a popular and inexpensive TAPR TNC 2 clone.

The Kantronics KPC-2 features built-in HF and VHF modems.

The GLB TNC2A is available as a kit or assembled and tested.

As in all other modes of Amateur Radio, packet allows you to "read the mail" or monitor channel activity. This is called the **monitor mode,** and looks like this:

WA6JPR>WB6YMH: HELLO SKIP, WHEN IS THE NEXT OSCAR 10 PASS? K

WB6YMH>WA6JPR: HANG ON WALLY, I'LL TAKE A LOOK.

To enable monitor mode, simply type: MON ON at the cmd: prompt. You may also need to type MFROM ALL. Consult the operating manual for your TNC to be sure. The call signs

Kantronics KPC-2. The KPC-2 is similar to the original Kantronics Packet Communicator. In addition to the standard RS-232-C serial port, it has a TTL interface (for computers like the Commodore C64™). Serial-port and radio-port data rates are software selectable.

MFJ-1270. This unit is preassembled and it provides a TTL interface (for Commodore 64 compatibility), as well as the standard RS-232-C serial port. Cosmetically it is very different from the other TNC 2 work-alikes, but it is functionally the same. (See the Product Review in September 1986 *QST.*)

Pac-Comm TNC-200. Another TNC 2 clone, the TNC-200 is available as either a kit or an assembled and tested unit. Other than cosmetic differences, Pac-Comm's versions of the TNC 2 are the same as the original.

Pac-Comm TNC-220. This TNC includes two radio ports; each port may be configured for 300-baud (HF) or 1200-baud (VHF) operation. Switching between ports is software controlled and an active bandpass filter is included for the HF port. The computer/terminal interface can be selected for TTL or RS-232-C compatibility.

GLB Electronics PK2. This TNC is compatible with both the TNC 2 and the unique GLB user interfaces. GLB's plug-in modem provides software-selectable HF and VHF operation.

Computer-Specific TNCs and Multimode Communications Terminals

AEA PK-64. The PAKRATT™ Model PK-64 provides the Commodore 64 and 128 computers with packet-radio capability; in addition Morse, RTTY (Baudot and ASCII) and AMTOR operation are possible. The PK-64 plugs into the user port of a C64 or C128. (See Product Review, *QST*, June 1986.)

AEA PK-232. The PAKRATT Model PK-232 provides Morse, Baudot, ASCII, AMTOR and packet radio capability for any computer equipped with an RS-232-C serial port. The unit is preassembled and includes internal modems for HF (200 Hz frequency shift) and VHF operation. (See the Product Review in January 1988 *QST.*)

Kantronics KAM™. The "Kantronics All Mode" does just that—Morse, Baudot, ASCII, AMTOR and packet radio. It may be connected to any computer or terminal with a standard RS-232-C serial port or simultaneous TTL interface. Two radio ports are included for HF and VHF operation. (See the Product Review in June 1989 *QST.*)

MFJ 1278. MFJ's latest all-mode data controller unit features Packet, Morse, Baudot, ASCII, AMTOR, SSTV and FAX capability. TTL and RS-232 ports are standard. (See the Product Review in September 1989 *QST.*)

RADIOS FOR PACKET OPERATION

VHF

Just about any VHF FM transceiver can be used on VHF packet. Even a small hand-held transceiver can be used, although an outside antenna is a good idea. If there is very little packet activity in your area, you may need a bit more power to reach the nearest dedicated digipeater. Many manufacturers sell amplifiers that increase the power from a hand-held transceiver to 10 or 25 watts. Most mobile FM transceivers are rated at 10 to 25 watts, and these radios make great packet rigs. For a list of available equipment, see the VHF FM section of this chapter.

Older crystal-controlled transceivers may be used on packet, but a few words of caution are in order. Many older rigs cannot switch from receive to transmit quickly enough for packet operation. Most TNCs have an adjustable "transmit delay" that causes the TNC to key the radio and then wait a second or so before actually sending the packet. Older radios may require this transmit delay. In addition, an older rig may have passed through the hands of several amateurs, and some of the internal settings may have been "adjusted." While this may not be a problem when you use the rig for voice communications, the rig may not work on packet. If you plan to use an older rig on packet, try to locate an experienced packet operator who can help you set up your packet station.

HF

Most SSB transceivers will work on HF packet. Again, your transceiver may require a bit of transmit delay from the TNC. Your TNC manual should cover wiring and operation guidelines in detail. On HF, packet operation uses lower sideband.

Antennas

Since many packet stations are put on the air by amateurs who have already been active on VHF FM, the same antennas are often used for both modes. As a result, vertically polarized antennas are the standard for VHF packet. Yagi antennas are usually not required, unless there is *very* little packet activity in your area or you have a particularly bad location. See the VHF FM section of this chapter for more antenna information.

of the stations involved appear as "FROM>TO" and the contents of the packet appear after the ":". In the monitor mode, your computer will display everything that is transmitted on the packet frequency, whether or not it is addressed to you (see Figure 10-11).

You can send a CQ by entering the converse mode of the TNC. You go to converse mode by typing:

CONV <CR>

(some TNCs allow you to type "K" instead of "CONV")

You can then type your CQ:

```
W1AW-4>K1CE:You have new mail, please kill after reading:
W1AW-4>K1CE:Msg# TR  Size  To    From   @ BBS  Date     Title
W1AW-4>K1CE:5807  N   420 K1CE   K1MON  W1AW   870308 HI AGAIN
W1AW-4>K1CE:K1CE de W1AW: at 2101z on 870308, 142 active msgs, last msg #5839
W1AW-4>K1CE:(A,B,D,H,I,J,K,L,R,S,T,U,W,X) >
W1AW-4>K1CE:Enter title for message:
W1AW-4>K1CE:Enter message, ^Z (CTL-Z) to end, it will be message 5840
WA3VIL>KA1MJP:Hi Leslie, are you going to the club meeting? KK
W1AW-4>K1CE:K1CE de W1AW: at 2101z on 870308, 143 active msgs, last msg #5840
W1AW-4>K1CE:(A,B,D,H,I,J,K,L,R,S,T,U,W,X) >
W1AW-4>KE3Z:Max. path length: 2 digis
W1AW-4>KE3Z:KE3Z de W1AW: at 2123z on 870308, 143 active msgs, last msg #5840
KA1MJP>WA3VIL:No, Larry, I have too much work. I have to go
W1AW-4>KE3Z:(A,B,D,H,I,J,K,L,R,S,T,U,W,X) >
W1AW-4>KE3Z:Enter title for message:
KA1MJP>WA3VIL:back in tonight. Maybe next month. KK
W1AW-4>KE3Z:Enter message, ^Z (CTL-Z) to end, it will be message 5841
```

Figure 10-11—In monitor mode, your TNC displays all the packet activity on the frequency, whether or not the packets are addressed to your station.

MIKE IN SAN DIEGO LOOKING FOR ANYONE IN SIMI VALLEY

Your TNC adds your call sign as the FROM address and CQ as the TO address. The receiving station's TNC adds these addresses to the front of the text when it is displayed.

MAKING A CONNECTION

You answer a CQ or establish a contact by using the CONNECT command. When two packet stations are **connected**, each station sends packets specifically addressed to the other station. When a station receives an error-free packet, it transmits an acknowledgment packet to let the sender know the packet has been received correctly.

To connect to another station, you type:

Connect K3OX <CR>

where K3OX is the call sign of the station you wish to contact. (Most TNCs let you use "C" as an abbreviation for "connect.")

If K3OX's packet-radio station is on the air and receives your connect request, your station and his will exchange packets to set up a connection. When the connection is completed, your terminal will display:

*** CONNECTED to K3OX

and your TNC automatically switches to the converse mode.

Now, everything you type into the terminal keyboard is sent to the other station. A packet is sent whenever you enter a carriage return. It's a good idea to use K, BK, O or > at the end of a thought to say "Okay, I'm done. It's your turn to transmit."

When you are finished conversing with the other station, return to the command mode by typing <CTRL-C> (hold down the CONTROL key and press the C key). When the command prompt (cmd:) is displayed, type:

Disconne <CR>

and your station will exchange packets with the other station to break the connection. (Most TNCs let you use "D" as an abbreviation.) When the connection is broken, your terminal will display:

*** DISCONNECTED

If, for some reason, the other station does not respond to your initial connect request, your TNC will resend the request until the number of attempts equals the internal retry counter. When the number of attempts exceeds the retry counter, your TNC will stop sending connect requests and your terminal will display:

*** retry count exceeded
*** DISCONNECTED

A TNC can reject a connect request if it is busy or if the operator has set CONOK (short for CONnect OK) off. If this happens when you try to connect, your TNC will display:

*** K3OX busy
*** DISCONNECTED

PACKET RADIO REPEATING

Sometimes terrain or propagation will prevent your signal from being received by the other station. Packet radio gets around this problem by using other packet-radio stations to relay your signal to the intended station. All you need to know is which on-the-air packet-radio stations can relay signals between your station and the station you want to contact. Once you know of a station that can relay your signals, type:

Connect K3OX Via W1AW-5 <CR>

where K3OX is the call sign of the station you want to connect to and W1AW-5 is the call sign of the station that will relay your packets. The "-5" following W1AW is a **secondary station identifier (SSID)**. The SSID permits up to 16 packet stations to operate with one call sign. For example, W1AW-5 is a 2-meter packet repeater and W1AW-6 is a 220-MHz packet repeater.

When W1AW-5 receives your connect request, it stores your request in memory until the frequency is quiet. It then retransmits your request to K3OX on the same frequency. This action is called *digipeating*, a contraction of "digital repeating." If K3OX's packet-radio station is on the air and receives the relayed connect request, your station and his will exchange packets through W1AW-5 to set up a connection. Once the connection is established, your terminal will display:

*** CONNECTED to K3OX VIA W1AW-5

W1AW-5 will continue to relay your packets until the connection is broken (see Figure 10-12).

Digital and voice repeaters both repeat, but the similarity ends there. Notice that digital repeaters differ from typical voice repeaters in a number of ways. A digital repeater **(digipeater)** usually receives and transmits on the same frequency (whereas a voice repeater receives and transmits on different frequencies). A digipeater does not receive and transmit at the same time (as compared to a voice repeater, which immediately transmits whatever it receives). Rather, a digipeater receives a packet, stores it temporarily until the frequency is clear, and then retransmits the packet. Also, a digipeater only repeats packets that are specifically sent to be repeated by that station (the address in the packet contains the call sign of the digipeater). A voice repeater repeats everything that it receives on its input frequency.

If one digipeater is insufficient to establish a connection, you can specify as many as eight stations in your connect request. Additional digipeaters are added to the connect command separated by commas. For example, typing:

Connect K3OX Via W1AW-5, WA2FTC-1 <CR>

after the command prompt (cmd:), causes your TNC to send the K3OX connect request to W1AW-5 which relays it to WA2FTC-1. Then, WA2FTC-1 relays it to K3OX.

Don't use more than one or two digipeaters at any one time, especially during the prime time operating hours (evenings and weekends). Each time you use a digipeater, you are competing with other stations attempting to use the same digipeater. Each station that you compete with has the potential of generating a packet that may collide with your packet (which causes your TNC to resend the packet). The more digipeaters you use, the more stations you compete with, greatly increasing the chance of a packet collision. As a result, it may be difficult to get one packet through multiple digipeaters, and your TNC will quickly reach its retry limit and disconnect the link.

Any packet-radio station can act as a digipeater. Most TNCs are set up to digipeat automatically without any

```
cmd:c aa2z v wlaw-5
*** CONNECTED to AA2Z via W1AW-5
Hi Bruce, what's up? KK
hi Mark. Just wondering if we can bring the desk over tonight. KK
No problem. Got the room all set up and everything. What time
do you think you'll be here? KK
Well, that's up to Rick. Probably around 6 or so. KK
OK, see you then. SK
See you at 6.
cmd:D
*** DISCONNECTED
```

Figure 10-12—Two packet stations *connect* to have a QSO. Everything you type is sent to the other station. Packets not addressed to your station are ignored by your TNC.

intervention by the operator of the station being used as a digipeater. You do not need his permission, only his co-operation, because he can disable his station's digipeater function. (In the spirit of Amateur Radio, most packet-radio operators leave the digipeater function on, disabling it only under special circumstances.)

Similar to VHF/UHF voice repeaters, some stations are set up as dedicated digipeaters. They are usually set up in good radio locations by packet-radio clubs. Besides location, the other advantage of a dedicated digipeater is that it is always there (barring a calamity). Stations do not have to depend on the whims of other packet-radio operators, who may or may not be on the air when their stations' digipeater functions are most needed.

Although you are not allowed to be the control operator of a voice repeater until you upgrade from the novice class, you *may* leave your TNC's digipeater function enabled. The FCC recognizes a distinction between digipeaters and voice repeaters in this case, and everyone realizes that an effective packet-radio system depends on having Novice digipeaters available.

Another form of digipeater is the *node*. To reach a distant station, first connect to the node. Then, instruct the node to connect you to the distant station. The node acknowledges packets sent from either station, then relays them to the other station. This has a number of advantages over a simple digipeater.

VHF/UHF PACKET OPERATING

Today, most amateur packet-radio activity occurs at VHF, on 2 meters. Rapid growth of packet on 220 MHz is expected, however. The most commonly used data rate on VHF is 1200 bauds, with frequency-modulated AFSK tones of 1200 and 2200 Hz. This is referred to as the "Bell 202" telephone modem standard.

Most TNCs are optimized for VHF/UHF FM operation, so getting one on the air is a simple matter of turning on your radio and tuning to your favorite packet-radio frequency. On 220 MHz, Novice packet activity centers on 223.4 MHz. If there is a voice repeater on that frequency in your area, ask around at a club meeting or on the repeater. Someone is bound to know where the packet activity is.

If you are conducting a direct connect (a contact without using a digipeater), move your contact to an unused simplex frequency. It is very inefficient to try to exchange packets on a frequency where other stations, especially digipeater stations, are also exchanging packets. The competition slows down your packets and, in return, you are also slowing down all of the other stations. You should use a frequency occupied by a digital repeater only if you are using that digital repeater. Consult the latest edition of *The ARRL Repeater Directory* for more information on recommended packet operating frequencies.

VHF/UHF packet-radio operation is similar to VHF/UHF FM voice operation. Under normal propagation conditions, you can communicate only with stations that are in line of sight of your station. By using a repeater (voice or digital), this limitation is reduced. If a repeater is within range of your station, you may communicate with other stations that are also within range of the same repeater, even though those stations may be beyond your own line of sight. Packet expands on this by permitting the simultaneous use of several digipeaters to relay a packet from one station within range of the first digipeater in the link to a station within radio coverage of the last digipeater in the chain.

HF PACKET OPERATING

HF packet radio is very different from VHF/UHF packet. An SSB transceiver is used to generate a 200-Hz-shift FSK signal, and 300 bauds is used rather than 1200 bauds.

As we mentioned, most TNCs are optimized for VHF/UHF FM operation, so getting on HF requires more effort than getting on VHF/UHF. For starters, the modems in some TNCs must be modified and recalibrated for HF operation. An external modem compatible with HF packet-radio operation may also be used. A few TNCs can be easily switched between VHF/UHF and HF modem requirements, either with a front-panel switch or in the software. Some include two modems, one designed for HF, the other for VHF/UHF.

Modem modification in most TNCs is simple. For example, the TAPR TNC 1 requires changing the values of a resistor and capacitor, followed by the recalibration of the modem. Adding an external HF modem (such as the AEA PM-1) is a simple matter with some TNCs and a major project for others. Consult your operating manual for instructions on using your TNC on HF.

Tuning is much more critical on HF packet than on VHF. Tune your receiver very slowly, in as small an increment as possible (10-Hz increments are good) until your terminal begins displaying packets. Do not change frequency until a whole packet is received. If you shift frequency in midpacket, that packet will not be received properly and will not be displayed on your terminal even if you were on the correct frequency before or after the frequency shift.

Some TNCs and external modems have tuning indicators that make tuning a lot easier. Kits are also available to allow you to add a tuning indicator to a TNC without one.

The same rule of thumb that applies to VHF/UHF packet-radio operation also applies to HF operation. Move to an unused frequency if there is other packet-radio activity on the frequency you are presently using. Often, one frequency is used as a calling frequency where stations transmit packets to attract the attention of other stations who may wish to contact them. Once a contact/connection is established, clear the calling frequency by moving off to another unused frequency.

PACKET BULLETIN-BOARD SYSTEMS (PBBS)

We mentioned RTTY mailboxes in the section on radioteletype, and some of you may be familiar with landline telephone bulletin-board systems. The packet-radio equivalent is the **Packet Bulletin Board System (PBBS)**. Most PBBS stations use a program written by Hank Oredson, WØRLI, for the Xerox® 820 computer. The PBBS computer allows packet stations to store messages for other amateurs, download and upload computer files, and even to link one packet station through a "gateway" to another band (see Figure 10-13).

```
cmd:c klce v wlaw-4
*** CONNECTED to K1CE
K1CE   BBS - West Hartford, CT
Hello Bruce, Last on 0257/870307, new 369 - 368, active 15.
KB1MW  de K1CE   BBS: (B,D,H,I,J,K,L,R,S,T,U,W,X) >
ll 5
 Msg# TR  Size To      From    @ BBS  Date    Title
  367  N    78 K1BA    K1CE    WA1RAJ 870307 Greetings
  366  N   652 ALL     KA1KRP         870306 FLEA MARKET
  360 TY   271 K1CE    WB1ASH         870305 QTC 1 K1CE
  358 BN   679 ALL     WA1OCK         870304 novice enhancement
  244 BN   319 ALL     K1CE           870228 NTS Traffic Handling Basics
KB1MW  de K1CE   BBS: (B,D,H,I,J,K,L,R,S,T,U,W,X) >
r 244
 Msg# TR  Size To      From    @ BBS  Date    Title
  244 BN   319 ALL     K1CE           870228 NTS Traffic Handling Basics
If you are interested in NTS traffic handling via packet radio,
and would like to know the basics, download a new file NTS.DOC for
information.  Good luck, and thanks for your help in handling NTS
traffic in Connecticut. If I can be of any assistance, please
leave a message.   73,   Rick K1CE NTSCT Section Node

KB1MW  de K1CE   BBS: (B,D,H,I,J,K,L,R,S,T,U,W,X) >
x
KB1MW  de K1CE   BBS: Date 870307 Time 0258 Last msg # 368, 15 active msgs
Messages: L - List, R - Read, S - Send, K - Kill
Files:    W - What, D - Download from, U - Upload to
B - Bye, G - GateWay, H - Help, T - Talk to Rick
P - Path for callsign, N - Enter/Change your name
I - Information, J - Calls heard,  X - Short/Long Menu >
b
KB1MW  Bruce de K1CE  : 73, CUL
*** DISCONNECTED
```

Figure 10-13—A packet bulletin board (PBBS) lets amateurs store messages for other amateurs. The messages may be forwarded automatically to other stations in the PBBS network.

PBBS computers have the ability to automatically forward messages from one computer to another, so you can store a message at one PBBS that is ultimately meant for an amateur thousands of miles away. The message will be forwarded from one PBBS to another until it reaches its ultimate destination. The term **network** is used to describe a system of packet stations that can interconnect to transmit data over long distances.

To use a PBBS, you must locate one. In any area with even a small amount of packet activity there is liable to be at least one PBBS. In addition, there are several HF PBBS stations, although many of the HF stations are set up for long-haul message traffic, not for individual-user connects. The *ARRL Operating Manual* contains more detailed information about using a PBBS.

THE DREADED BEACON

All TNCs have a beacon function. This function allows a station to send an unconnected packet at regular intervals. These unconnected packets usually contain a message to the effect that the station originating the beacon is on the air and is ready, willing and able to carry on a packet-radio contact.

The purpose of the beacon function is to generate activity when there is none. This purpose was legitimate when there was little packet-radio activity. Back in the early 1980s, it was a rare occurrence when a new packet-radio station appeared on the air. Without beacons, that new operator might believe that his packet-radio station was the only one active in the area. Similarly, packet-radio stations already on the air would not be aware of the new station's existence. It would be very discouraging to build a TNC (they were all kits in the early days), get on the air and find no one to conact. The beacon function was the solution to the problem. It let people know that a new packet station was on the air.

Today, beacons are usually unnecessary. There is absolutely no need to resort to beaconing in order to make your existence known. On HF and 220 MHz, there is plenty of activity in most areas. If you are getting on the air for the first time, monitor 223.4 MHz for a few minutes and you will quickly have a list of other stations that are on the air. When one of these stations disconnects, send a connect request to that station. After a few connections, your existence on the air will be known.

Instead of sending beacons, leave a message announcing your existence on your local PBBS. This is more effective than sending beacons because your message will be read even when your station is off the air (you can't send beacons while you are off the air).

Beacons only add congestion to already crowded packet-radio channels, so do the packet-radio community a favor and disable your TNC's beacon function by typing:

Beacon Every 0 <CR>

WHAT IF I STILL HAVE QUESTIONS?

The material presented here is only a very basic sketch of packet-radio operation. More material is presented in the *ARRL Operating Manual. QST* and the other ham magazines often have articles about packet radio. The ARRL also publishes *Gateway*, the packet-radio newsletter, 25 times a year. *Your Gateway to Packet Radio*, by Stan Horzepa, WA1LOU, is an ARRL publication that provides plenty of details about packet-radio operation. Finally, try your local radio club! There are sure to be a few "packeteers" in your local club. For a list of clubs in your area, write to the Field Services Department at ARRL HQ.

[Now turn to Chapter 12 and study questions 2B-5-1.1, 2B-5-1.2, 2B-5-2.1 and 2B-5-2.2. If you have difficulty with any of these questions, review this section.]

ON-THE-AIR ACTIVITIES

Once your license is in hand, you'll join thousands of other active Novices. You'll discover that a whole new world of on-the-air operating activities awaits you. At first, you'll probably just make contacts. Maybe you'll ragchew (chat) with other hams. After a while, however, you'll want to try your hand at specialized activities. These include handling traffic, contesting or DXing. You will probably want to exchange QSL cards with the hams you work. You may want to go after some of the various awards available to hams.

You really won't have to confine your interests to a narrow range. You'll find that certain aspects of Amateur Radio seem more appealing than others. You should at least try some of the activities. You might find you enjoy them very much.

HANDLING TRAFFIC

Passing "radiogram" messages is a vital part of an amateur's emergency communications training. A scheduled sequence of traffic **nets**, under the banner of the ARRL National Traffic System, provides an orderly path for getting messages from their origins to their destinations. These can be anywhere in the US and Canada, and some points overseas. Traffic nets furnish an opportunity to improve your code speed. They also serve as a training ground for

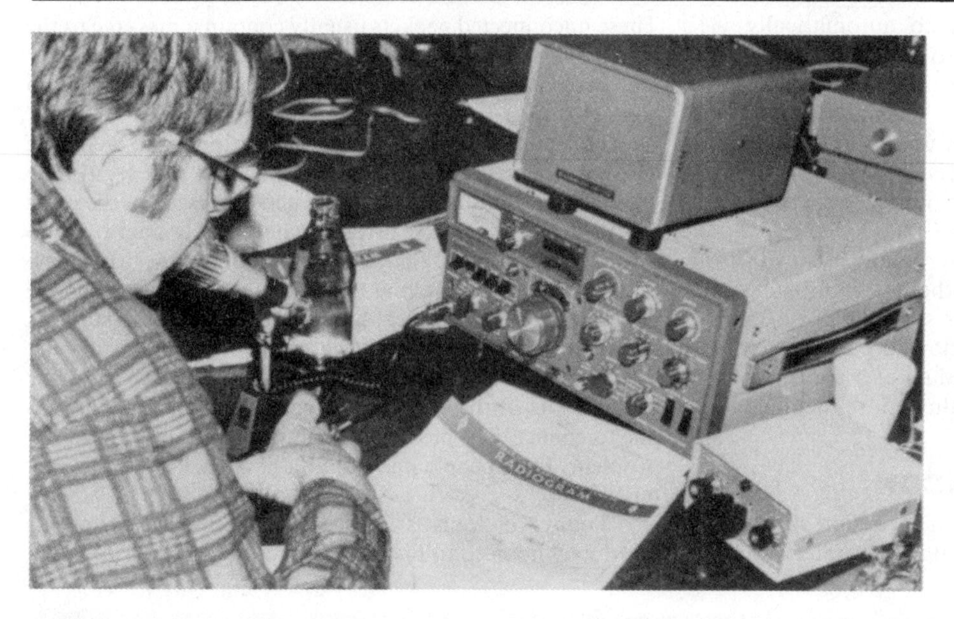

Each year, volunteer hams like this one handle tens of thousands of messages. They use the National Traffic System. Sometimes these messages are fresh from disaster areas. Hams help those seeking news of friends and relatives. Other messages are from state fairs and exhibitions. Still others contain birthday greetings. You can get involved in handling traffic. Contact the Field Services Department at ARRL Headquarters for more information.

amateurs who want to enhance their on-the-air operating abilities. Novice and slow-speed nets are particularly popular among new amateurs.

Getting started in traffic handling is a simple matter. Simply familiarize yourself with net procedures, and then check in. The times and frequencies of most traffic nets are listed in the *ARRL Net Directory*. The *Directory* also includes background information and instructions for typical net operations. Emergency and traffic operating are addressed in more detail in the *Public Service Communications Manual (PSCM)*. The *Net Directory* and *PSCM* are available from ARRL Headquarters for $1 each. Another source of information is *The ARRL Operating Manual*, which can be purchased from your local radio dealer or direct from ARRL.

Whether your interest is in emergency preparedness, handling messages for your friends in the community, or improving your code speed and operating skills, you will find traffic handling and net operation an enjoyable and educational aspect of Amateur Radio.

DX

When the DX bug bites, it bites hard, with little warning. One day as you call a casual CQ, a weak station answers. You are expecting to hear from someone in the next county or state. He sends his call sign. "Hey, that's a strange sounding call sign. Let's check the international call signs allocations list to see what country this guy might be in. Z2. That's Zimbabwe in southeastern Africa! This guy is halfway around the world!" Before you know it, you have succumbed and become a DXer.

DX is one of those loosely defined terms that means different things to different people. The term DX (distance) is generally understood to include all stations that are not in your own country. It can be a two-mile contact across the border or a 10,000-mile QSO two continents away. A foreign border or ocean often separates a DX QSO from a domestic QSO.

DX and DXing (working fellow amateurs in foreign lands) is a fascinating and absorbing aspect of Amateur Radio. You could easily spend an entire ham radio "career" solely in pursuit of DX. And indeed, some hams do.

Fact: Any amateur station is capable of working DX. It is true that bigger antennas, linear amplifiers (when you get your General or higher class license) and refinements in operating techniques can make the process easier. But even the newest Novice with the most basic station is capable of snagging his share of DX.

What is the ultimate goal of the DXer? The most common answer would probably be to "work 'em all." Contact and confirm by QSL card as many of the countries on the ARRL DXCC countries list as possible (presently about 315 in number). That will ensure your high standing among fellow DXers pursuing that same goal. The **DX Century Club (DXCC)** is a prestigious award issued by the ARRL. It is recognized worldwide as the premier award for DXing achievement in all of Amateur Radio. For the basic DXCC award, you'll need to work 100 different countries from the DXCC list. QSLs are required from each of them. The cards are then sent to ARRL for verification. An official application form is required. (The ARRL DXCC list of countries is available from ARRL Headquarters for $1. It's also included in the current edition of *The ARRL Operating Manual*.) After you obtain the basic award, the challenge continues. A series of endorsements for working and confirming still more DXCC countries awaits you. DXCC membership is open to amateurs worldwide (although US and Canadian amateurs must be ARRL or CRRL members to participate). It is the basis for a friendly worldwide competition. Everyone tries to work more countries than their friends across town (or across the seas).

Working all the countries on the DXCC list is not a weekend affair. Complexities of international politics and the absence of amateur activity in some of the less-populated countries work against it. Some hams with 20 years or more of DX experience are still missing quite a few DXCC countries. Some relative newcomers, on the other hand, lack only a couple of countries.

There are many resources available to the avid DXer. At the top of the list is personal on-the-air experience. Spend lots of listening time on the bands. Get to know the peculiarities of radio-wave propagation. Get to know when and where DX stations are likely to hang out. Listen to successful DXers. Copy their good operating habits. You can develop effective strategies for working DX just by monitoring the bands.

Working DX requires a different set of operating skills. This is true from looking for DX QSOs, to sending and receiving QSL cards.

Much has been printed on the art and science of working DX. Among the more useful printed materials are DX bulletins and newsletters. These bring subscribers current and helpful information on some of the "rarer" DX stations. The monthly "How's DX?" column in *QST* is chock full of useful DX news. A highly recommended treatment of DXing techniques can be found in *The ARRL Operating Manual*. For the CW operator, Bob Locher, W9KNI, wrote a book called *The Complete DXer* (available from ARRL for $10). Bob's book gives both beginner and experienced operators a "feel" for how to chase DX on CW. Whatever the source of your latest and hottest DX tip, success comes only to the station that joins in the fray. After all, that's the name of the game. Good luck and good DXing.

CONTESTS AND CONTESTING

Even the most casual, passive ham has a competitive spirit to some degree. We all like to test our skills against others. In ham radio, this is an excellent way to learn good operating techniques. Contesting provides the vehicle for us as amateurs to satisfy our desire to compete. In contests, hams go "head-to-head"—and may the best ham win.

Contesting is a diverse facet of Amateur Radio. Most amateurs join in at one time or another to satisfy their competitive appetites. You might totally immerse yourself for the full 48 hours of a big contest. Those who do may come away with 2000 or more contest QSOs and the prize for first place! You can also spend but a few minutes to achieve a personal goal. Maybe you'd like to work a new DX country. Or you might try to make the most QSOs you ever made in a given time period.

Generally, a contest involves working as many different

stations as possible in a given period of time. A premium is placed on working specific types of stations for "multipliers." Multipliers may be different states (or provinces) or counties in a North American contest. In international (DX) contests, multipliers may be different countries or International Telecommunication Union (ITU) Zones.

Contests come in all shapes and sizes, each tailored to one of the many different types of amateurs. There is a Novice Roundup designed to help beginning amateurs. There are DX contests for DXers. VHF contests exist for those whose enjoy working the bands above 30 MHz. These are only a few of many types of contests... but you get the idea. "There is a contest for any taste."

"Winning" in Amateur Radio contesting, is a relative term. One operator might turn in the number one score among thousands of participants. Another might significantly improve his or her operating techniques. All contests are learning experiences. Both hams are "winners." Sharpened operating skills and greater station efficiency result from contesting. That's true for serious contenders and casual participants.

The contest operator knows from experience that short transmissions aid in efficient and courteous operating. A good contest operator wastes neither words nor motions. Contesters are likely to have some of the best signals on the band. They may not have the most elaborate stations, however. Their crisp signals come from effective use of available station equipment. Contests encourage operators to use their equipment and skills to the best advantage.

Even if you've never entered a contest before, don't be afraid to try your hand! There is a wealth of good information available. Contact your local Amateur Radio club. Every club has at least a few avid contesters. They'll be happy to talk about their favorite Amateur Radio activity. The "Contest Corral" column in *QST*, and contest sections in most other Amateur Radio periodicals, list hundreds of contests each year. The basic rules and entry requirements are also listed.

Reading and talking about contesting have their place. But nothing can beat actual on-the-air experience. Don't be gun-shy. Choose an event you wish to participate in. Read and understand the rules. Then, jump right in and give it a shot. Don't worry about letting your inexperience show. Every top-notch contester was a beginner at one time. You'll be surprised how helpful your fellow contesters are.

FIELD DAY

Field Day is a special event. It's fun, challenging, frustrating, exciting, exhausting, and satisfying. Field Day is Amateur Radio from A to Z.

Ask 10 amateurs "What is Field Day?" and you are likely to get 12 different answers. The premier operating event of the year is held on the fourth full weekend in June. It is sponsored by the American Radio Relay League. But that's where any and all agreement ends. Field Day is a chance to take to the hills and fields. It's a chance to test our ability to communicate in adverse conditions.

Field Day is more than just a contest. Field Day is a chance to spend a weekend in the outdoors with family and friends. Field Day is a 24-hour excuse for enjoying Amateur Radio. Field Day is an experience not to be missed. Mark your calendar now.

Twenty-four hours spent at even a modest-sized Field Day setup is a crash course in Amateur Radio. It's probably equivalent to a year's worth of "normal hamming." The experience draws heavily on ingenuity and resourcefulness. Even the newcomer is quickly pressed into service. Everyone must deal with the problems that occur in remote locations.

If Field Day sounds like your "cup of tea," we can help. Write ARRL for the name of a radio club in your area. Although you can operate Field Day by yourself, you'll miss out on a lot of fun. There's a lot to be learned in working with a group.

QSL?

"Please QSL." Once you've made a thousand QSOs, you will have probably heard this a thousand times. If you've yet to make your first QSO, you will undoubtedly copy this request as part of your first contact. (Or you will want to make the request!) There's an unwritten commandment in Hamdom. It says "You shall QSL when asked. The time will certainly come when you shall be in need." By the way, QSL sent as a "Q signal" on CW means "I acknowledge receipt," or "I will confirm this contact." QSL? asks "Can you acknowledge receipt?" or "Can you confirm this QSO for me?" QSL is an important "word" in amateur jargon. It is understood to mean a written or printed confirmation of contact.

Okay, you're just starting out in Amateur Radio. You're looking for a supply of QSL cards. You'll need to send QSLs to the stations whose call signs will soon fill your logbook. There are a few things to keep in mind as you select, fill out and, send your QSLs.

QSLs come in all sizes, shapes, colors and textures. QSLs have been silk-screened on T-shirts, and sent as commercial telegrams. They've been molded in plastic and typed on tissue. Usually, however, they're printed on postcard sized card stock. (Remember, the US Postal Service will not accept cards smaller than 3½ by 5½ inches.)

Your QSLs can be printed professionally, or the cards can be homemade. (See the *Ham Ads* section of *QST* for suppliers.) The final design is left up to you. Keep the card's recipient in mind, however. He or she will form a lasting impression of you from your QSL card. Your QSL must also include certain information about each QSO.

What do thousands of hams do on the fourth weekend in June every year? It's something they swear they'll never do again, but they usually do. You guessed it. It's Field Day!

Although points are awarded for various categories, it's not really a contest. Field Day is actually an emergency preparedness exercise. Many groups of amateurs literally operate out in a field. Others try the nearest mountaintop, beach or backyard. The idea is to test your readiness to operate under adverse conditions. It's a group effort, and a great deal of fun. You may be in charge of untangling a longwire antenna at 3 AM. Or you may begin operating at sunrise after being up the entire night! It's an experience you're not likely to forget. It shows what the Amateur Radio Service can do in emergencies.

Shortly after you begin your on-the-air operation, you will experience the excitement of collecting QSL cards from all over the world.

The ultimate use of a QSL will be as an awards submission. There are many Amateur Radio awards. These include the DX Century Club (DXCC), Worked All States (WAS) and Worked All Continents (WAC). Most societies that issue the awards will honor your QSL as valid confirmation only if it contains the following essential information: your call sign, your specific location (QTH), confirmation that a two-way QSO has taken place, the call sign of the station worked, date of the QSO, time of the QSO (in UTC, please), frequency band used, mode (CW, SSB, RTTY, FM, and so on) used, and signal report given. When choosing the design for your QSL, be sure the card has the proper layout for recording all QSO information.

There are conventions to follow when filling out QSLs. Always record times in Coordinated Universal Time (UTC). It is frustrating to receive a QSL marked in local time. You'll have to convert some obscure "local" time to UTC. Otherwise, you won't be able to find the QSO in your logbook. Amateurs around the world use UTC. There's a good rule to follow when recording the QSO date on your QSL. The date is written in Arabic numerals. The month is written in Roman numerals. The year is written in Arabic numerals. Example: July 10, 1989 would be written as 10 VII 1989. Most DX (non-US or Canadian) amateurs record the date in this manner. Standardization makes life easier for all involved. An alternative is to abbreviate the month, but use the same format (10 JUL 1989). Do not use the method of just writing Arabic numerals, such as 10/7/89 or 7/10/89. (Is that July 10 or October 7?)

When filling out your QSL card, use regular ink or permanent marker. Pencil and most felt-tipped markers tend to fade. They smudge easily and deteriorate over the years. A rainy day is all it takes to "wipe out" a hard-earned contact. If you make a mistake when filling out a QSL card, throw it away. Do not try to correct it by writing over the mistake. Destroy the card and start again. Most organizations will not accept "altered" QSLs as valid awards confirmations. There is no way to determine who altered the card. It cannot be counted for credit.

Once the QSL is made out, you must get it to the operator of the station you've worked. There are many ways to send QSLs. They depend on its destination and how quickly you want it to arrive. Domestic QSLs (US, Canada or Mexico) are sent as postcards. They can also be mailed in envelopes, as First Class mail. Hams in rarer states or provinces receive many QSL requests. They appreciate receiving an SASE (self-addressed, stamped envelope) with your QSL. That helps defray postage costs and saves time in addressing many cards.

QSLs going to DX stations do not have to be sent directly to the DX station's address. They can be sent via a DX *QSL bureau*. A QSL bureau is a system developed to help amateurs exchange QSL cards with other amateurs in foreign countries. Hams send QSLs by regular mail to one central bureau location in their country. When enough QSLs for a foreign country accumulate, a bureau staff member sends them all to that country. Once the cards arrive, someone working in an incoming QSL bureau sorts and mails them to individual amateurs. Using a QSL bureau means that each amateur does not have to send single cards overseas.

The ARRL sponsors such a system. The outgoing DX QSL Bureau (for ARRL members only) sends your QSLs to bureaus around the world. The incoming DX QSL Bureau (for use by all US and Canadian amateurs) forwards DX QSLs to their final destinations in the US and Canada. Details on the use of the ARRL DX QSL Bureau system are printed in *QST*. When mailing DX QSLs directly overseas, remember an important rule. United States stamps on your SASE will not be accepted by foreign postal authorities. Instead of using a US stamp on the return envelope, enclose one or two International Reply Coupons. IRCs are available at US Post Offices. They can be exchanged for the proper stamps in another country. As an added bonus, you'll start receiving beautiful foreign stamps with your DX QSLs! Many successful stamp collectors trace their "collecting" roots back to Amateur Radio DX. *The ARRL Operating Manual* has more information about sending QSLs to, and receiving them from, DX stations.

Collecting QSL cards is a satisfying hobby within a hobby. You may collect QSLs for the fun of it, or in an attempt to earn an award.

QSLs, like money, are more fun to receive than to give. In receiving a QSL, there is an implied responsibility to return the courtesy. QSLing is a worthwhile chore—do it cheerfully and promptly.

AWARDS

Working toward operating awards is a popular aspect of Amateur Radio. The ARRL sponsors many coveted awards. These include Worked All States, DX Century Club and the A-1 Operator Club. In addition, the League also administers the Worked All Continents award. This is sponsored by the International Amateur Radio Union. They're all available to licensed Amateur Radio operators throughout the world. Hams in North America must be League members to apply for most ARRL awards. Foreign amateurs need not be League members.

Perhaps the first award you'll earn is the *Rag Chewers' Club* award. To qualify, you need to maintain a contact for a solid half-hour. Like most new Novices, your first contact will be quite exciting. Your 5-word-per-minute "conversation" will make the first half hour fly by. And, you'll have earned your RCC to boot! Just report your QSO to ARRL Headquarters, and we'll send you your award.

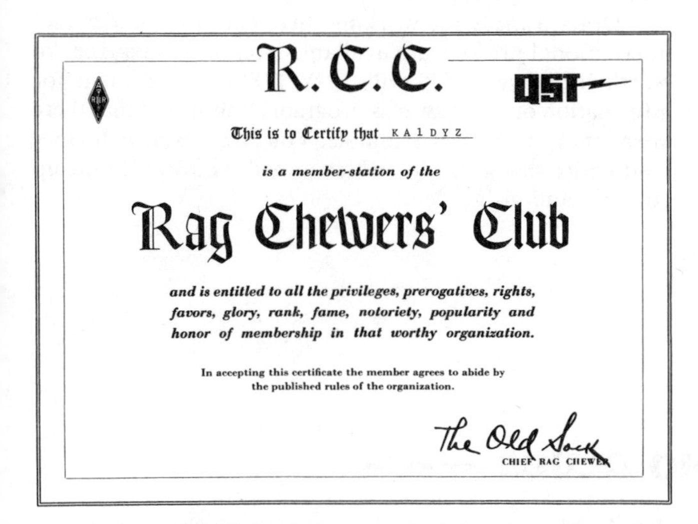

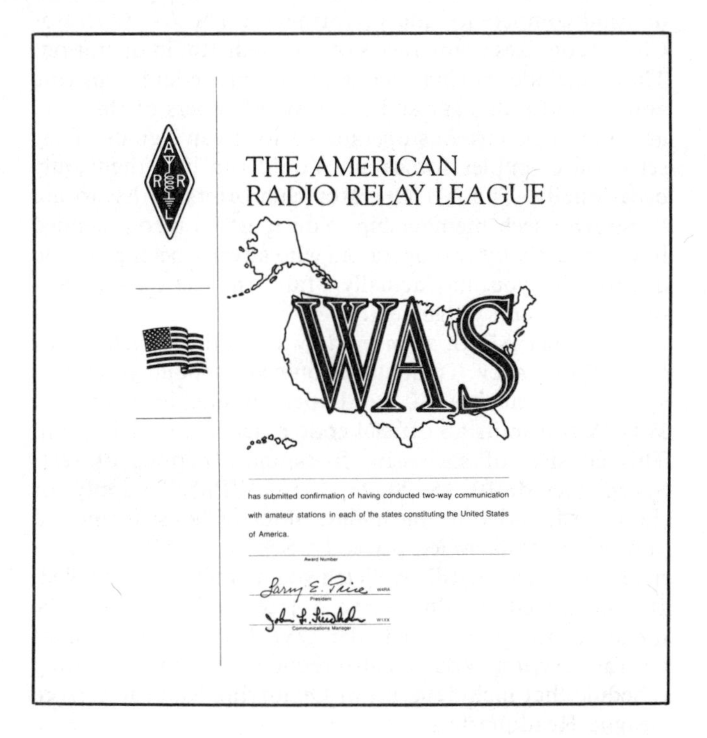

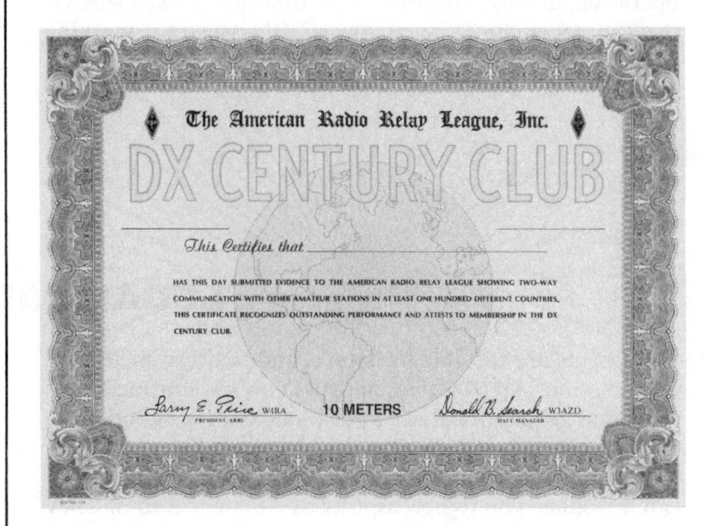

The *Worked All States* (WAS) award is probably the most popular achievement award. This is especially true among US hams. Eligibility requirements are simple. You must establish two-way Amateur Radio contact with other hams in all 50 states. Each contact must be confirmed by a QSL card or other written verification. As a Novice, WAS is well within your reach. Many Novices qualify. Send an SASE along with your request to ARRL Headquarters.

The *DX Century Club* (DXCC) is a prestigious operating achievement award. You must work and collect QSL cards from licensed hams in at least 100 different countries to qualify for this award. ("Countries" as defined by the DXCC rules, that is. Alaska and Hawaii, for example, each count as separate "countries.") Although DXCC is somewhat difficult to achieve in the Novice bands, it is possible. Years of high solar activity make it easier. Your patience and perseverance will pay off in experience and satisfaction. All DX contacts made as a Novice will count toward DXCC. You don't lose anything when you upgrade.

The *Worked All Continents* (WAC) award is sponsored by the International Amateur Radio Union. It's the first DX award many new hams get. To qualify, you must work and get QSL cards from one ham in each of the continents. Continents include North America, South America, Europe, Africa, Asia and Oceania. Write to ARRL Headquarters and request a WAC application form and complete rules. Please include an SASE with your request.

All amateurs should strive for membership in the *A-1 Operator Club.* Membership in this elite group attests to unusual competence and performance. The A-1 Operator Club recognizes many facets of Amateur Radio operation. These include keying, modulation, procedure, copying ability, and judgment and courtesy. Members of this club are smooth, courteous operators with clean signals. They set good examples for us all. You'll run into them only occasionally. The A-1 Operator club offers no list to aid those who seek membership. You must be recommended independently by two operators who already belong. If you ask to join, you may actually "fail" the test on the basis of courtesy and judgement!

Another ARRL-sponsored award is the *Certificate of Code Proficiency*. To qualify, you have to build your code speed up to at least 10 words per minute. Each month, W1AW transmits an official code practice qualifying run. This consists of successive five-minute periods of text. Speeds include 10, 15, 20, 25, 30, 35 WPM. To apply for this award, underline one minute of text. Choose the period you believe you copied perfectly. Send it to ARRL Headquarters with an SASE. We'll tell you if you passed or failed. If you passed, you'll receive a free certificate. W1AW qualifying runs are announced in *QST*. Look in the "Contest Corral" section. You can also request a W1AW operating schedule that includes a list of Qualifying Run times from League Headquarters.

Many other organizations besides the ARRL sponsor operating awards. Many Novices' first step toward WAS is the *Ten American Districts Award*. It's issued for working

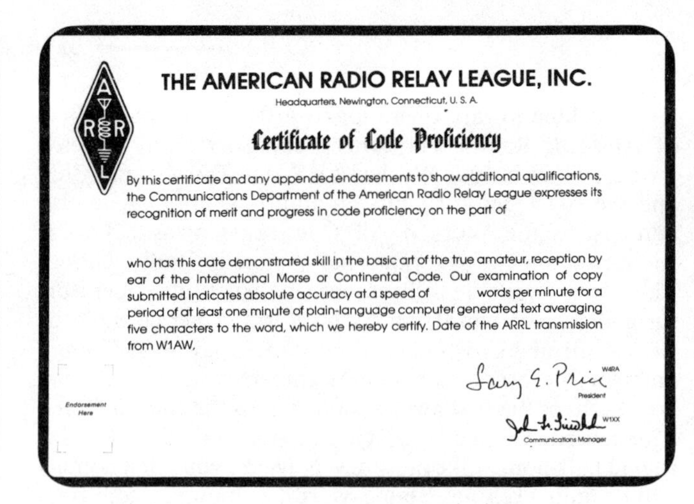
THE AMERICAN RADIO RELAY LEAGUE, INC.

Headquarters, Newington, Connecticut, U. S. A.

Certificate of Code Proficiency

By this certificate and any appended endorsements to show additional qualifications, the Communications Department of the American Radio Relay League expresses its recognition of merit and progress in code proficiency on the part of

who has this date demonstrated skill in the basic art of the true amateur, reception by ear of the International Morse or Continental Code. Our examination of copy submitted indicates absolute accuracy at a speed of words per minute for a period of at least one minute of plain-language computer generated text averaging five characters to the word, which we hereby certify. Date of the ARRL transmission from W1AW,

Endorsement Here

President

Communications Manager

and confirming contacts in all 10 US call areas. Apply to the Lockheed Employees Recreation Club, 2814 Empire Blvd, Burbank, CA 91504. Enclose $1 with your application to cover postage and handling.

Other awards for working different US counties and international prefixes are available from *CQ* magazine, 76 North Broadway, Hicksville, NY 11801. Write them for information on their awards program. You'll find that there are a lot of certificates available. You could even wallpaper your entire shack. Decide which awards you want, tune up your rig and go get 'em!

LOCAL RADIO CLUBS

People are sociable by nature, and radio amateurs are no exception. All over the country, hams have formed local radio clubs. Your local Amateur Radio club is a good place to meet other hams. At club meetings, you'll meet others with similar interests. They can help you get on the air. They can introduce you to new aspects of Amateur Radio. They can also amuse you with tales of the "good old days." Through membership in your local club, you'll make fast, lifelong friendships.

Most radio clubs sponsor a wide range of activities for their members. These range from banquets to auctions, from flea markets to community service projects, and from training classes to Field Day efforts. By participating, you'll add enjoyment and dimension to your amateur experience.

Special-interest clubs may interest you as well. Some examples of these are contest clubs, DX clubs and VHF clubs. Other groups, such as Explorer posts, canoers, motorcyclists and marathoners use Amateur Radio as an integral part of their activities. If you develop a special interest within Amateur Radio, you'll want to investigate —or even help start—a local special-interest club. Often, special-interest club members are the most knowledgeable hams in your area. Many are glad to help a beginner learn the ropes.

How do you find out about these clubs? More than 2000 are affiliated with the ARRL. To find the club (or clubs) in your area, write the Field Services Department at League Headquarters. We'll send you the information you need.

If there is no club in your area, why not start your own? If you supply the members, ARRL will supply the information. Often, hobbies are more fun if you have someone to share them with.

FIND YOURSELF A CLUB

There are 2000-plus Amateur Radio clubs scattered throughout the US and Canada. The Sheboygan (Wisconsin) ARC is one of the most active. One popular activity is helping American Field Service foreign exchange students from Latin America. The club helps the students contact their families, thousands of miles away. It's a thrill for all involved. It's rewarding to bring together homesick students and their anxious families.

The club also sponsors licensing classes and a Novice buddy program. That's where senior members of the club assist newcomers. They help in setting up stations and antennas. Big or small, problems are solved on a one-to-one basis. Long-lasting friendships are begun.

As a special project, the club visited a nearby Coast Guard station. Members learned more about the operation of marine radio and radar. Some club members were able to see the equipment in action. They traveled aboard a Coast Guard cutter (see photo).

Whether you're looking for a Novice transceiver or just want to learn more about Amateur Radio, a local club is for you. Write to the Field Services Department, ARRL, 225 Main St, Newington, CT 06111, for the name of one nearby.

THE ARRL

In addition to participating locally, you should also take part on the national level. The American Radio Relay League (ARRL) is the membership organization of Amateur Radio operators in the United States. There are many reasons why you, as a ham, should join the League.

Each month, the ARRL sends its official journal, *QST*, to every member. In *QST* you'll find articles dealing with nearly every aspect of Amateur Radio. As a new ham, you'll appreciate the beginners' articles. They explain how transmitters, receivers, antennas and accessories work. There are easy-to-build construction projects. In addition, *QST* carries information on operating events, activities and contests. In the back of *QST*, you'll find page after page of informative ads. These can help you choose the right equipment and accessories. *QST* also brings you the very latest news from the world of Amateur Radio.

QST isn't the only reason you should join the ARRL, though. The League represents the interests of Amateur Radio in Washington. ARRL makes sure the FCC and other government agencies know where US hams stand on current issues. ARRL also answers many questions from members. The staff at League Headquarters exists to help its members. DXers can take advantage of the ARRL Outgoing QSL Bureau. It saves considerable money in forwarding your cards to the right place. These are only a few of the services available to ARRL members.

The ARRL is much more than an organization in Newington, Connecticut. The ARRL is each local member across the country. The ARRL Field Organization depends on local volunteers to carry out various programs. The work of Amateur Radio begins at the local level by amateurs working in official elected or appointed capacities. There are jobs for those interested in many aspects of public service. These include emergency communications, communications for community events and handling routine messages. There are jobs for public relations people, government liaisons and those who will provide technical assistance. Share some of your talent and enthusiasm by becoming involved with The American Radio Relay League.

By supporting your League, you're supporting Amateur Radio. Write to us today.

The ARRL's monthly journal, *QST*, is crammed full of projects and information geared to the beginner. Whether you're into DX, simple construction projects or late-breaking news that affects all amateurs, *QST* is indispensable.

KEY WORDS

AC power-line filter—A filter connected in the power line to an amateur transmitter. It keeps RF energy from entering the power line and radiating from power lines near a house. You can also connect a line filter in the power line to other electronic devices. This will keep unwanted RF energy out of those devices.

Attenuate—To decrease or lessen in strength. Filters attenuate undesired signals.

Chirp—A slight shift in transmitter frequency each time you key the transmitter. Chirp can be caused by poor voltage regulation in the transmitter power supply. The transmitter oscillator is sensitive to current variations in the frequency-determining elements.

Fundamental frequency—The basic desired operating frequency of an oscillator.

Harmonics—Signals from a transmitter or oscillator occurring on whole-number multiples of the desired operating frequency.

High-pass filter—A filter designed to allow high-frequency signals to go through, while blocking lower-frequency signals. You could connect a high-pass filter in the antenna feed line to a television receiver. This would allow the television signal to enter the receiver and block lower-frequency amateur signals.

Key clicks—A click or thump at the beginning or end of a CW signal. Clicks occur when a transmitted signal has excessively fast turn-on or turn-off times.

Low-pass filter—A filter designed to allow low-frequency signals to go through, while blocking higher-frequency signals. You could connect a low-pass filter in the feed line from your amateur transmitter. This would allow the amateur signal to pass to the antenna and block any harmonics of that signal.

Neutralization—Using negative feedback in an amplifier to eliminate the effects of coupling between stages. Neutralization prevents oscillation in an amplifier stage.

Parasitics—Oscillations in a transmitter amplifier that have no relation to the frequencies intended to be amplified.

Radio-frequency interference (RFI)—Interference to electronic equipment caused by undesired radio-frequency energy getting into the equipment.

Receiver overload—Interference to electronic equipment caused by a very strong RF signal getting into the circuits in the equipment. Because the signal usually overloads the receiver front end, we sometimes call this front-end overload.

Spurious emissions—Signals from a transmitter on frequencies other than the desired operating frequency.

Superimposed hum—A low-pitched buzzing sound added to a radio signal. A poor filter circuit in a transmitter or receiver power supply causes superimposed hum.

Television interference (TVI)—Interference to a television receiver.

Chapter 11

But What if I Have Trouble?

In this chapter, you'll learn about some of the more common problems you may encounter when you get on the air. Although you may never have any trouble, you should be aware of possible problems and learn how to cure them. We'll show you how to identify interference to consumer electronics equipment. We'll also suggest ways to cure some of these interference problems, often caused by overload and harmonics. We will discuss some common transmitter problems: key clicks, chirp and superimposed hum. After you finish this chapter you should be able to identify the cause of and cure for these problems.

SPURIOUS SIGNALS

An ideal transmitter would emit a signal only on the operating frequency and nowhere else. Real-world transmitters radiate undesired signals, or **spurious emissions**, as well. Using good design and construction practices, transmitter builders can reduce spurious emissions to the point that they cause no problems. You can, however, run into trouble caused by these undesired signals. Spurious emissions fall into two general categories: harmonics and parasitics. As a Novice, you are most likely to run into problems with harmonics.

HARMONICS

Harmonics are multiples of a given frequency. For example, the second harmonic of 100 Hz is 200 Hz. The fifth harmonic of 100 Hz is 500 Hz. Every oscillator generates harmonics in addition to a signal at the resonant, or

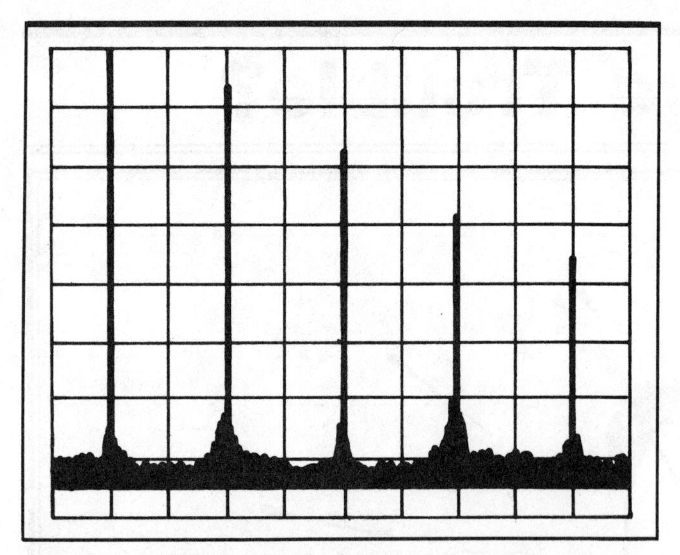

Figure 11-1—Harmonics are signals that appear at integral multiples of the resonant, or fundamental, frequency. This drawing represents a spectrum analyzer display screen, and shows a 2-MHz signal and some of its harmonics. A spectrum analyzer is an instrument that allows you to look at energy radiated over a wide range of frequencies. Here, the analyzer is adjusted to display any RF energy between 1 MHz and 11 MHz. Each vertical line in the background grid denotes an increment of 1 MHz. The first thick black vertical "pip" represents energy from the fundamental signal of a 2-MHz oscillator. The next pip, two vertical divisions later, is the second harmonic at 4 MHz (twice the fundamental frequency). The pip at the center of the photo is the third harmonic at 6 MHz (three times the fundamental frequency). This figure shows the second, third, fourth and fifth harmonics.

fundamental, frequency. For example, consider an oscillator tuned to 7.125 MHz in the 40-meter band. It will also generate signals at 14.250 MHz (second harmonic), 21.375 MHz (third harmonic), 28.500 MHz (fourth harmonic, and so on. Figure 11-1 shows the output of an oscillator that has many harmonics.

Harmonics can cause interference to other amateurs or to other users of the radio spectrum. The second through fourth harmonics of a 40-meter transmitter fall in the 20, 15 and 10-meter amateur bands. Imagine the interference (QRM) that would result if everyone transmitted two, four, six or more harmonics besides the desired signal!

To prevent possible chaos, FCC regulations specify limits for harmonic and other spurious radiation. For example, if you are operating a 100-W-output transmitter, your harmonic signals can be no greater than 10 milliwatts. A transmitter that complies with the rules still generates some harmonic energy. That energy is so small that it is not likely to cause problems, however.

Good transmitter engineering calls for tuned circuits between the oscillator and amplifier stages in amateur transmitters. These circuits reduce or eliminate spurious signals such as harmonics. The tuned circuits allow signals at the desired frequency to pass, but they **attenuate** (reduce) harmonics.

Transmitters with vacuum tubes in the final amplifier often use a pi network impedance-matching circuit at the final amplifier output. The plate-tuning and antenna-loading controls vary the value of the pi-network capacitors. Careful adjustment of these controls allows maximum power to pass at the frequency of operation and reduces harmonics.

Transmitters with solid-state final amplifiers usually operate over a wide range of frequencies and have no external tuning controls. In these transmitters, the band switch activates a separate tuned *band-pass filter* for each band. As the name implies, this filter allows energy to pass at the desired frequency and attenuates spurious signals above and below.

How can you be sure that your transmitter does not generate excessive harmonics? The FCC requires all transmitter manufacturers to prove that their equipment complies with regulations. This means that commercially manufactured equipment usually produces clean signals. If you build a transmitter from a magazine or book article, check for information on harmonic radiation. ARRL requires that all transmitter projects published in *QST* and in their technical books meet the FCC specifications for commercial equipment.

PARASITICS

Parasitics are oscillations in the final amplifier on frequencies that have no relation to those intended to be amplified. These spurious signals are not harmonics. They result from an unintentional tuned feedback path in the final-amplifier circuit. Most transmitters have suppressors built into the final-amplifier circuit to remove parasitics.

NEUTRALIZATION

Spurious signals may result from improper **neutralization** of the transmitter. An oscillator is no more than an amplifier that has some of the output signal fed back to the input. This is fine for an oscillator. We do *not* want feedback in an amplifier, however. Feedback in an audio amplifier causes a howling noise. You have probably heard this noise from a public-address system with the gain set too high.

We don't want feedback in our radio amplifiers, either. This is what happens in an improperly neutralized transmitter: some of the output feeds back to the input. This undesired feedback causes problems. The amplifier can oscillate at frequencies where we don't want output. These oscillations can be transmitted as spurious signals, and they can sometimes cause damage to the transmitter. Neutralization is simply the process of eliminating this feedback, or neutralizing it. If you suspect that you need to neutralize your transmitter, you may have to adjust an internal control.

You will need to operate your transmitter to check for proper neutralization, and for other transmitter adjustments and tests. You should do this without transmitting a signal on the air. The best way to adjust your transmitter is to use a relatively inexpensive station accessory called a *dummy load*. This is simply a special resistor that replaces your antenna when you want to operate your transmitter without radiating a signal. See Chapter 9 for more information about dummy loads.

The procedure of neutralizing transmitters varies from unit to unit. If you have a manual for your transmitter, follow its instructions. If no manual is available, ask your instructor or a local ham to show you how to neutralize your transmitter. Some transmitters are "self neutralized," and need no additional adjustment.

Remember that you have exposed the high-voltage components of your rig when you are neutralizing it. Be extra careful to avoid the possibility of electric shock.

[Now turn to Chapter 12 and study questions 2H-7.1 and 2H-7.2. Review this section if you have any difficulty with those questions.]

INTERFERENCE WITH OTHER SERVICES

Radio frequency interference (RFI) has been a headache for amateurs for years. It can occur whenever an electronic device is surrounded by a field of RF energy. Your transmitter sends a powerful field of RF energy into the atmosphere each time you transmit. This RF energy may interfere with your own or your neighbor's television set (causing **television interference—TVI**). You may also have problems with a stereo system, electronic organ, video cassette recorder or any other piece of consumer electronic equipment.

Remember that you may be a more visible member of your community once you get your license. If you have a very obvious antenna in your yard, your neighbors may blame you for any interference they experience, even when you're not on the air! If you have any problems with interference to your own equipment, it's a good bet that your neighbors do too. On the other hand, if you can show your neighbors that you don't interfere with *your* television, they may be more open to your suggestions for curing problems.

So what should you do if someone tells you that you are causing interference? The first thing to do is to make sure that your equipment is operating properly. Check to see if you cause interference to your own TV. If you do, stop operating! You should try to cure the problem before you continue to operate. You may have a serious problem with your rig.

If you are sure your equipment is operating properly and you don't cause interference to your own TV, don't stop there. It's tempting just to tell your neighbors that you're not at fault. This can cause even more problems, however. The best thing to do is to work with your neighbors to determine whether you are actually causing the problem. If you are, try to help them solve the problem. When you do this, a more-experienced ham can be a great help. If you don't know any other hams in your area, write to ARRL HQ. We'll try to help you find someone who can help.

[Now go back to Chapter 12 and study those questions with numbers that begin 2D-8-3. Review this section if you have any problems.]

RECEIVER OVERLOAD

Receiver overload is one very common cause of TV and FM interference. This happens most often when the consumer electronic equipment is near the amateur station.

Figure 11-2—There has been a boom in home electronic devices since the 1950s. Many more CB and amateur antennas are close to a growing number of home-entertainment devices. The result: an RFI problem that shows little sign of disappearing.

It occurs when the intense RF signal from the amateur transmitter (at the fundamental frequency) enters the receiver. This overloads one or more of the receiver circuits, usually the front end (first RF amplifier stage after the antenna). For this reason, we sometimes call this *front-end overload.*

In the presence of an intense RF field, spurious signals may be produced in the receiver. These unwanted signals can cause interference. Receiver overload can cause problems with equipment in your house or equipment owned by neighbors. This type of interference can happen no matter what ham band you're using.

Receiver overload usually has a dramatic effect on the television picture. Whenever you key your transmitter, the video on the TV screen will be completely wiped out. The TV screen may go black, or it might just become light with traces of color. The sound (audio) portion of the TV signal will probably be blotted out also. In an FM receiver, the audio will be blocked each time you send a code character. More often than not, overload affects only TV channels 2 through 13. In cases of severe interference, however, it may also affect the UHF channels.

The objective in curing receiver overload is to prevent the amateur signal from entering the front end of the

Figure 11-3—A high-pass filter can prevent fundamental energy from an amateur signal from entering a television set. This type of high-pass filter goes in the 300-ohm feed line that connects the television with the antenna.

entertainment receiver. The first step is to have the equipment owner or a qualified service technician install a **high-pass filter**. See Figure 11-3.

The filter should be installed at the TV or FM receiver input. The best location is the point where the antenna feed line enters the tuner of the set. It is *not* a good idea for an Amateur Radio operator to install a filter on a neighbor's entertainment equipment. Only the owner or a qualified technician should install the filter. If you install the filter yourself you might be blamed for other problems with the TV set. A high-pass filter is a tuned circuit that passes high frequencies (TV channels start at 54 MHz). The filter blocks low frequencies (the Novice bands are in the range of 3-30 MHz).

[Now study the questions in Chapter 12 with numbers that begin 2D-8-1. Review this section as needed.]

Harmonic Interference

Another problem for hams is harmonic interference to entertainment equipment. As we learned before, harmonics are multiples of a given frequency. Your transmitter radiates undesired harmonics along with your signal. Your transmitting frequency is much lower than the TV or FM channels. Some of your harmonics will fall within the home entertainment bands, however. The entertainment receiver cannot distinguish between the TV or FM signals (desired signals) and your harmonics, (undesirable intruders). If your harmonic radiation is strong enough, it can seriously interfere with the received signal.

Figure 11-4—"Crosshatching," caused by harmonics radiated from an amateur transmitter.

You can usually tell harmonic interference when you see it on a TV set. This type of interference shows up as crosshatch or a herringbone pattern on the TV screen. See Figure 11-4.

Unlike receiver overload, interference from radiated harmonics seldom affects all channels. Rather, it may bother the one channel that has a harmonic relationship to the band you're on. Generally, harmonics from amateur transmitters operating below 30 MHz affect the lower TV channels (2 through 6). Ten-meter transmitters usually bother channels 2 and 6, and channels 3 and 6 experience trouble from 15-meter transmitters.

Harmonic interference must be cured at your transmitter. As a licensed amateur, you must take steps to see that harmonics from your transmitter do not interfere with other services. All harmonics generated by your transmitter must be attenuated well below the strength of the fundamental frequency. If harmonics from your transmitting equipment exceed these limits, you are at fault.

There are several possible cures for harmonic interference. We will discuss a few of them here. Try each step in the order that we introduce them—chances are good that your problem will be solved quickly.

The first step you should take is to install a **low-pass filter** like the one shown in Figure 11-5. The filter goes in the transmission line between your transmitter and antenna. As the name implies, a low-pass filter is the opposite of a high-pass filter. A low-pass filter allows RF energy in the amateur bands to pass freely. It blocks very high frequency harmonics that can fall in the TV and FM bands. Low-pass filters usually have a specified cutoff frequency, often 40 MHz, above which they severely attenuate the passage of RF energy.

Even if your transmitter is working well within FCC specifications, you may need additional attenuation to reduce harmonics. Remember, your goal is to eliminate

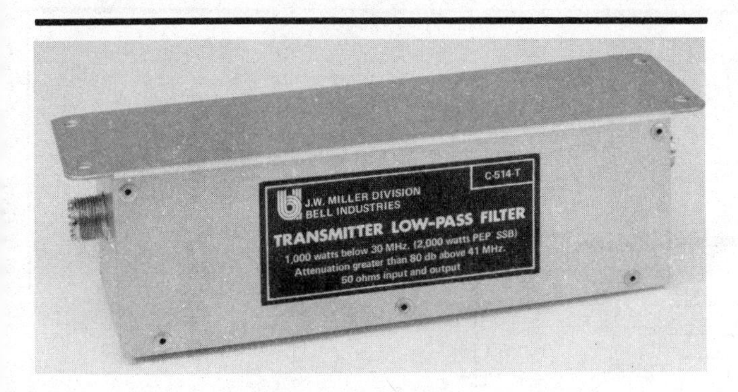

Figure 11-5—This is a low-pass filter. It goes in the coaxial cable feed line between an amateur transmitter and the antenna. A low-pass filter can reduce the strength of harmonics generated by the transmitter.

interference. Good-quality low-pass filters often attenuate signals falling in the entertainment bands by 70 or 80 dB. This is significantly better than the 40 to 50 dB typical of amateur transmitters. A decibel (dB) is a number (the logarithm of a ratio) used to describe how effective the filter is. Larger numbers indicate better filtering. Figure 11-6 shows the output of a transmitter before and after filtering.

Another source of interference is RF energy from your transmitter that enters the ac power lines. The **ac power-line filter** is another kind of low-pass network. It prevents RF energy from entering the ac line and radiating from power lines near your house.

You can also run into trouble if you use a multiband antenna. If your antenna works on two or three different bands, it will radiate any harmonics present on those frequencies. After all, we *want* the antenna to radiate energy at a given frequency. It cannot tell the difference between desired signal energy and unwanted harmonic energy. This problem does not usually affect home entertainment equipment. It may cause interference to other amateurs or to other radio services operating near the amateur bands, however.

For example, consider a multiband dipole antenna that covers 80 and 40 meters. With this antenna, you may radiate 40-meter energy while you operate on 80 meters. Remember that the second harmonic of your 80-meter signal (3.7 MHz) falls just above the 40-meter amateur band at 7.4 MHz. If your transmitter is free of excessive harmonic output, you will probably not have a problem. If your transmitter radiates strong harmonics, you should use an external low-pass filter to attenuate them.

Proper shielding and grounding are essential in your efforts to reduce harmonic radiation. The only place you want RF to leave your transmitter is through the antenna connection. Your transmitter must be fully enclosed in a metal cabinet. The various shields that make up the metal cabinet should be securely screwed or welded together at the seams. You must also connect the transmitter to a good ground.

Remember: A low-pass filter will only block harmonics from reaching your antenna. It will do nothing for a poorly

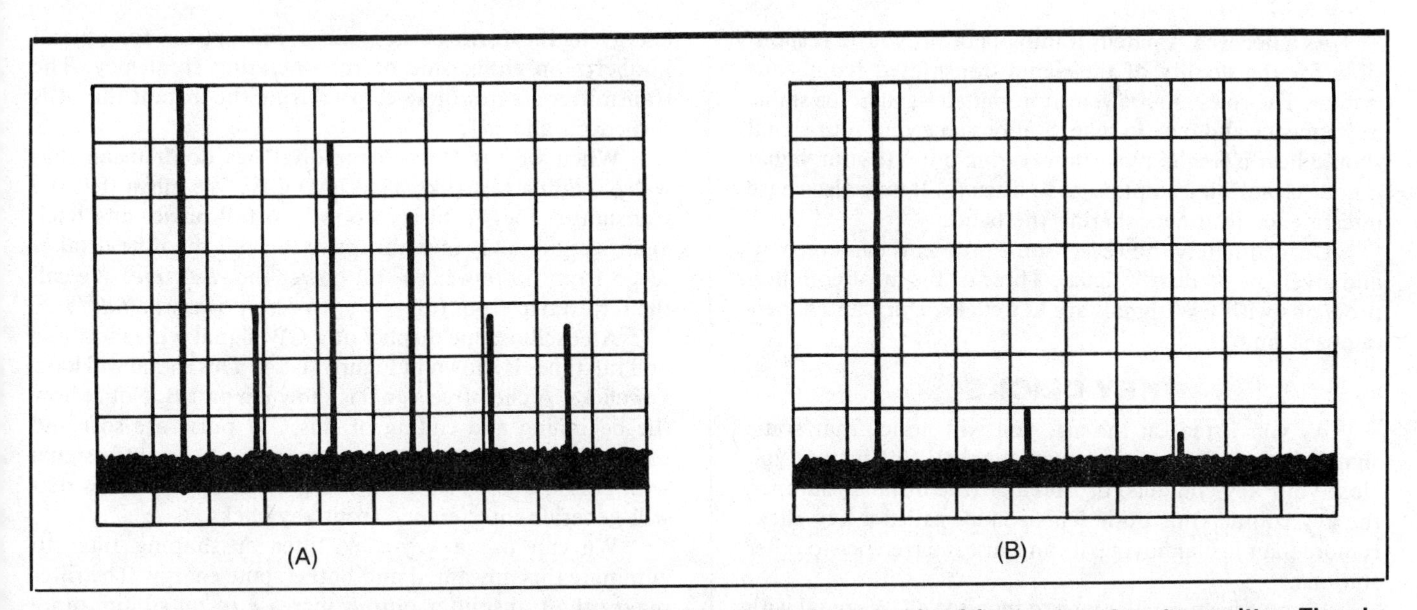

Figure 11-6—The drawing at A is a spectrum-analyzer display of the signals emitted from an amateur transmitter. The pip at the left of the display (the one that extends to the top horizontal line) is the fundamental. All other pips represent harmonics. This particular transmitter generates several harmonics. On an analyzer display, stronger signals create taller pips. Here, the harmonic signals are quite strong. In fact, the third harmonic is about one-tenth as strong as the fundamental. The fundamental signal is 100 W, so the transmitter is radiating a potent 10-W signal at the third harmonic. Most of these harmonics will cause interference to other services. The drawing at B is the output signal from the same transmitter, operating at the same power level at the same frequency, after installation of a low-pass filter in the antenna lead. Notice that the harmonics have all but disappeared from the display. The stronger of the two remaining harmonics is only about 100 microwatts—weak enough that it is unlikely to cause interference.

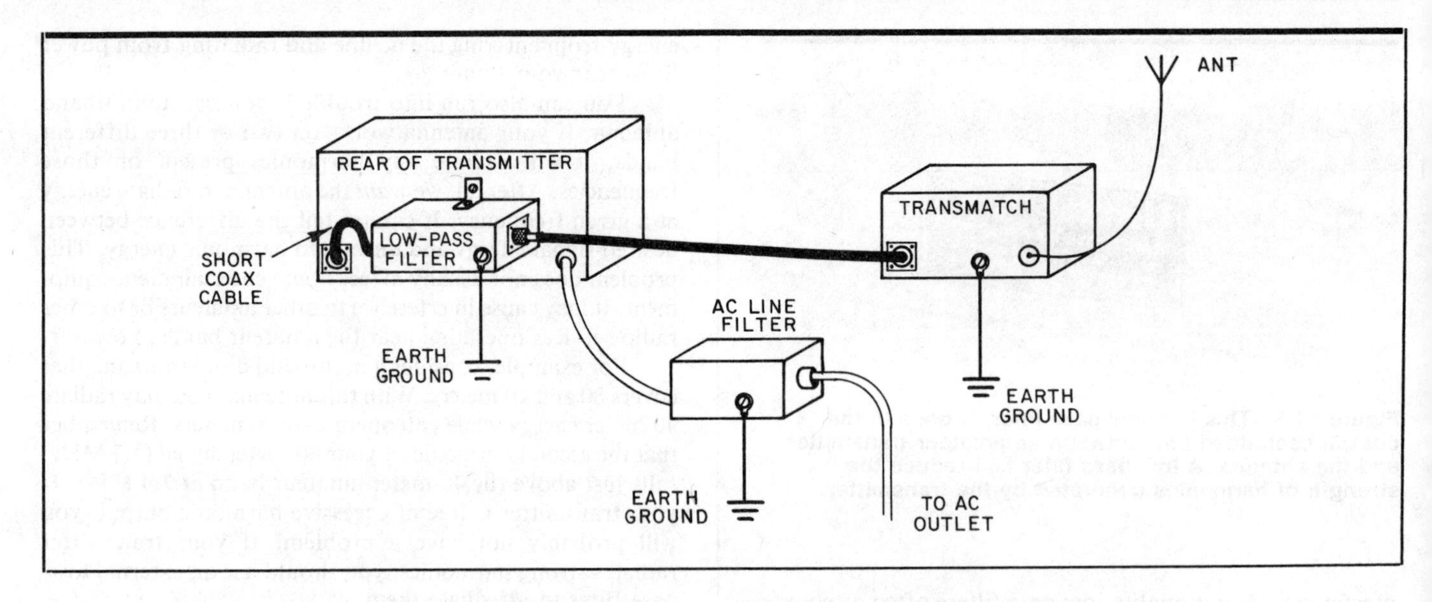

Figure 11-7—Suggested techniques for filtering harmonic energy from the leads of an amateur transmitter.

shielded transmitter that leaks stray RF from places other than the antenna connector. Figure 11-7 summarizes the steps you can take to reduce harmonic radiation from your station.

[Now turn to Chapter 12 and study the questions that begin 2H-5, question 2H-7.3, and the questions that begin 2D-8-2. Review this section as necessary.]

SIGNAL PURITY

As a licensed Amateur Radio operator, you're responsible for the quality of the signal transmitted from your station. The rules require your transmitted signal to be stable in frequency and pure in tone. Stations receiving your signal should hear a single, pure, unwavering note. If your signal is not "clean," it is unpleasant to listen to. It may also cause interference to others sharing the band.

Unfortunately, however, some problems can creep up and give you a "dirty" signal. Three of the most common problems with CW signals are **key clicks, chirp** and **superimposed hum**.

KEY CLICKS

As you listen on the air, you will notice that some signals have a click or thump on "make" (the instant you close your key contacts) or "break" (the instant you open the key contacts) or both. This sound, called a **key click**, is more than just annoying; it can cause interference to other stations.

If you use an oscilloscope to monitor a CW signal with key clicks, you will see excessively square CW keyed waveforms. Imagine a good, stable transmitter sending a CW signal on 3.720 MHz. What happens when we turn the transmitter on and off with a telegraph key? You'd think that the only frequency the output energy could have would be 3.720 MHz. If you turn the transmitter output on and off rapidly, however, something else happens. Unwanted energy in the form of key clicks will appear for several kilohertz on either side of the operating frequency. The transmitter creates these clicks during the instant that it is turned on and off.

When we say "keyed rapidly," we don't mean that we're sending fast (like 35 WPM) CW. We mean that the transmitter goes from zero power to full power and back again very quickly and abruptly. We call the time it takes to go from no power to full power the *rise time*. We call the return trip from full power to zero power the *fall time*.

An oscilloscope display of a CW signal with short rise and fall times is shown in Figure 11-8A. This signal will have key clicks. A click-free signal is shown in part B. Notice how the beginning and ending of this CW pulse are soft and round. Compare this with the square shoulders of the signal with clicks. Part C shows a string of dots and dashes that will generate interference from key clicks.

We can use a key-click filter or shaping filter to eliminate this unwanted and bothersome energy. The filter makes the transmitter output increase to maximum more slowly on make, and fall from full output more slowly on break. The keying waveform is softer, limiting the output energy to a few hundred hertz on either side of the operating frequency.

[Turn to Chapter 12 now and study the questions with numbers that begin 2H-2. Review this section if any of those questions confuse you.]

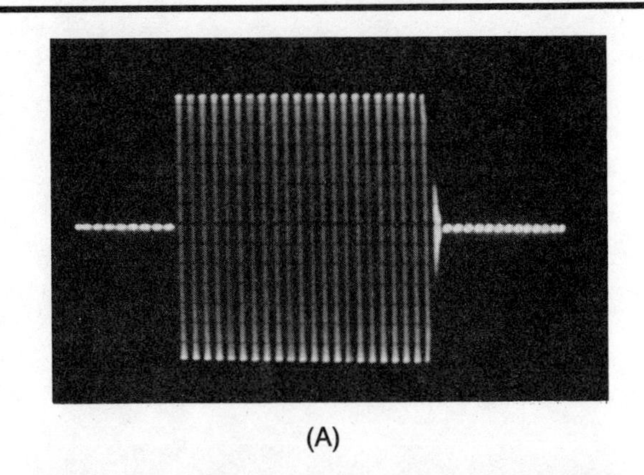

(A)

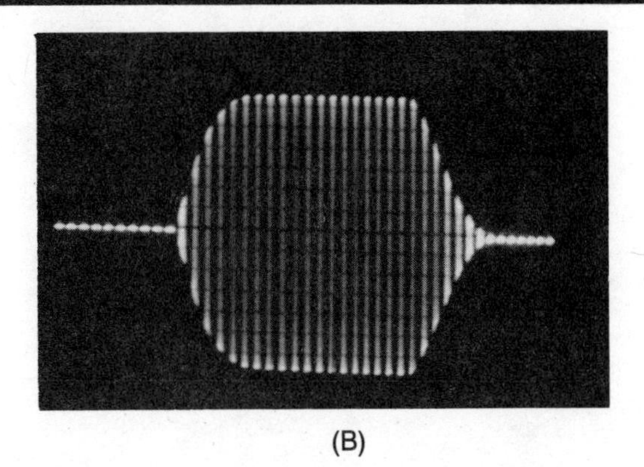

(B)

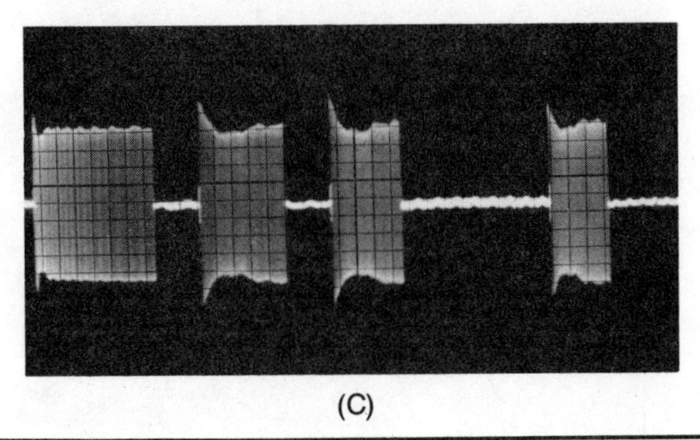

(C)

Figure 11-8—These photographs show the shape of CW signals as viewed on an oscilloscope. The photo at A shows a CW dot with no intentional wave shaping. This type of signal will cause key clicks. At B is a similar dot, but this signal has been shaped to eliminate key clicks. Notice how the signal at B builds and decays gradually. Compare this with the abrupt on-and-off characteristics of the one at A. The photo at C shows a string of unshaped CW characters. Notice the sharp spikes at the beginning of each pulse. These spikes are audible in a receiver as key clicks. Such spikes can appear at either end of a CW character.

CHIRP

Chirp is another common transmitter problem. This occurs when the oscillator in your transmitter shifts frequency slightly whenever you close your telegraph key. The result is that other stations receive your transmitted signal as a chirping sound rather than as a pure tone. Your "dahdidahdit" will sound like "whoopwhiwhoopwhip." It isn't very much fun to copy a chirpy CW signal!

Chirp usually happens when the oscillator power supply voltage changes as you transmit. Your transmitter may also chirp if the load conditions on the oscillator change as you key it. If the supply voltage is changing, you must improve the voltage regulation. With better regulation, the voltage won't shift when you key your transmitter.

If it's not a voltage problem, then what? You may need a better buffer (isolation) or driver stage between the oscillator and the next stage in your transmitter. Some oscillators are sensitive to temperature changes. If there is too much current through the frequency-determining components, their temperature may increase and the resonant frequency will change.

[Time for another trip to the Question Pool. Study the questions in Chapter 12 with numbers that begin 2H-3. Review this section if you have any problems.]

SUPERIMPOSED HUM

The last kind of signal problem we will discuss here is superimposed hum. Power supplies contain filters that remove the ac present in the rectifier output. The result is a pure, filtered dc output. If a filter capacitor fails, then the filtering action will be incomplete.

If the filter doesn't work properly, ac will be present in the power supply output. This ac will also work its way into the transmitter output. Instead of hearing a pure tone, stations receiving your signal would hear a low-pitched hum as well. If there is a lot of ac present, the signal will have a raspy tone. It may even have a loud buzz as if you were keying an electric razor! To cure a hum problem, you should check the power-supply filter circuit. A bad filter capacitor can cause superimposed hum.

[Now study the questions in Chapter 12 with numbers beginning 2H-4. Review this section as necessary.]

HOW TO GET HELP

Transmitting a signal with chirp, key clicks or hum is not a good idea. It's a violation of both the letter and spirit of the regulations governing Amateur Radio. Your best way of finding out if your signal has a problem is from the stations you work. If another operator tells you about a problem with your signal, don't be offended. He cares about the image of the Amateur Radio Service and only wants to help you. You might never be aware of a problem otherwise.

Don't let this chapter scare you. Spurious signals, unwanted harmonics, poor neutralization, key clicks, parasitic oscillations, television interference (TVI) or other equipment difficulties can all be solved. Two ARRL publications will be helpful: *The ARRL Handbook for the Radio Amateur* and *Radio Frequency Interference*. Check with your Amateur Radio instructor or any experienced ham for help. ARRL HQ can give you the address of the nearest radio club. There's no substitute for experience. Another ham may know just how to solve your problem and return you to the air as soon as possible.

Chapter 12

Novice Question Pool —With Answers

DON'T START HERE!

Before you read the questions and answers printed in this chapter, be sure to read the text in the previous chapters. Use these questions as review exercises, when the text tells you to study them. Don't try to memorize all 372 questions and answers. This book has been carefully written and prepared to guide you step by step as you learn about Amateur Radio. By understanding the electronics principles and Amateur Radio concepts in this book, you will enjoy our hobby more. You will also better appreciate the privileges granted by an Amateur Radio license.

This chapter contains the complete question pool for the Novice license exam, Element 2. Your Element 2 exam will consist of 30 of these questions, selected by the Volunteer Examiners giving you the test (see the sidebar "A Note to Those Preparing for the Novice Exam"). Most VECs have agreed to use the multiple-choice answers and distractors presented here. The ARRL/VEC and most other VECs have agreed to use these multiple-choice answers. Some VECs may use the questions printed here with different answers and/or distractors. You may take your exam at a session coordinated by a Volunteer Examiner Coordinator. If so, check with that VEC to find out if they use these multiple-choice answers.

You might also take your exam from two Volunteer Examiners with General class or higher licenses, but not part of the VEC program. These examiners must still use the questions exactly as printed. They may choose other answer formats, however. Ask your examiners what answer format they will use before you go for the exam.

Volunteer Examiners for Novice licenses can contact the Educational Activities Branch at ARRL Headquarters and request a Novice exam. They should include their call signs, license classes and expiration dates so the staff can verify that they are eligible to give the exam. If your examiners select the questions for your exam, we recommend the multiple-choice answers printed in this chapter. They have been carefully worded to provide the best answer and good distractors (the incorrect choices) for each question. The examiners are free to choose another answer format, such as fill-in-the-blank or even to give an oral exam, however.

The question pool is divided into nine sections, called subelements. (A subelement is a portion of the exam element, in this case Element 2.) The FCC specifies how many questions from each section must be on your test. For example, there must be ten questions from the Commission's Rules section, Subelement 2A. There must also be two questions from the Operating Procedures section, Subelement 2B and so on. Table 12-1 summarizes the number of questions from each subelement that make up an exam. The number of questions on an exam appears at the beginning of each subelement in the question pool, too. The subelements are broken into smaller groups, with one question specified to come from each smaller group. These sections are clearly shown in the question pool.

Table 12-1
Novice Exam Content

Subelement	*Topic*	*Number of questions*
2A	Commission's Rules	10
2B	Operating Procedures	2
2C	Radio-Wave Propagation	1
2D	Amateur Radio Practices	4
2E	Electrical Principles	4
2F	Circuit Components	2
2G	Practical Circuits	2
2H	Signals and Emissions	2
2I	Antennas and Feed Lines	3

The question numbers used in this question pool relate to the syllabus or study guide printed at the end of Chapter 1. The syllabus is an outline of topics covered by the exam. The first part of each question number (up to the decimal point) lists the syllabus point covered by that question. The number after the decimal point identifies the individual questions that go with that point. For example, question 2C-4.2 is the second question in the series about syllabus point 2C-4, sunspot cycle.

Good luck with your studies.

A Note to Those Preparing for the Novice Exam

The VEC Question Pool Committee released a supplement to Element 2 in October 1989. This supplement contains questions and answers that have been revised because of the changes to Part 97, the FCC Rules governing Amateur Radio.

Please note: The revised questions and answers contained in the supplement should not be used on any Novice written exams before July 1, 1990.

- If you take your exam *before* July 1, 1990, you need not be concerned with the supplement. (Your test *may* include the questions that will be revised when the supplement goes into effect on July 1, 1990. We are suggesting to VEs that these questions not be used, however.)
- If you take your exam *on or after* July 1, 1990, the questions in the supplement will be part of the pool from which your exam is prepared. Be sure to study the supplement questions instead of the like-numbered questions in the regular pool.

A Note to VEs

For Element 2 exams administered before July 1, 1990, the ARRL recommends that examiners *not* include the questions that will be revised by the supplement, since these questions are now out of date.

For Element 2 exams administered on or after July 1, 1990, substitute the questions, answers and distractors in the supplement for the like-numbered questions, answers and distractors in the Element 2 Question Pool.

Subelement 2A—Commission's Rules (10 Questions)

One question must be from the following:

2A-1.1 What are the five principles that express the fundamental purpose for which the Amateur Radio Service rules are designed?
- A. Recognition of emergency communications, advancement of the radio art, improvement of communication and technical skills, increase in the number of trained radio operators and electronics experts, and the enhancement of international goodwill
- B. Recognition of business communications, advancement of the radio art, improvement of communication and business skills, increase in the number of trained radio operators and electronics experts, and the enhancement of international goodwill
- C. Recognition of emergency communications, preservation of the earliest radio techniques, improvement of communication and technical skills, maintain a pool of people familiar with early tube-type equipment, and the enhancement of international goodwill
- D. Recognition of emergency communications, advancement of the radio art, improvement of communication and technical skills, increase in the number of trained radio operators and electronics experts, and the enhancement of a sense of patriotism

2A-1.2 Which of the following is ***not*** one of the basic principles for which the Amateur Radio Service rules are defined?
- A. Providing emergency communications
- B. Improvement of communication and technical skills
- C. Advancement of the radio art
- D. Enhancement of a sense of patriotism and nationalism

2A-1.3 The Amateur Radio Service rules were defined to provide a radio communications service that meets five fundamental purposes. Which of the following is ***not*** one of those principles?
- A. Improvement of communication and technical skills
- B. Enhancement of international goodwill
- C. Increase the number of trained radio operators and electronics experts
- D. Preserving the history of radio communications

2A-1.4 The Amateur Radio Service rules were defined to provide a radio communications service that meets five fundamental purposes. What are those principles?
- A. Recognition of business communications, advancement of the radio art, improvement of communication and business skills, increase in the number of trained radio operators and electronics experts, and the enhancement of international goodwill
- B. Recognition of emergency communications, advancement of the radio art, improvement of communication and technical skills, increase in the number of trained radio operators and electronics experts, and the enhancement of international goodwill
- C. Recognition of emergency communications, preservation of the earliest radio techniques, improvement of communication and technical skills, maintain a pool of people familiar with early tube-type equipment, and the enhancement of international goodwill
- D. Recognition of emergency communications, advancement of the radio art, improvement of communication and technical skills, increase in the number of trained radio operators and electronics experts, and the enhancement of a sense of patriotism

2A-2.1 What is the *Amateur Radio Service*?
- A. A private radio service used for personal gain and public benefit
- B. A public radio service used for public service communications
- C. A radio communication service for self-training and technical experimentation
- D. A private radio service intended for the furtherance of commercial radio interests

2A-2.2 What name is given to the radio communication service that is designed for self-training and technical experimentation?
- A. The Amateur Radio Service
- B. The Citizen's Radio Service
- C. The Experimenter's Radio Service
- D. The Maritime Radio Service

2A-3.1 What is *Amateur Radio communication*?
- A. Non-commercial radio communication between Amateur Radio stations with a personal aim and without pecuniary interest
- B. Commercial radio communications between radio stations licensed to non-profit organizations and businesses
- C. Experimental or educational radio transmissions controlled by student operators
- D. Non-commercial radio communications intended for the education and benefit of the general public

2A-3.2 What is the term used to describe non-commercial radio communications conducted with a personal aim and without pecuniary interest?
- A. Experimental Radio communications
- B. Personal Radio communications
- C. Non-commercial Radio communications
- D. Amateur Radio communications

2A-4.1 Who is an *Amateur Radio operator*?
A. A person who has not received any training in radio operations
B. Someone who performs communications in the Amateur Radio Service
C. A person who performs private radio communications for hire
D. A trainee in a commercial radio station

2A-4.2 What is the term used to describe someone who performs communications in the Amateur Radio Service?
A. A Citizen Radio operator
B. A Personal Radio operator
C. A Radio Service operator
D. An Amateur Radio operator

One question must be from the following:

2A-5.1 What is the portion of an Amateur Radio license that conveys operator privileges?
A. The verification section
B. Form 610
C. The operator license
D. The station license

2A-5.2 What authority is derived from an Amateur Radio operator license?
A. The authority to operate any shortwave radio station
B. The authority to operate an Amateur Radio station
C. The authority to have an Amateur Radio station at a particular location
D. The authority to transmit on either amateur or Class D citizen's band frequencies

2A-6.1 What authority is derived from an Amateur Radio station license?
A. The authority to use specified operating frequencies
B. The authority to have an Amateur Radio station at a particular location
C. The authority to enforce FCC Rules when violations are noted on the part of other operators
D. The authority to transmit on either amateur or Class D citizen's band frequencies

2A-6.2 What part of your Amateur Radio license gives you authority to have an Amateur Radio station at a particular location?
A. The operator license
B. The FCC Form 610
C. The station license
D. An Amateur Radio license does not specify a station location

2A-7.1 What is an *Amateur Radio station*?
A. A licensed radio station engaged in broadcasting to the public in a limited and well-defined area
B. A radio station used to further commercial radio interests
C. A private radio service used for personal gain and public service
D. A radio station operated by a person interested in self-training, intercommunication and technical investigation

2A-8.1 Who is a *control operator*?
A. A licensed operator designated to be responsible for the emissions of a particular station
B. A person, either licensed or not, who controls the emissions of an Amateur Radio Station
C. An unlicensed person who is speaking over an Amateur Radio Station's microphone while a licensed person is present
D. A government official who comes to an Amateur Radio Station to take control for test purposes

2A-8.2 As an Amateur Radio station licensee, you may designate another Amateur Radio operator to be responsible for the emissions from your station. What is this other operator called?
A. Auxiliary operator
B. Operations coordinator
C. Third party
D. Control operator

2A-9.1 List the five United States Amateur Radio license classes in order of increasing privileges.
A. Novice, General, Technician, Advanced, Amateur Extra
B. Novice, Technician, General, Advanced, Digital
C. Novice, Technician, General, Amateur, Extra
D. Novice, Technician, General, Advanced, Amateur Extra

2A-9.2 Which US Amateur Radio operator license is considered to be the "entry level" or "beginner's" license?
A. The Novice class license
B. The CB license
C. The Technician class license
D. The Amateur class License

2A-9.3 What is the license class immediately above Novice class?
A. The Digital class license
B. The Technician class license
C. The General class license
D. The Experimenter's class license

One question must be from the following:

2A-10.1 What frequencies may a Novice control operator use in the amateur 80-meter band?
A. 3500 to 4000 kHz
B. 3700 to 3750 kHz
C. 7100 to 7150 kHz
D. 7000 to 7300 kHz

2A-10.2 What frequencies may a Novice control operator use in the amateur 40-meter band?
A. 3500 to 4000 kHz
B. 3700 to 3750 kHz
C. 7100 to 7150 kHz
D. 7000 to 7300 kHz

2A-10.3 What frequencies may a Novice control operator use in the amateur 15-meter band?
A. 21.100 to 21.200 MHz
B. 21.000 to 21.450 MHz
C. 28.000 to 29.700 MHz
D. 28.100 to 28.200 MHz

2A-10.4 What frequencies may a Novice control operator use in the amateur 10-meter band?
A. 28.000 to 29.700 MHz
B. 28.100 to 28.300 MHz
C. 28.100 to 28.500 MHz
D. 28.300 to 28.500 MHz

2A-10.5 What frequencies may a Novice control operator use in the amateur 220-MHz band?
A. 225.0 to 230.5 MHz
B. 222.1 to 223.91 MHz
C. 224.1 to 225.1 MHz
D. 222.2 to 224.0 MHz

2A-10.6 What frequencies may a Novice control operator use in the amateur 1270-MHz band?
A. 1260 to 1270 MHz
B. 1240 to 1300 MHz
C. 1270 to 1295 MHz
D. 1240 to 1246 MHz

2A-10.7 If you are operating your Amateur Radio station on 3725 kHz, in what meter band are you operating?
A. 80 meters
B. 40 meters
C. 15 meters
D. 10 meters

2A-10.8 If you are operating your Amateur Radio station on 7125 kHz, in what meter band are you operating?
A. 80 meters
B. 40 meters
C. 15 meters
D. 10 meters

2A-10.9 If you are operating your Amateur Radio station on 21,150 kHz, in what meter band are you operating?
A. 80 meters
B. 40 meters
C. 15 meters
D. 10 meters

2A-10.10 If you are operating your Amateur Radio station on 28,150 kHz, in what meter band are you operating?
A. 80 meters
B. 40 meters
C. 15 meters
D. 10 meters

One question must be from the following:

2A-11.1 Who is eligible to obtain a US Amateur Radio *operator* license?
A. Anyone except a representative of a foreign government
B. Only a citizen of the United States
C. Anyone
D. Anyone except an employee of the United States Government

2A-11.2 Who is ***not*** eligible to obtain a US Amateur Radio *operator* license?
A. Any citizen of a country other than the United States
B. A representative of a foreign government
C. No one
D. An employee of the United States Government

2A-12.1 What FCC examination elements are required for a Novice class license?
A. Elements 1(A) and 2(A)
B. Elements 1(B) and 2
C. Elements 1(A) and 2
D. Elements 1 and 2

2A-12.2 What is an FCC Element 1(A) examination intended to prove?
A. The applicant's ability to send and receive Morse code at 5 words per minute
B. The applicant's ability to send and receive Morse code at 13 words per minute
C. The applicant's knowledge of Novice class theory and regulations
D. The applicant's ability to recognize Novice frequency assignments and operating modes

2A-12.3 What is an FCC Element 2 examination?
A. A test of the applicant's ability to send and receive Morse code at 5 words per minute
B. The written examination for the Technician class operator license
C. A test of the applicant's ablility to recognize Novice frequency assignments
D. The written examination for the Novice class operator license

2A-13.1 Who is eligible to obtain a US Amateur Radio *station* license?
A. A licensed Amateur Radio operator
B. Any unlicensed person, except an agent of a foreign government
C. Any unlicensed person, except an employee of the United States Government
D. Any unlicensed United States Citizen

2A-14.1 Why is an Amateur Radio operator required to furnish the FCC with a current mailing address?
A. So the FCC has a record of the location of each Amateur Radio station
B. So the FCC can direct correspondence to the licensee
C. So the FCC can send license-renewal notices
D. So the FCC can compile a list for use in a call sign directory

2A-15.1 Which one of the following call signs is a valid US amateur call?
A. UA4HAK
B. KBL7766
C. KA9OLS
D. BY7HY

2A-15.2 Which one of the following call signs is a valid US amateur call?
A. CE2FTF
B. G3GVA
C. UA1ZAM
D. AA2Z

2A-15.3 Which one of the following call signs is ***not*** a valid US amateur call?
A. KDV5653
B. WA1DVU
C. KA5BUG
D. NTØZ

2A-15.4 What letters may be used for the first letter in a valid US amateur call sign?
- A. K, N, U and W
- B. A, K, N and W
- C. A, B, C and D
- D. A, N, V and W

2A-15.5 Excluding special-event call signs that may be issued by the FCC, what numbers may be used in a valid US call sign?
- A. Any double-digit number, 10 through 99
- B. Any double-digit number, 22 through 45
- C. Any single digit, 1 though 9
- D. A single digit, 0 through 9

2A-16.1 Your Novice license was issued on November 1, 1988. When will it expire?
- A. November 1, 1998
- B. November 30, 1998
- C. November 1, 1993
- D. November 1, 1990

One question must be from the following:

2A-17.1 What does the term *emission* mean?
- A. RF signals transmitted from a radio station
- B. Signals refracted by the E layer
- C. Filter out the carrier of a received signal
- D. Baud rate

2A-17.2 What emission types are Novice control operators permitted to use on the amateur 80-meter band?
- A. A1A only
- B. F1B only
- C. A80A only
- D. A1A and J3E only

2A-17.3 What emission types are Novice control operators permitted to use in the 40-meter band?
- A. A1A only
- B. F1B only
- C. A40A only
- D. A1A and J3E only

2A-17.4 What emission types are Novice control operators permitted to use in the 15-meter band?
- A. A1A only
- B. F1B only
- C. A15A only
- D. A1A and J3E only

2A-17.5 What emission types are Novice control operators permitted to use from 3700 to 3750 kHz?
- A. A1A and F1B only
- B. A1A and J3E only
- C. All amateur emission privileges authorized for use on those frequencies
- D. A1A only

2A-17.6 What emission types are Novice control operators permitted to use from 7100 to 7150 kHz?
- A. A1A and F1B only
- B. A1A and J3E only
- C. All amateur emission privileges authorized for use on those frequencies
- D. A1A only

2A-17.7 What emission types are Novice control operators permitted to use on frequencies from 21.1 to 21.2 MHz?
- A. A1A and F1B only
- B. A1A and J3E only
- C. All amateur emission privileges authorized for use on those frequencies
- D. A1A only

2A-17.8 What emission types are Novice control operators permitted to use on frequencies from 28.1 to 28.3 MHz?
- A. All authorized amateur emission privileges
- B. F1B and J3E
- C. A1A and F1B
- D. A1A and J3E

2A-17.9 What emission types are Novice control operators permitted to use on frequencies from 28.3 to 28.5 MHz?
- A. All authorized amateur emission privileges
- B. A1A and F1B
- C. A1A and J3E
- D. A1A and F3E

2A-17.10 What emission types are Novice control operators permitted to use on the amateur 220-MHz band?
- A. F1B and J3E
- B. A1A and F1B
- C. A1A and J3E
- D. All amateur emission privileges authorized for use on 220 MHz

2A-17.11 What emission types are Novice control operators permitted to use on the amateur 1270-MHz band?
- A. F1B and J3E
- B. A1A and F1B
- C. A1A and J3E
- D. All amateur emission privileges authorized for use on 1270 MHz

2A-17.12 On what frequencies may a Novice control operator operate single-sideband voice?
- A. 3700 to 3750 kHz
- B. 7100 to 7150 kHz
- C. 21,100 to 21,200 kHz
- D. 28,300 to 28,500 kHz

2A-17.13 On what frequencies may a Novice control operator operate FM voice?
- A. 28.3 to 28.5 MHz
- B. 144.0 to 148.0 MHz
- C. 222.1 to 223.91 MHz
- D. 1240 to 1270 MHz

One question must be from the following:

2A-18.1 What amount of output transmitting power may a Novice class control operator use when operating below 30 MHz?
- A. 200 watts input
- B. 250 watts output
- C. 1500 watts PEP output
- D. The minimum legal power necessary to maintain reliable communications

2A-18.2 What is the maximum transmitting power ever permitted to be used by an amateur station transmitting in the 80, 40 and 15-meter Novice bands?
A. 75 watts PEP output
B. 100 watts PEP output
C. 200 watts PEP output
D. 1500 watts PEP output

2A-18.3 What is the maximum transmitting power permitted an amateur station transmitting on 3725 kHz?
A. 75 watts PEP output
B. 100 watts PEP output
C. 200 watts PEP output
D. 1500 watts PEP output

2A-18.4 What is the maximum transmitting power permitted an amateur station transmitting on 7125 kHz?
A. 75 watts PEP output
B. 100 watts PEP output
C. 200 watts PEP output
D. 1500 watts PEP output

2A-18.5 What is the maximum transmitting power permitted an amateur station transmitting on 21.125 MHz?
A. 75 watts PEP output
B. 100 watts PEP output
C. 200 watts PEP output
D. 1500 watts PEP output

2A-19.1 What is the maximum transmitting power permitted an amateur station with a Novice control operator transmitting on 28.125 MHz?
A. 75 watts PEP output
B. 100 watts PEP output
C. 200 watts PEP output
D. 1500 watts PEP output

2A-19.2 What is the maximum transmitting power permitted an amateur station with a Novice control operator transmitting in the amateur 10-meter band?
A. 25 watts PEP output
B. 200 watts PEP output
C. 1000 watts PEP output
D. 1500 watts PEP output

2A-19.3 What is the maximum transmitting power permitted an amateur station with a Novice control operator transmitting in the amateur 220-MHz band?
A. 5 watts PEP output
B. 10 watts PEP output
C. 25 watts PEP output
D. 200 watts PEP output

2A-19.4 What is the maximum transmitting power permitted an amateur station with a Novice control operator transmitting in the amateur 1270-MHz band?
A. 5 milliwatts PEP output
B. 500 milliwatts PEP output
C. 1 watt PEP output
D. 5 watts PEP output

2A-19.5 What amount of transmitting power may an amateur station with a Novice control operator use in the amateur 220-MHz band?
A. Not less than 5 watts PEP output
B. The minimum legal power necessary to maintain reliable communications
C. Not more than 50 watts PEP output
D. Not more than 200 watts PEP output

2A-20.1 What term is used to describe amateur communications intended to be received and printed automatically?
A. Teleport communications
B. Direct communications
C. Digital communications
D. Third-party communications

2A-20.2 What term is used to describe amateur communications for the direct transfer of information between computers?
A. Teleport communications
B. Direct communications
C. Digital communications
D. Third-party communications

2A-20.3 On what frequencies in the 10-meter band are Novice control operators permitted to transmit emission F1B RTTY?
A. 28.1 to 28.5 MHz
B. 28.0 to 29.7 MHz
C. 28.1 to 28.2 MHz
D. 28.1 to 28.3 MHz

One question must be from the following:

2A-21.1 Who is held responsible for the proper operation of an amateur station?
A. Only the control operator
B. Only the station licensee
C. Both the control operator and the station licensee
D. The person who owns the property where the station is located

2A-21.2 You allow another Amateur Radio operator to use your amateur station. What are your responsibilities, as the station licensee?
A. You are responsible for the proper operation of your station
B. Only the control operator is responsible for the proper operation of the station
C. As the station licensee, you must be at the control point of your station whenever it is operated
D. You must notify the FCC when another amateur will be the control operator of your station

2A-21.3 What is your primary responsibility as the station licensee?
A. You must permit any licensed amateur radio operator to operate your station at any time upon request
B. You must be present whenever the station is operated
C. You must notify the FCC in writing whenever another Amateur Radio operator will act as the control operator
D. You are responsible for the proper operation of the station for which you are licensed

2A-21.4 You are the licensee of an Amateur Radio station. When are you ***not*** responsible for its proper operation?
A. Only when another licensed amateur is the control operator
B. The licensee is responsible for the proper operation of the station for which he or she is licensed
C. Only after notifying the FCC in writing that another licensed amateur will assume responsibility for the proper operation of your station
D. Only when your station is in repeater operation

2A-22.1 When must an amateur station have a control operator?
A. A control operator is only required for training purposes
B. Whenever the station receiver is operated
C. Whenever the transmitter is operated, except when the station is under automatic control
D. A control operator is not required

2A-22.2 Another amateur gives you permission to use her Amateur Radio station. What are your responsibilities, as the control operator?
A. You are responsible for the proper operation of the station
B. Only the station licensee is responsible for the proper operation of the station
C. You must be certain the station licensee has given proper FCC notice that you will be the control operator
D. You must inspect all antennas and related equipment to ensure they are working properly

2A-23.1 Who may be the control operator of an amateur station?
A. Any person over 21 years of age
B. Any licensed Amateur Radio operator
C. Any licensed Amateur Radio operator with an Advanced class license or higher
D. Any person over 21 years of age with a General class license or higher

2A-24.1 Where must an Amateur Radio operator be when he or she is performing the duties of control operator?
A. Anywhere in the same building as the transmitter
B. At the control point of the Amateur Radio station
C. At the station entrance, to control entry to the room
D. Within sight of the station monitor, to view the output spectrum of the transmitter

2A-25.1 Where must you keep your Amateur Radio operator license when you are operating a station?
A. Your original operator license must always be posted in plain view
B. Your original operator license must always be taped to the inside front cover of your station log
C. You must have the original or a photocopy of your operator license in your possession
D. You must have the original or a photocopy of your operator license posted at your primary station location. You need not have the original license nor a copy in your possession to operate another station

2A-26.1 Where must you keep your Amateur Radio station license when your station is being operated?
A. Your original station license must always be taped to the inside front cover of your station log
B. Your original station license must always be posted in plain view
C. You must post the original or a photocopy of your station license at the main entrance to the transmitter building
D. You must post the original or a photocopy of your station license near your station or keep it in the personal possession of the licensed operator

One question must be from the following:

2A-27.1 How often must an amateur station be identified?
A. At the beginning of the contact and at least every ten minutes during a contact
B. At least once during each transmission
C. At least every ten minutes during a contact and at the end of the contact
D. Every 15 minutes during a contact and at the end of the contact

2A-27.2 As an Amateur Radio operator, how should you correctly identify your station communications?
A. With the name and location of the control operator
B. With the station call sign
C. With the call of the control operator, even when he or she is visiting another radio amateur's station
D. With the name and location of the station licensee, followed by the two-letter designation of the nearest FCC Field Office

2A-27.3 What station identification, if any, is required at the beginning of a QSO?
A. The operator originating the contact must transmit both call signs
B. No identification is required at the beginning of the contact
C. Both operators must transmit their own call signs
D. Both operators must transmit both call signs

2A-27.4 What station identification, if any, is required at the end of a QSO?
A. Both operators must transmit their own call sign
B. No identification is required at the end of the contact
C. The operator originating the contact must always transmit both call signs
D. Both operators must transmit their own call sign followed by a two-letter designator for the nearest FCC field office

2A-27.5 What do the FCC Rules for amateur station identification require?
A. Each Amateur Radio station shall give its call sign at the beginning of each communication, and every ten minutes or less during a communication
B. Each Amateur Radio station shall give its call sign at the end of each communication, and every ten minutes or less during a communication
C. Each Amateur Radio station shall give its call sign at the beginning of each communication, and every five minutes or less during a communication
D. Each Amateur Radio station shall give its call sign at the end of each communication, and every five minutes or less during a communication

2A-27.6 What is the fewest number of times you must transmit your Amateur Radio station identification during a 25 minute QSO?
A. 1
B. 2
C. 3
D. 4

2A-27.7 What is the longest period of time during a QSO that an amateur station does not need to transmit its station identification?
A. 5 minutes
B. 10 minutes
C. 15 minutes
D. 20 minutes

2A-28.1 With which amateur stations may an FCC-licensed amateur station communicate?
A. All amateur stations
B. All public noncommercial radio stations unless prohibited by the station's government
C. Only with US amateur stations
D. All amateur stations, unless prohibited by the amateur's government

2A-28.2 With which non-Amateur Radio stations may an FCC-licensed amateur station communicate?
A. No non-amateur stations
B. All such stations
C. Only those authorized by the FCC
D. Only those who use the International Morse code

2A-29.1 When must the licensee of an Amateur Radio station in portable or mobile operation notify the FCC?
A. One week in advance if the operation will last for more than 24 hours
B. FCC notification is not required for portable or mobile operation
C. One week in advance if the operation will last for more than a week
D. One month in advance of any portable or mobile operation

2A-29.2 When may you operate your Amateur Radio station at a location within the United States, its territories or possessions other than the one listed on your station license?
A. Only during times of emergency
B. Only after giving proper notice to the FCC
C. During an emergency or an FCC-approved emergency preparedness drill
D. Whenever you want to

2A-30.1 When are business communications permitted in the Amateur Radio Service?
A. Only when the immediate safety of life of individuals or property is threatened
B. There are no rules against conducting business communications in the Amateur Radio Service
C. No business communications of any kind are ever permitted in the Amateur Radio Service
D. Business communications are permitted between the hours of 9 AM to 5 PM, only on weekdays

2A-30.2 You wish to obtain an application for membership in the American Radio Relay League. When would you be permitted to send an Amateur Radio message requesting the application?
A. At any time, since the ARRL is a not-for-profit organization
B. Never. Business communications are not permitted in the Amateur Radio service
C. Only during normal business hours, between 9 AM and 5 PM
D. At any time, since there are no rules against conducting business communications in the Amateur Radio Service

2A-30.3 On your way home from work you decide to order pizza for dinner. When would you be permitted to use the autopatch on your radio club repeater to order the pizza?
A. At any time, since you will not profit from the communications
B. Only during normal business hours, between 9 AM and 5 PM
C. At any time, since there are no rules against conducting business communications in the Amateur Radio Service
D. Never. Business communications are not permitted in the Amateur Radio Service

One question must be from the following:

2A-31.1 When may an FCC-licensed Amateur Radio operator communicate with an Amateur Radio operator in a foreign country?
- A. Only when the foreign operator uses English as his primary language
- B. All the time, except on 28.600 to 29.700 MHz
- C. Only when a third-party agreement exists between the US and the foreign country
- D. At any time unless prohibited by either the US or the foreign government

2A-32.1 When may an Amateur Radio station be used to transmit messages for hire?
- A. Under no circumstances may an Amateur Radio station be hired to transmit messages
- B. Modest payment from a non-profit charitable organization is permissible
- C. No money may change hands, but a radio amateur may be compensated for services rendered with gifts of equipment or services rendered as a returned favor
- D. All payments received in return for transmitting messages by Amateur Radio must be reported to the IRS

2A-32.2 When may the control operator be paid to transmit messages from an Amateur Radio station?
- A. The control operator may be paid if he or she works for a public service agency such as the Red Cross
- B. The control operator may not be paid under any circumstances
- C. The control operator may be paid if he or she reports all income earned from operating an Amateur Radio Station to the IRS as receipt of tax-deductible contributions
- D. The control operator may be paid if he or she works for an Amateur Radio Station that operates primarily to broadcast telegraphy practice and news bulletins for Radio Amateurs

2A-33.1 When is an Amateur Radio operator permitted to broadcast information intended for the general public?
- A. Amateur Radio operators are not permitted to broadcast information intended for the general public
- B. Only when the operator is being paid to transmit the information
- C. Only when such transmissions last less than 1 hour in any 24-hour period
- D. Only when such transmissions last longer than 15 minutes

2A-34.1 What is *third-party traffic*?
- A. A message passed by one Amateur Radio control operator to another Amateur Radio control operator on behalf of another person
- B. Public service communications handled on behalf of a minor political party
- C. Only messages that are formally handled through Amateur Radio channels
- D. A report of highway conditions transmitted over a local repeater

2A-34.2 Who is a *third-party* in Amateur Radio communications?
- A. The Amateur Radio station that breaks into a two-way contact between two other Amateur Radio stations
- B. Any person passing a message through Amateur Radio communication channels other than the control operators of the two stations handling the message
- C. A shortwave listener monitoring a two-way Amateur Radio communication
- D. The control operator present when an unlicensed person communicates over an Amateur Radio station

2A-34.3 When is an Amateur Radio operator permitted to transmit a message to a foreign country for a third party?
- A. Anytime
- B. Never
- C. Anytime, unless there is a third-party traffic agreement between the US and the foreign government
- D. Only if there is a third-party traffic agreement between the US and the foreign government

2A-35.1 When is an Amateur Radio operator permitted to transmit music?
- A. The transmission of music is not permitted in the Amateur Radio Service
- B. When the music played produces no disonances or spurious emissions
- C. When it is used to jam an illegal transmission
- D. Only above 1280 MHz

2A-36.1 When is the transmission by radio of messages in codes or ciphers permitted in domestic and international communications between Amateur Radio stations?
- A. Codes and ciphers are permitted during ARRL-sponsored contests
- B. Codes and ciphers are permitted during nationally declared emergencies
- C. The transmission of codes and ciphers is not permitted in domestic or international Amateur Radio communications
- D. Codes and ciphers are permitted above 1280 MHz

2A-36.2 When is an Amateur Radio operator permitted to use abbreviations that are intended to obscure the meaning of the message?
- A. Only during ARRL-sponsored contests
- B. Only on frequencies above 222.5 MHz
- C. Only during a declared communications emergency
- D. Abbreviations that are intended to obscure the meaning of the message may never be used

One question must be from the following:

2A-37.1 Under what circumstances, if any, may the control operator cause *false or deceptive signals or communications* to be transmitted?
A. Under no circumstances
B. When operating a beacon transmitter in a "fox hunt" exercise
C. When playing a harmless "practical joke" without causing interference to other stations that are not involved
D. When you need to obscure the meaning of transmitted information to ensure secrecy

2A-37.2 If an Amateur Radio operator transmits the word "MAYDAY" when no actual emergency has occurred, what is this called?
A. A traditional greeting in May
B. An Emergency Action System test transmission
C. False or deceptive signals
D. "MAYDAY" has no significance in an emergency situation

2A-38.1 When may an Amateur Radio operator transmit unidentified communications?
A. A transmission need not be identified if it is restricted to brief tests not intended for reception by other parties
B. A transmission need not be identified when conducted on a clear frequency or "dead band" where interference will not occur
C. An amateur operator may never transmit unidentified communications
D. A transmission need not be identified unless two-way communications or third-party traffic handling are involved

2A-38.2 What is the meaning of the term *unidentified radio communications or signals*?
A. Radio communications in which the transmitting station's call sign is transmitted in modes other than CW and voice
B. Radio communications approaching a receiving station from an unknown direction
C. Radio communications in which the operator fails to transmit his or her name and QTH
D. Radio communications in which the transmitting station's call sign is not transmitted

2A-38.3 What is the term used to describe a transmission from an Amateur Radio station without the required station identification?
A. Unidentified transmission
B. Reluctance modulation
C. NØN emission
D. Tactical communication

2A-39.1 When may an Amateur Radio operator willfully or maliciously interfere with a radio communication or signal?
A. You may jam another person's transmissions if that person is not operating in a legal manner
B. You may interfere with another station's signals if that station begins transmitting on a frequency already occupied by your station
C. You may never intentionally interfere with another station's transmissions
D. You may expect, and cause, deliberate interference because it is unavoidable during crowded band conditions

2A-39.2 What is the meaning of the term *malicious interference*?
A. Accidental interference
B. Intentional interference
C. Mild interference
D. Occasional interference

2A-39.3 What is the term used to describe an Amateur Radio transmission that is intended to disrupt other communications in progress?
A. Interrupted CW
B. Malicious interference
C. Transponded signals
D. Unidentified transmissions

2A-40.1 As an Amateur Radio operator, you receive an *Official Notice of Violation* from the FCC. How promptly must you respond?
A. Within 90 days
B. Within 30 days
C. Within 10 days
D. The next day

2A-40.2 As an Amateur Radio operator, you receive an *Official Notice of Violation* from the FCC. To whom must you respond?
A. Any office of the FCC
B. The Gettysburg, PA office of the FCC
C. The Washington, DC office of the FCC
D. The FCC office that originated the notice

2A-40.3 As an Amateur Radio operator, you receive an *Official Notice of Violation* from the FCC relating to a violation that may be due to the physical or electrical characteristics of your transmitter. What information must be included in your response?
A. The make and model of the apparatus
B. The steps taken to guarantee future violations
C. The date that the apparatus was returned to the manufacturer
D. The steps taken to prevent future violations

Subelement 2B—Operating Procedures (2 Questions)

One question must be from the following:

2B-1-1.1 What is the most important factor to consider when selecting a transmitting frequency within your authorized subband?
- A. The frequency should not be in use by other amateurs
- B. You should be able to hear other stations on the frequency to ensure that someone will be able to hear you
- C. Your antenna should be resonant at the selected frequency
- D. You should ensure that the SWR on the antenna feed line is high enough at the selected frequency

2B-1-1.2 You wish to contact an Amateur Radio station more than 1500 miles away on a summer afternoon. Which band is most likely to provide a successful contact?
- A. The 80- or 40-meter bands
- B. The 40- or 15-meter bands
- C. The 15- or 10-meter bands
- D. The 1¼ meter or 23-centimeter bands

2B-1-1.3 How can on-the-air transmitter tune-up be kept as short as possible?
- A. By using a random wire antenna
- B. By tuning up on 40 meters first, then switching to the desired band
- C. By tuning the transmitter into a dummy load
- D. By using twin lead instead of coaxial-cable feed lines

2B-1-2.1 You are having a QSO with your uncle in Pittsburgh when you hear an emergency call for help on the frequency you are using. What should you do?
- A. Inform the station that the frequency is in use
- B. Direct the station to the nearest emergency net frequency
- C. Call your local Civil Preparedness Office and inform them of the emergency
- D. Immediately stand by to copy the emergency communication

2B-2-1.1 What is the format of a standard Morse code CQ call?
- A. Transmit the procedural signal "CQ" three times, followed by the procedural signal "DE," followed by your call three times
- B. Transmit the procedural signal "CQ" three times, followed by the procedural signal "DE," followed by your call one time
- C. Transmit the procedural signal "CQ" ten times, followed by the procedural signal "DE," followed by your call one time
- D. Transmit the procedural signal "CQ" continuously until someone answers your call

2B-2-1.2 How should you answer a Morse code CQ call?
- A. Send your call sign four times
- B. Send the other station's call sign twice, followed by the procedural signal "DE," followed by your call sign twice
- C. Send the other station's call sign once, followed by the procedural signal "DE," followed by your call sign four times
- D. Send your call sign followed by your name, station location and a signal report

2B-2-2.1 At what telegraphy speed should a "CQ" message be transmitted?
- A. Only speeds below five WPM
- B. The highest speed your keyer will operate
- C. Any speed at which you can reliably receive
- D. The highest speed at which you can control the keyer

2B-2-3.1 What is the meaning of the Morse code character $\overline{\text{AR}}$?
- A. Only the called station transmit
- B. All received correctly
- C. End of transmission
- D. Best regards

2B-2-3.2 What is the meaning of the Morse code character $\overline{\text{SK}}$?
- A. Received some correctly
- B. Best regards
- C. Wait
- D. End of contact

2B-2-3.3 What is the meaning of the Morse code character $\overline{\text{BT}}$?
- A. Double dash "="
- B. Fraction bar "/"
- C. End of contact
- D. Back to you

2B-2-3.4 What is the meaning of the Morse code character $\overline{\text{DN}}$?
- A. Double dash "="
- B. Fraction bar "/"
- C. Done now (end of contact)
- D. Called station only transmit

2B-2-3.5 What is the meaning of the Morse code character $\overline{\text{KN}}$?
- A. Fraction bar "/"
- B. End of contact
- C. Called station only transmit
- D. Key now (go ahead to transmit)

2B-2-4.1 What is the procedural signal "CQ" used for?
- A. To notify another station that you will call on the quarter hour
- B. To indicate that you are testing a new antenna and are not listening for another station to answer
- C. To indicate that only the called station should transmit
- D. A general call when you are trying to make a contact

2B-2-4.2 What is the procedural signal "DE" used for?
- A. To mean "from" or "this is," as in "W9NGT de N9BTT"
- B. To indicate directional emissions from your antenna
- C. To indicate "received all correctly"
- D. To mean "calling any station"

2B-2-4.3 What is the procedural signal "K" used for?
A. To mean "any station transmit"
B. To mean "all received correctly"
C. To mean "end of message"
D. To mean "called station only transmit"

2B-2-5.1 What does the *R* in the RST signal report mean?
A. The recovery of the signal
B. The resonance of the CW tone
C. The rate of signal flutter
D. The readability of the signal

2B-2-5.2 What does the *S* in the RST signal report mean?
A. The scintillation of a signal
B. The strength of the signal
C. The signal quality
D. The speed of the CW transmission

2B-2-5.3 What does the *T* in the RST signal report mean?
A. The tone of the signal
B. The closeness of the signal to "telephone" quality
C. The timing of the signal dot to dash ratio
D. The tempo of the signal

2B-2-6.1 What is one meaning of the Q signal "QRS"?
A. Interference from static
B. Send more slowly
C. Send RST report
D. Radio station location is

2B-2-6.2 What is one meaning of the Q signal "QRT"?
A. The correct time is
B. Send RST report
C. Stop sending
D. Send more slowly

2B-2-6.3 What is one meaning of the Q signal "QTH"?
A. Time here is
B. My name is
C. Stop sending
D. My location is . . .

2B-2-6.4 What is one meaning of the Q signal "QRZ," when it is followed with a question mark?
A. Who is calling me?
B. What is your radio zone?
C. What time zone are you in?
D. Is this frequency in use?

2B-2-6.5 What is one meaning of the Q signal "QSL," when it is followed with a question mark?
A. Shall I send you my log?
B. Can you acknowledge receipt (of my message)?
C. Shall I send more slowly?
D. Who is calling me?

2B-3-1.1 What is the format of a standard radiotelephone CQ call?
A. Transmit the phrase "CQ" at least ten times, followed by "this is," followed by your call sign at least two times
B. Transmit the phrase "CQ" at least five times, followed by "this is," followed by your call sign once
C. Transmit the phrase "CQ" three times, followed by "this is," followed by your call sign three times
D. Transmit the phrase "CQ" at least ten times, followed by "this is," followed by your call sign once

2B-3-1.2 How should you answer a radiotelephone CQ call?
A. Transmit the other station's call sign at least ten times, followed by "this is," followed by your call sign at least twice
B. Transmit the other station's call sign at least five times phonetically, followed by "this is," followed by your call sign at least once
C. Transmit the other station's call sign at least three times, followed by "this is," followed by your call sign at least five times phonetically
D. Transmit the other station's call sign once, followed by "this is," followed by your call sign given phonetically

2B-3-2.1 How is the call sign "KA3BGQ" stated in Standard International Phonetics?
A. Kilo Alfa Three Bravo Golf Quebec
B. King America Three Bravo Golf Quebec
C. Kilowatt Alfa Three Bravo George Queen
D. Kilo America Three Baker Golf Quebec

2B-3-2.2 How is the call sign "WE5TZD" stated phonetically?
A. Whiskey Echo Foxtrot Tango Zulu Delta
B. Washington England Five Tokyo Zanzibar Denmark
C. Whiskey Echo Five Tango Zulu Delta
D. Whiskey Easy Five Tear Zebra Dog

2B-3-2.3 How is the call sign "KC4HRM" stated phonetically?
A. Kilo Charlie Four Hotel Romeo Mike
B. Kilowatt Charlie Four Hotel Roger Mexico
C. Kentucky Canada Four Honolulu Radio Mexico
D. King Charlie Foxtrot Hotel Roger Mary

2B-3-2.4 How is the call sign "AF6PSQ" stated phonetically?
A. America Florida Six Portugal Spain Quebec
B. Adam Frank Six Peter Sugar Queen
C. Alfa Fox Sierra Papa Santiago Queen
D. Alfa Foxtrot Six Papa Sierra Quebec

2B-3-2.5 How is the call sign "NB8LXG" stated phonetically?
A. November Bravo Eight Lima Xray Golf
B. Nancy Baker Eight Love Xray George
C. Norway Boston Eight London Xray Germany
D. November Bravo Eight London Xray Germany

2B-3-2.6 How is the call sign "KJ1UOI" stated phonetically?
A. King John One Uncle Oboe Ida
B. Kilowatt George India Uncle Oscar India
C. Kilo Juliette One Uniform Oscar India
D. Kentucky Juliette One United Ontario Indiana

2B-3-2.7 How is the call sign "WV2BPZ" stated phonetically?
A. Whiskey Victor Two Bravo Papa Zulu
B. Willie Victor Two Baker Papa Zebra
C. Whiskey Victor Tango Bravo Papa Zulu
D. Willie Virginia Two Boston Peter Zanzibar

2B-3-2.8 How is the call sign "NY3CTJ" stated phonetically?
A. Norway Yokohama Three California Tokyo Japan
B. Nancy Yankee Three Cat Texas Jackrabbit
C. Norway Yesterday Three Charlie Texas Juliette
D. November Yankee Three Charlie Tango Juliette

2B-3-2.9 How is the call sign "KG7DRV" stated phonetically?
A. Kilo Golf Seven Denver Radio Venezuela
B. Kilo Golf Seven Delta Romeo Victor
C. King John Seven Dog Radio Victor
D. Kilowatt George Seven Delta Romeo Video

2B-3-2.10 How is the call sign "WX9HKS" stated phonetically?
A. Whiskey Xray Nine Hotel Kilo Sierra
B. Willie Xray November Hotel King Sierra
C. Washington Xray Nine Honolulu Kentucky Santiago
D. Whiskey Xray Nine Henry King Sugar

2B-3-2.11 How is the call sign "AEØLQY" stated phonetically?
A. Able Easy Zero Lima Quebec Yankee
B. Arizona Equador Zero London Queen Yesterday
C. Alfa Echo Zero Lima Quebec Yankee
D. Able Easy Zero Love Queen Yoke

One question must be from the following:

2B-4-1.1 What is the format of a standard RTTY CQ call?
A. Transmit the phrase "CQ" three times, followed by "DE," followed by your call sign two times
B. Transmit the phrase "CQ" three to six times, followed by "DE," followed by your call sign three times
C. Transmit the phrase "CQ" ten times, followed by the procedural signal "DE," followed by your call one time
D. Transmit the phrase "CQ" continuously until someone answers your call

2B-4-2.1 You receive an RTTY CQ call at 45 bauds. At what speed should you respond?
A. 22½ bauds
B. 45 bauds
C. 90 bauds
D. Any speed, since radioteletype systems adjust to any signal rate

2B-5-1.1 What does the term *connected* mean in a packet-radio link?
A. A telephone link has been established between two amateurs
B. An Amateur Radio message has reached the station for local delivery
C. The transmitting station is sending data specifically addressed to the receiving station, and the receiving station is acknowledging that the data has been received correctly
D. The transmitting station and a receiving station are using a certain digipeater, so no other contacts can take place until they are finished

2B-5-1.2 What does the term *monitoring* mean on a frequency used for packet radio?
A. The FCC is copying all messages to determind their content
B. A member of the Amateur Auxilliary to the FCC's Field Operations Bureau is copying all messages to determine their content
C. The receiving station's video monitor is displaying all messages intended for that station, and is acknowledging correct receipt of the data
D. The receiving station is displaying information that may not be addressed to that station, and is not acknowledging correct receipt of the data

2B-5-2.1 What is a *digipeater*?
A. A packet-radio station used to retransmit data that is specifically addressed to be retransmitted by that station
B. An Amateur Radio repeater designed to retransmit all audio signals in a digital form
C. An Amateur Radio repeater designed using only digital electronics components
D. A packet-radio station that retransmits any signals it receives

2B-5-2.2 What is the meaning of the term *network* in packet radio?
A. A system of telephone lines interconnecting packet-radio stations to transfer data
B. A method of interconnecting packet-radio stations so that data can be transferred over long distances
C. The interlaced wiring on a terminal-node-controller board
D. The terminal-node-controller function that automatically rejects another caller when the station is connected

2B-6-1.1 What is a good way to establish a contact on a repeater?
A. Give the call sign of the station you want to contact three times
B. Call the other operator by name and then give your call sign three times
C. Call the desired station and then identify your own station
D. Say, "Breaker, breaker," and then give your call sign

2B-6-2.1 What is the main purpose of a repeater?
A. To provide a station that makes local information available 24 hours a day
B. To provide a means of linking Amateur Radio stations with the telephone system
C. To retransmit NOAA weather information during severe storm warnings
D. Repeaters extend the operating range of portable and mobile stations

2B-6-3.1 What does it mean to say that a repeater has an *input* and an *output* frequency?
A. The repeater receives on one frequency and transmits on another
B. All repeaters offer a choice of operating frequency, in case one is busy
C. One frequency is used to control repeater functions and the other frequency is the one used to retransmit received signals
D. Repeaters require an access code to be transmitted on one frequency while your voice is transmitted on the other

2B-6-4.1 When should simplex operation be used instead of using a repeater?

A. Whenever greater communications reliability is needed
B. Whenever a contact is possible without using a repeater
C. Whenever you need someone to make an emergency telephone call
D. Whenever you are traveling and need some local information

2B-6-5.1 What is an *autopatch*?

A. A repeater feature that automatically selects the strongest signal to be repeated
B. An automatic system of connecting a mobile station to the next repeater as it moves out of range of the first
C. A device that allows repeater users to make telephone calls from their portable or mobile stations
D. A system that automatically locks other stations out of the repeater when there is a QSO in progress

2B-6-5.2 What is the purpose of a repeater *time-out timer*?

A. It allows the repeater to have a rest period after heavy use
B. It logs repeater transmit time to determine when the repeater mean time between failure rating is exceeded
C. It limits repeater transmission time to no more than ten minutes
D. It limits repeater transmission time to no more than three minutes

Subelement 2C—Radio-Wave Propagation (1 Question)

One question must be from the following:

2C-1.1 What type of radio-wave propagation occurs when the signal travels in a straight line from the transmitting antenna to the receiving antenna?

A. Line-of-sight propagation
B. Straight-line propagation
C. Knife-edge diffraction
D. Tunnel propagation

2C-1.2 What path do radio waves usually follow from a transmitting antenna to a receiving antenna at VHF and higher frequencies?

A. A bent path through the ionosphere
B. A straight line
C. A great circle path over either the north or south pole
D. A circular path going either east or west from the transmitter

2C-2.1 What type of propagation involves radio signals that travel along the surface of the Earth?

A. Sky-wave propagation
B. Knife-edge diffraction
C. E-layer propagation
D. Ground-wave propagation

2C-2.2 What is the meaning of the term *ground-wave propagation*?

A. Signals that travel along seismic fault lines
B. Signals that travel along the surface of the earth
C. Signals that are radiated from a ground-plane antenna
D. Signals that are radiated from a ground station to a satellite

2C-2.3 Two Amateur Radio stations a few miles apart and separated by a low hill blocking their line-of-sight path are communicating on 3.725 MHz. What type of propagation is probably being used?

A. Tropospheric ducting
B. Ground wave
C. Meteor scatter
D. Sporadic E

2C-2.4 When compared to sky-wave propagation, what is the usual effective range of ground-wave propagation?

A. Much smaller
B. Much greater
C. The same
D. Dependent on the weather

2C-3.1 What type of propagation uses radio signals refracted back to earth by the ionosphere?

A. Sky wave
B. Earth-moon-earth
C. Ground wave
D. Tropospheric

2C-3.2 What is the meaning of the term *sky-wave propagation*?

A. Signals reflected from the moon
B. Signals refracted by the ionosphere
C. Signals refracted by water-dense cloud formations
D. Signals retransmitted by a repeater

2C-3.3 What does the term *skip* mean?
A. Signals are reflected from the moon
B. Signals are refracted by water-dense cloud formations
C. Signals are retransmitted by repeaters
D. Signals are refracted by the ionosphere

2C-3.4 What is the area of weak signals between the ranges of ground waves and the first hop called?
A. The skip zone
B. The hysteresis zone
C. The monitor zone
D. The transequatorial zone

2C-3.5 What is the meaning of the term *skip zone*?
A. An area covered by skip propagation
B. The area where a satellite comes close to the earth, and skips off the ionosphere
C. An area that is too far for ground-wave propagation, but too close for skip propagation
D. The area in the atmosphere that causes skip propagation

2C-3.6 What type of radio-wave propagation makes it possible for amateur stations to communicate long distances?
A. Direct-inductive propagation
B. Knife-edge diffraction
C. Ground-wave propagation
D. Sky-wave propagation

2C-4.1 How long is an average *sunspot cycle*?
A. 2 years
B. 5 years
C. 11 years
D. 17 years

2C-4.2 What is the term used to describe the long-term variation in the number of visible sunspots?
A. The 11-year cycle
B. The Solar magnetic flux cycle
C. The hysteresis count
D. The sunspot cycle

2C-5.1 What affect does the number of sunspots have on the maximum usable frequency (MUF)?
A. The more sunspots there are, the higher the MUF will be
B. The more sunspots there are, the lower the MUF will be
C. The MUF is equal to the square of the number of sunspots
D. The number of sunspots effects the lowest usable frequency (LUF) but not the MUF

2C-5.2 What affect does the number of sunspots have on the ionization level in the atmosphere?
A. The more sunspots there are, the lower the ionization level will be
B. The more sunspots there are, the higher the ionization level will be
C. The ionization level of the ionosphere is equal to the square root of the number of sunspots
D. The ionization level of the ionosphere is equal to the square of the number of sunspots

2C-6.1 Why can a VHF or UHF radio signal that is transmitted toward a mountain often be received at some distant point in a different direction?
A. You can never tell what direction a radio wave is traveling in
B. These radio signals are easily bent by the ionosphere
C. These radio signals are easily reflected by objects in their path
D. These radio signals are sometimes scattered in the ectosphere

2C-6.2 Why can the direction that a VHF or UHF radio signal is traveling be changed if there is a tall building in the way?
A. You can never tell what direction a radio wave is traveling in
B. These radio signals are easily bent by the ionosphere
C. These radio signals are easily reflected by objects in their path
D. These radio signals are sometimes scattered in the ectosphere

Subelement 2D—Amateur Radio Practice (4 Questions)

One question must be from the following:

2D-1.1 How can you prevent the use of your amateur station by unauthorized persons?
- A. Install a carrier-operated relay in the main power line
- B. Install a key-operated "ON/OFF" switch in the main power line
- C. Post a "Danger - High Voltage" sign in the station
- D. Install ac line fuses in the main power line

2D-1.2 What is the purpose of a key-operated "ON/OFF" switch in the main power line?
- A. To prevent the use of your station by unauthorized persons
- B. To provide an easy method for the FCC to put your station off the air
- C. To prevent the power company from inadvertently turning off your electricity during an emergency
- D. As a safety feature, to kill all power to the station in the event of an emergency

2D-2.1 Why should all antenna and rotator cables be grounded when an amateur station is not in use?
- A. To lock the antenna system in one position
- B. To avoid radio frequency interference
- C. To save electricity
- D. To protect the station and building from damage due to a nearby lightning strike

2D-2.2 How can an antenna system be protected from damage caused by a nearby lightning strike?
- A. Install a balun at the antenna feed point
- B. Install an RF choke in the feed line
- C. Ground all antennas when they are not in use
- D. Install a line fuse in the antenna wire

2D-2.3 How can amateur station equipment be protected from damage caused by voltage induced in the power lines by a nearby lightning strike?
- A. Use heavy insulation on the wiring
- B. Keep the equipment on constantly
- C. Disconnect the ground system
- D. Disconnect all equipment after use, either by unplugging or by using a main disconnect switch

2D-2.4 For proper protection from lightning strikes, what equipment should be grounded in an amateur station?
- A. The power supply primary
- B. All station equipment
- C. The feed line center conductors
- D. The ac power mains

2D-3.1 What is a convenient indoor grounding point for an amateur station?
- A. A metallic cold water pipe
- B. PVC plumbing
- C. A window screen
- D. A natural gas pipe

2D-3.2 To protect against electrical shock hazards, what should you connect the chassis of each piece of your equipment to?
- A. Insulated shock mounts
- B. The antenna
- C. A good ground connection
- D. A circuit breaker

2D-3.3 What type of material should a driven ground rod be made of?
- A. Ceramic or other good insulator
- B. Copper or copper-clad steel
- C. Iron or steel
- D. Fiberglass

2D-3.4 What is the shortest ground rod you should consider installing for your amateur station RF ground?
- A. 4 foot
- B. 6 foot
- C. 8 foot
- D. 10 foot

One question must be from the following:

2D-4.1 What precautions should you take when working with 1270-MHz waveguide?
- A. Make sure that the RF leakage filters are installed at both ends of the waveguide
- B. Never look into the open end of a waveguide when RF is applied
- C. Minimize the standing-wave ratio before you test the waveguide
- D. Never have both ends of the waveguide open at the same time when RF is applied

2D-4.2 What precautions should you take when you mount a UHF antenna in a permanent location?
- A. Make sure that no one can be near the antenna when you are transmitting
- B. Make sure that the RF field screens are in place
- C. Make sure that the antenna is near the ground to maximize directional effect
- D. Make sure you connect an RF leakage filter at the antenna feed point

2D-4.3 What precautions should you take before removing the shielding on a UHF power amplifier?
- A. Make sure all RF screens are in place at the antenna
- B. Make sure the feed line is properly grounded
- C. Make sure the amplifier cannot be accidentally energized
- D. Make sure that the RF leakage filters are connected

2D-4.4 Why should you use only good-quality, well-constructed coaxial cable and connectors for a UHF antenna system?
- A. To minimize RF leakage
- B. To reduce parasitic oscillations
- C. To maximize the directional characteristics of your antenna
- D. To maximize the standing-wave ratio of the antenna system

2D-4.5 Why should you be careful to position the antenna of your 220-MHz hand-held transceiver away from your head when you are transmitting?
- A. To take advantage of the directional effect
- B. To minimize RF exposure
- C. To use your body to reflect the signal, improving the directional characteristics of the antenna
- D. To minimize static discharges

2D-4.6 Which of the following types of radiation produce health risks most like the risks produced by radio frequency radiation?
A. Microwave oven radiation and ultraviolet radiation
B. Microwave oven radiation and radiation from an electric space heater
C. Radiation from Uranium or Radium and ultraviolet radiation
D. Sunlight and radiation from an electric space heater

2D-5.1 Why is there a switch that turns off the power to a high-voltage power supply if the cabinet is opened?
A. To prevent RF from escaping from the supply
B. To prevent RF from entering the supply through the open cabinet
C. To provide a way to turn the power supply on and off
D. To reduce the danger of electrical shock

2D-5.2 What purpose does a safety interlock on an amateur transmitter serve?
A. It reduces the danger that the operator will come in contact with dangerous high voltages when the cabinet is opened while the power is on
B. It prevents the transmitter from being turned on accidentally
C. It prevents RF energy from leaking out of the transmitter cabinet
D. It provides a way for the station licensee to ensure that only authorized operators can turn the transmitter on

2D-6.1 What type of safety equipment should you wear when you are working at the top of an antenna tower?
A. A grounding chain
B. A reflective vest
C. Loose clothing
D. A carefully inspected safety belt

2D-6.2 Why should you wear a safety belt when you are working at the top of an antenna tower?
A. To provide a way to safely hold your tools so they don't fall and injure someone on the ground
B. To maintain a balanced load on the tower while you are working
C. To provide a way to safely bring tools up and down the tower
D. To prevent an accidental fall

2D-6.3 For safety purposes, how high should you locate all portions of your horizontal wire antenna?
A. High enough so that a person cannot touch them from the ground
B. Higher than chest level
C. Above knee level
D. Above electrical lines

2D-6.4 What type of safety equipment should you wear when you are on the ground assisting someone who is working on an antenna tower?
A. A reflective vest
B. A safety belt
C. A grounding chain
D. A hard hat

2D-6.5 Why should you wear a hard hat when you are on the ground assisting someone who is working on an antenna tower?
A. To avoid injury from tools dropped from the tower
B. To provide an RF shield during antenna testing
C. To avoid injury if the tower should accidentally collapse
D. To avoid injury from walking into tower guy wires

One question must be from the following:

2D-7-1.1 What accessory is used to measure standing wave ratio?
A. An ohm meter
B. An ammeter
C. An SWR meter
D. A current bridge

2D-7-1.2 What instrument is used to indicate the relative impedance match between a transmitter and antenna?
A. An ammeter
B. An ohmmeter
C. A voltmeter
D. An SWR meter

2D-7-2.1 What does an SWR-meter reading of 1:1 indicate?
A. An antenna designed for use on another frequency band is probably connected
B. An optimum impedance match has been attained
C. No power is being transferred to the antenna
D. An SWR meter never indicates 1:1 unless it is defective

2D-7-2.2 What does an SWR-meter reading of less than 1.5:1 indicate?
A. An unacceptably low reading
B. An unacceptably high reading
C. An acceptable impedance match
D. An antenna gain of 1.5

2D-7-2.3 What does an SWR-meter reading of 4:1 indicate?
A. An unacceptably low reading
B. An acceptable impedance match
C. An antenna gain of 4
D. An impedance mismatch, which is not acceptable; it indicates problems with the antenna system

2D-7-2.4 What does an SWR-meter reading of 5:1 indicate?
A. The antenna will make a 10-watt signal as strong as a 50-watt signal
B. Maximum power is being delivered to the antenna
C. An unacceptable mismatch is indicated
D. A very desirable impedance match has been attained

2D-7-3.1 What kind of SWR-meter reading may indicate poor electrical contact between parts of an antenna system?
A. An erratic reading
B. An unusually low reading
C. No reading at all
D. A negative reading

2D-7-3.2 What does an unusually high SWR-meter reading indicate?

A. That the antenna is not the correct length, or that there is an open or shorted connection somewhere in the feed line
B. That the signals arriving at the antenna are unusually strong, indicating good radio conditions
C. That the transmitter is producing more power than normal, probably indicating that the final amplifier tubes or transistors are about to go bad
D. That there is an unusually large amount of solar white-noise radiation, indicating very poor radio conditions

2D-7-3.3 The SWR-meter reading at the low-frequency end of an amateur band is 2.5:1, and the SWR-meter reading at the high-frequency end of the same band is 5:1. What does this indicate about your antenna?

A. The antenna is broadbanded
B. The antenna is too long for operation on this band
C. The antenna is too short for operation on this band
D. The antenna has been optimized for operation on this band

2D-7-3.4 The SWR-meter reading at the low-frequency end of an amateur band is 5:1, and the SWR-meter reading at the high-frequency end of the same band is 2.5:1. What does this indicate about your antenna?

A. The antenna is broadbanded
B. The antenna is too long for operation on this band
C. The antenna is too short for operation on this band
D. The antenna has been optimized for operation on this band

One question must be from the following:

2D-8-1.1 What is meant by *receiver overload*?

A. Interference caused by transmitter harmonics
B. Interference caused by overcrowded band conditions
C. Interference caused by strong signals from a nearby transmitter
D. Interference caused by turning the receiver volume too high

2D-8-1.2 What is a likely indication that radio-frequency interference to a receiver is caused by front-end overload?

A. A low pass filter at the transmitter reduces interference sharply
B. The interference is independent of frequency
C. A high pass filter at the receiver reduces interference little or not at all
D. Grounding the receiver makes the problem worse

2D-8-1.3 Your neighbor reports interference to his television whenever you are transmitting from your amateur station. This interference occurs regardless of your transmitter frequency. What is likely to be the cause of the interference?

A. Inadequate transmitter harmonic suppression
B. Receiver VR tube discharge
C. Receiver overload
D. Incorrect antenna length

2D-8-1.4 What type of filter should be installed on a TV receiver as the first step in preventing RF overload from an amateur HF station transmission?

A. Low pass
B. High pass
C. Band pass
D. Notch

2D-8-2.1 What is meant by *harmonic radiation*?

A. Transmission of signals at whole number multiples of the fundamental (desired) frequency
B. Transmission of signals that include a superimposed 60-Hz hum
C. Transmission of signals caused by sympathetic vibrations from a nearby transmitter
D. Transmission of signals to produce a stimulated emission in the air to enhance skip propagation

2D-8-2.2 Why is harmonic radiation from an amateur station undesirable?

A. It will cause interference to other stations and may result in out-of-band signal radiation
B. It uses large amounts of electric power
C. It will cause sympathetic vibrations in nearby transmitters
D. It will produce stimulated emission in the air above the transmitter, thus causing aurora

2D-8-2.3 What type of interference may radiate from a multi-band antenna connected to an improperly tuned transmitter?

A. Harmonic radiation
B. Auroral distortion
C. Parasitic excitation
D. Intermodulation

2D-8-2.4 What is the purpose of shielding in a transmitter?

A. It gives the low pass filter structural stability
B. It enhances the microphonic tendencies of radiotelephone transmitters
C. It prevents unwanted RF radiation
D. It helps maintain a sufficiently high operating temperature in circuit components

2D-8-2.5 Your neighbor reports interference on one or two channels of her television when you are transmitting from your amateur station. This interference only occurs when you are operating on 15 meters. What is likely to be the cause of the interference?

A. Excessive low-pass filtering on the transmitter
B. Sporadic E de-ionization near your neighbor's TV antenna
C. TV Receiver front-end overload
D. Harmonic radiation from your transmitter

2D-8-2.6 What type of filter should be installed on an amateur transmitter as the first step in reducing harmonic radiation?

A. Key click filter
B. Low pass filter
C. High pass filter
D. CW filter

2D-8-3.1 If you are notified that your amateur station is causing television interference, what should you do first?
A. Make sure that your amateur equipment is operating properly, and that it does not cause interference to your own television
B. Immediately turn off your transmitter and contact the nearest FCC office for assistance
C. Install a high-pass filter at the transmitter output and a low-pass filter at the antenna-input terminals of the TV
D. Continue operating normally, since you have no legal obligation to reduce or eliminate the interference

2D-8-3.2 Your neighbor informs you that you are causing television interference, but you are sure your amateur equipment is operating properly and you cause no interference to your own TV. What should you do?
A. Immediately turn off your transmitter and contact the nearest FCC office for assistance
B. work with your neighbor to determine that you are actually the cause of the interference
C. Install a high-pass filter at the transmitter output and a low-pass filter at the antenna-input terminals of the TV
D. Continue operating normally, since you have no legal obligation to reduce or eliminate the interference

Subelement 2E—Electrical Principles (4 questions)

One question must be from the following:

2E-1-1.1 Your receiver dial is calibrated in megahertz and shows a signal at 1200 MHz. At what frequency would a dial calibrated in gigahertz show the signal?
A. 1,200,000 GHz
B. 12 GHz
C. 1.2 GHz
D. 0.0012 GHz

2E-1-2.1 Your receiver dial is calibrated in kilohertz and shows a signal at 7125 kHz. At what frequency would a dial calibrated in megahertz show the signal?
A. 0.007125 MHz
B. 7.125 MHz
C. 71.25 MHz
D. 7,125,000 MHz

2E-1-2.2 Your receiver dial is calibrated in gigahertz and shows a signal at 1.2 GHz. At what frequency would a dial calibrated in megahertz show the same signal?
A. 1.2 MHz
B. 12 MHz
C. 120 MHz
D. 1200 MHz

2E-1-3.1 Your receiver dial is calibrated in megahertz and shows a signal at 3.525 MHz. At what frequency would a dial calibrated in kilohertz show the signal?
A. 0.003525 kHz
B. 3525 kHz
C. 35.25 kHz
D. 3,525,000 kHz

2E-1-3.2 Your receiver dial is calibrated in kilohertz and shows a signal at 3725 kHz. At what frequency would a dial calibrated in Hertz show the same signal?
A. 3,725 Hz
B. 3.725 Hz
C. 37.25 Hz
D. 3,725,000 Hz

2E-1-4.1 How long (in meters) is an antenna that is 400 centimeters long?
A. 0.0004 meters
B. 4 meters
C. 40 meters
D. 40,000 meters

2E-1-5.1 What reading will be displayed on a meter calibrated in amperes when it is being used to measure a 3000-milliampere current?
A. 0.003 amperes
B. 0.3 amperes
C. 3 amperes
D. 3,000,000 amperes

2E-1-5.2 What reading will be displayed on a meter calibrated in volts when it is being used to measure a 3500-millivolt potential?
A. 350 volts
B. 35 volts
C. 3.5 volts
D. 0.35 volts

2E-1-6.1 How many farads is 500,000 microfarads?
A. 0.0005 farads
B. 0.5 farads
C. 500 farads
D. 500,000,000 farads

2E-1-7.1 How many microfarads is 1,000,000 picofarads?
A. 0.001 microfarads
B. 1 microfarad
C. 1,000 microfarads
D. 1,000,000,000 microfarads

One question must be from the following:

2E-2-1.1 What is the term used to describe the flow of electrons in an electric circuit?
A. Voltage
B. Resistance
C. Capacitance
D. Current

2E-2-2.1 What is the basic unit of electric current?
A. The volt
B. The watt
C. The ampere
D. The ohm

2E-3-1.1 What supplies the force that will cause electrons to flow through a circuit?
A. Electromotive force, or voltage
B. Magnetomotive force, or inductance
C. Farad force, or capacitance
D. Thermodynamic force, or entropy

2E-3-1.2 The pressure in a water pipe is comparable to what force in an electrical circuit?
A. Current
B. Resistance
C. Gravitation
D. Voltage

2E-3-1.3 An electric circuit must connect to two terminals of a voltage source. What are these two terminals called?
A. The north and south poles
B. The positive and neutral terminals
C. The positive and negative terminals
D. The entrance and exit terminals

2E-3-2.1 What is the basic unit of voltage?
A. The volt
B. The watt
C. The ampere
D. The ohm

2E-4.1 List at least three good electrical conductors.
A. Copper, gold, mica
B. Gold, silver, wood
C. Gold, silver, aluminum
D. Copper, aluminum, paper

2E-5.1 List at least four good electrical insulators.
A. Glass, air, plastic, porcelain
B. Glass, wood, copper, porcelain
C. Paper, glass, air, aluminum
D. Plastic, rubber, wood, carbon

2E-6-1.1 There is a limit to the electric current that can pass through any material. What is this current limiting called?
A. Fusing
B. Reactance
C. Saturation
D. Resistance

2E-6-1.2 What is an electrical component called that opposes electron movement through a circuit?
A. A resistor
B. A reactor
C. A fuse
D. An oersted

2E-6-2.1 What is the basic unit of resistance?
A. The volt
B. The watt
C. The ampere
D. The ohm

One question must be from the following:

2E-7.1 What electrical principle relates voltage, current and resistance in an electric circuit?
A. Ampere's Law
B. Kirchhoff's Law
C. Ohm's Law
D. Tesla's Law

2E-7.2 There is a 2-amp current through a 50-ohm resistor. What is the applied voltage?
A. 0.04 volts
B. 52 volts
C. 100 volts
D. 200 volts

2E-7.3 If 200 volts is applied to a 100-ohm resistor, what is the current through the resistor?
A. 0.5 amps
B. 2 amps
C. 50 amps
D. 20000 amps

2E-7.4 There is a 3-amp current through a resistor and we know that the applied voltage is 90 volts. What is the value of the resistor?
A. 0.03 ohms
B. 10 ohms
C. 30 ohms
D. 2700 ohms

2E-8.1 What is the term used to describe the ability to do work?
A. Voltage
B. Power
C. Inertia
D. Energy

2E-8.2 What is converted to heat and light in an electric light bulb?
A. Electrical energy
B. Electrical voltage
C. Electrical power
D. Electrical current

2E-9-1.1 What term is used to describe the rate of energy consumption?
A. Energy
B. Current
C. Power
D. Voltage

2E-9-1.2 You have two lamps with different wattage light bulbs in them. How can you determine which bulb uses electrical energy faster?
A. The bulb that operates from the higher voltage will consume energy faster
B. The physically larger bulb will consume energy faster
C. The bulb with the higher wattage rating will consume energy faster
D. The bulb with the lower wattage rating will consume energy faster

2E-9-2.1 What is the basic unit of electrical power?
A. Ohm
B. Watt
C. Volt
D. Ampere

2E-10.1 What is the term for an electrical circuit in which there can be no current?
A. A closed circuit
B. A short circuit
C. An open circuit
D. A hyper circuit

2E-11.1 What is the term for a failure in an electrical circuit that causes excessively high current?
A. An open circuit
B. A dead circuit
C. A closed circuit
D. A short circuit

One question must be from the following:

2E-12-1.1 What is the term used to describe a current that flows only in one direction?
A. Alternating current
B. Direct current
C. Periodic current
D. Pulsating current

2E-12-2.1 What is the term used to describe a current that flows first in one direction, then in the opposite direction, over and over?
A. Alternating current
B. Direct current
C. Negative current
D. Positive current

2E-12-3.1 What is the term for the number of complete cycles of an alternating waveform that occur in one second?
A. Pulse repetition rate
B. Hertz
C. Frequency per wavelength
D. Frequency

2E-12-3.2 A certain ac signal makes 2000 complete cycles in one second. What property of the signal does this number describe?
A. The frequency of the signal
B. The pulse repetition rate of the signal
C. The wavelength of the signal
D. The hertz per second of the signal

2E-12-3.3 What is the basic unit of frequency?
A. The hertz
B. The cycle
C. The kilohertz
D. The megahertz

2E-12-4.1 What range of frequencies are usually called *audio frequencies*?
A. 0 to 20 Hz
B. 20 to 20,000 Hz
C. 200 to 200,000 Hz
D. 10,000 to 30,000 Hz

2E-12-4.2 A signal at 725 Hz is in what frequency range?
A. Audio frequency
B. Intermediate frequency
C. Microwave frequency
D. Radio frequency

2E-12-4.3 Why do we call signals in the range 20 Hz to 20,000 Hz *audio frequencies*?
A. Because the human ear rejects signals in this frequency range
B. Because the human ear responds to sounds in this frequency range
C. Because frequencies in this range are too low for a radio to detect
D. Because a radio converts signals in this range directly to sounds the human ear responds to

2E-12-5.1 Signals above what frequency are usually called *radio-frequency* signals?
A. 20 Hz
B. 2000 Hz
C. 20,000 Hz
D. 1,000,000 Hz

2E-12-5.2 A signal at 7125 kHz is in what frequency range?
A. Audio frequency
B. Radio frequency
C. Hyper-frequency
D. Super-high frequency

2E-13.1 What is the term for the distance an ac signal travels during one complete cycle?
A. Wave velocity
B. Velocity factor
C. Wavelength
D. Wavelength per meter

2E-13.2 In the time it takes a certain radio signal to pass your antenna, the leading edge of the wave travels 12 meters. What property of the signal does this number refer to?
A. The signal frequency
B. The wave velocity
C. The velocity factor
D. The signal wavelength

Subelement 2F—Circuit Components (2 Questions)

One question must be from the following:

2F-1.1 **What is the symbol used on schematic diagrams to represent a resistor?**

A. 

B.

C. 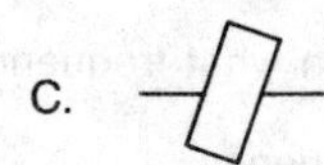

D.

2F-1.2 **What is the symbol used on schematic diagrams to represent a variable resistor or potentiometer?**

A. 

B.

C. 

D.

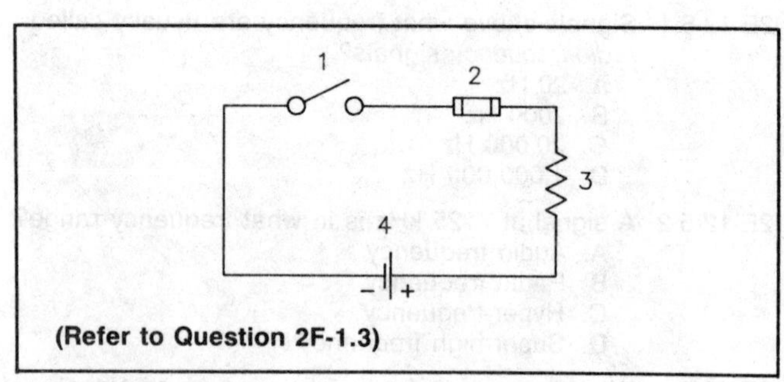

(Refer to Question 2F-1.3)

Figure 2F-1

2F-1.3 **In diagram 2F-1, which component is a resistor?**
A. 1
B. 2
C. 3
D. 4

2F-2.1 **What is the symbol used on schematic diagrams to represent a single-pole, single-throw switch?**

A. 

B.

C. 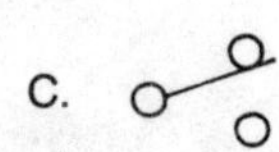

D.

2F-2.2 **What is the symbol used on schematic diagrams to represent a single-pole, double-throw switch?**

A.

B.

C.

D.

2F-2.3 **What is the symbol used on schematic diagrams to represent a double-pole, double-throw switch?**

A.

B.

C.

D.

2F-2.4 **What is the symbol used on schematic diagrams to represent a single-pole 5-position switch?**

A.

B. 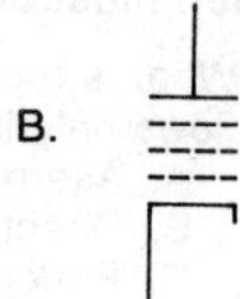

C.

D. 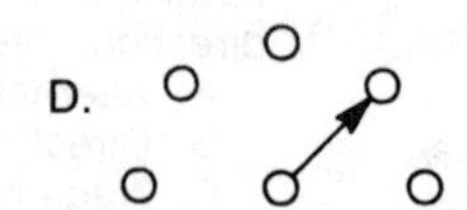

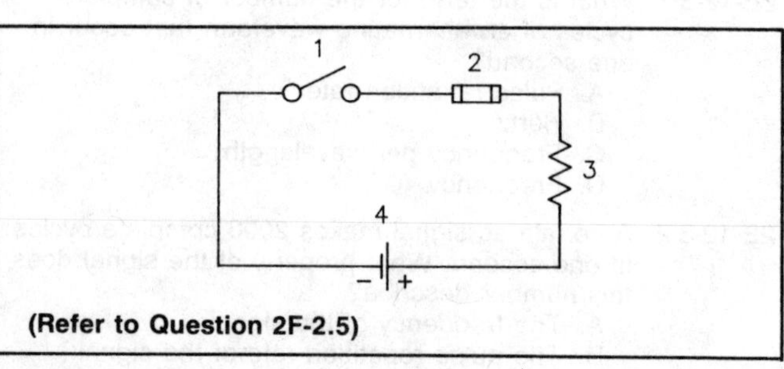

(Refer to Question 2F-2.5)

Figure 2F-2

2F-2.5 **In diagram 2F-2, which component is a switch?**
A. 1
B. 2
C. 3
D. 4

2F-3.1 What is the symbol used on schematic diagrams to represent a fuse?

A. 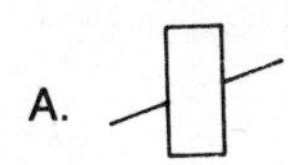

B.

C.

D.

2F-4.1 What is the symbol used on schematic diagrams to represent a single-cell battery?

A. 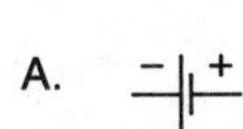

B.

C.

D.

2F-4.2 What is the symbol used on schematic diagrams to represent a multiple-cell battery?

A. 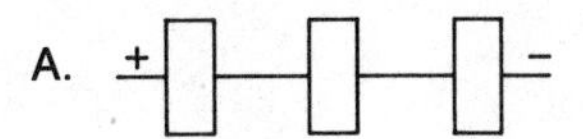

B.

C.

D.

One question must be from the following:

2F-5.1 What is the symbol normally used to represent an earth-ground connection on schematic diagrams?

A. 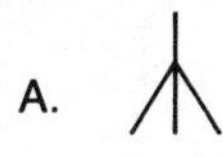

B.

C.

D.

2F-5.2 What is the symbol normally used to represent a chassis-ground connection on schematic diagrams?

A.

B.

C.

D.

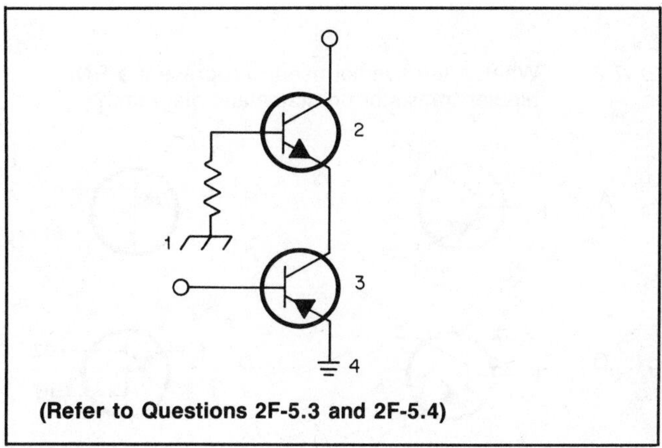

(Refer to Questions 2F-5.3 and 2F-5.4)

Figure 2F-5

2F-5.3 In diagram 2F-5, which symbol represents a chassis ground connection?
A. 1
B. 2
C. 3
D. 4

2F-5.4 In diagram 2F-5, which symbol represents an earth ground connection?
A. 1
B. 2
C. 3
D. 4

2F-6.1 What is the symbol used to represent an antenna on schematic diagrams?

A.

B.

C.

D.

2F-7.1 What is the symbol used to represent an NPN bipolar transistor on schematic diagrams?

A.
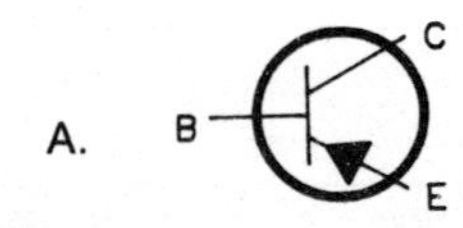

B.
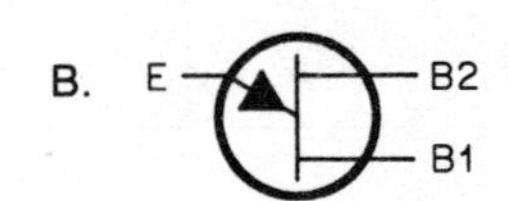

C. B C E

D.
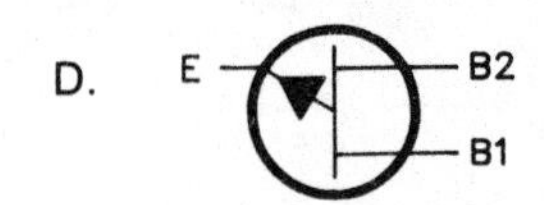

2F-7.2 What is the symbol used to represent a PNP bipolar transistor on schematic diagrams?

A.
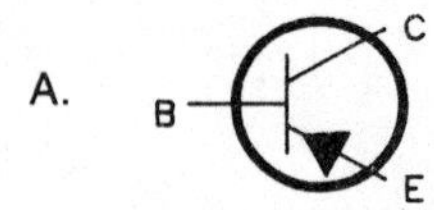

B.

C.
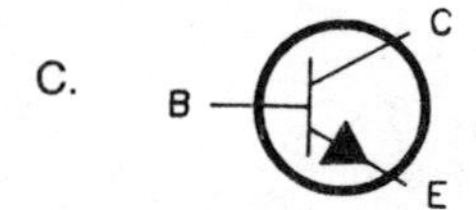

D.
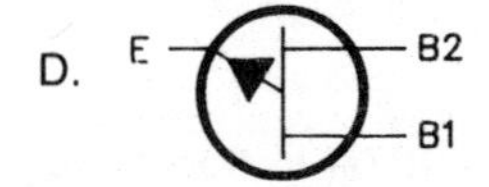

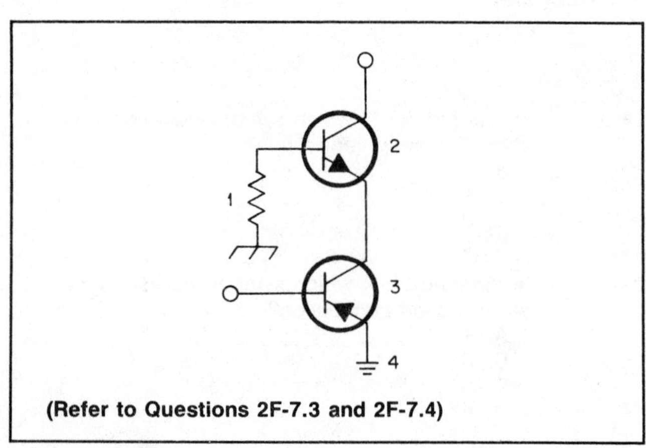

(Refer to Questions 2F-7.3 and 2F-7.4)

Figure 2F-7

2F-7.3 In diagram 2F-7, which symbol represents a PNP bipolar transistor?

A. 1
B. 2
C. 3
D. 4

2F-7.4 In diagram 2F-7, which symbol represents an NPN bipolar transistor?

A. 1
B. 2
C. 3
D. 4

2F-8.1 What is the symbol used to represent a triode vacuum tube on schematic diagrams?

A. PLATE GRID CATHODE

B.
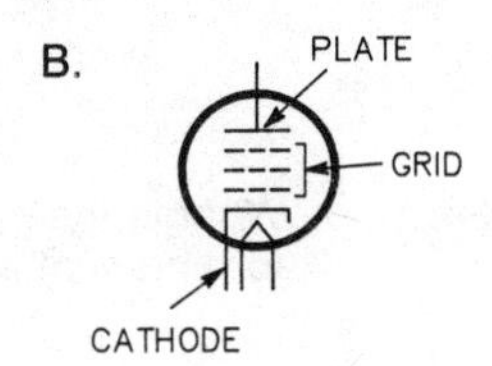

C.
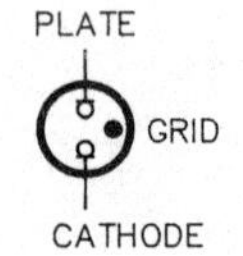

D.
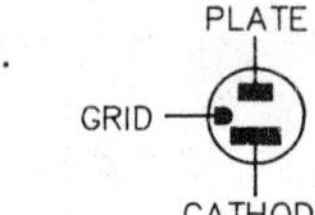

Subelement 2G—Practical Circuits (2 Questions)

One question must be from the following:

2G-1-1.1 What is the unlabeled block (?) in this diagram?

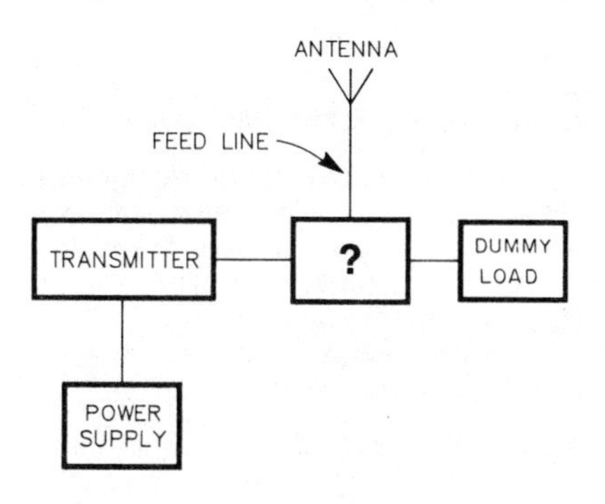

A. A terminal-node controller
B. An antenna switch
C. A telegraph key
D. A TR switch

2G-1-1.2 What is the unlabeled block (?) in this diagram?

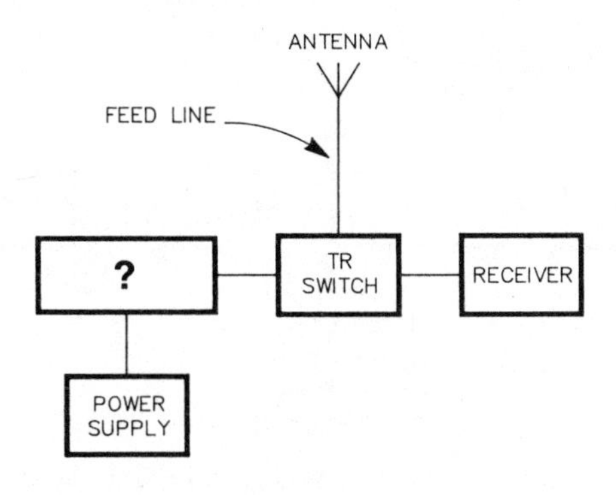

A. A microphone
B. A receiver
C. A transmitter
D. An SWR meter

2G-1-1.3 What is the unlabeled block (?) in this diagram?

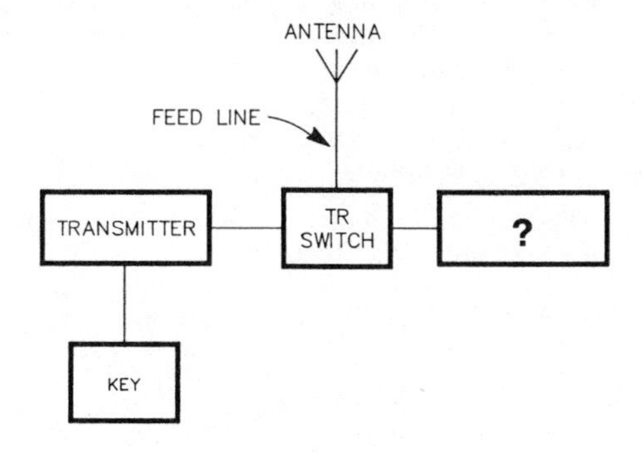

A. A key click filter
B. An antenna tuner
C. A power supply
D. A receiver

2G-1-1.4 What is the unlabeled block (?) in this diagram?

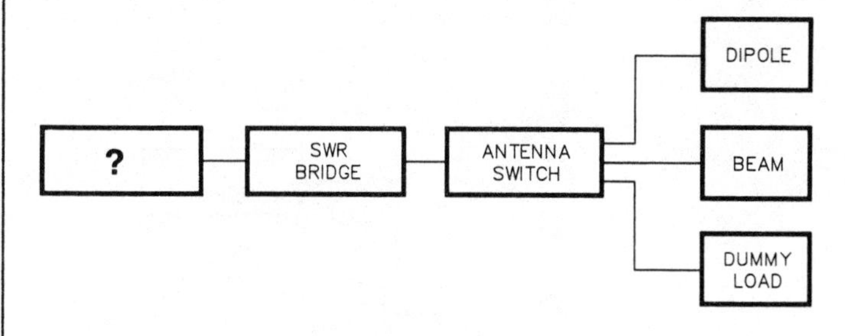

A. A transceiver
B. A TR switch
C. An antenna tuner
D. A modem

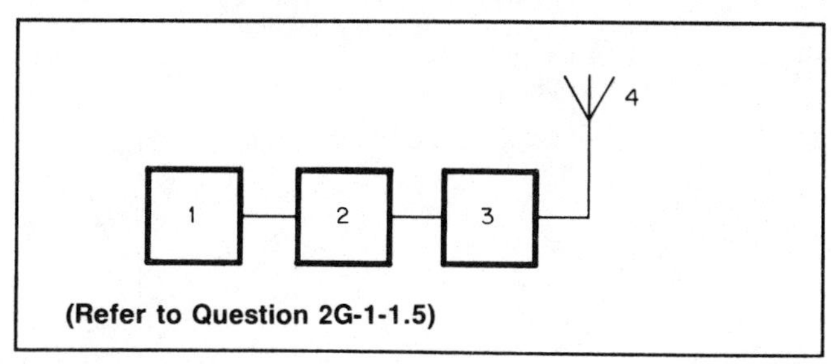

(Refer to Question 2G-1-1.5)

Figure 2G-1

2G-1-1.5 In block diagram 2G-1, which symbol represents an antenna?
A. 1
B. 2
C. 3
D. 4

2G-1-2.1 What is the unlabeled block (?) in this diagram?

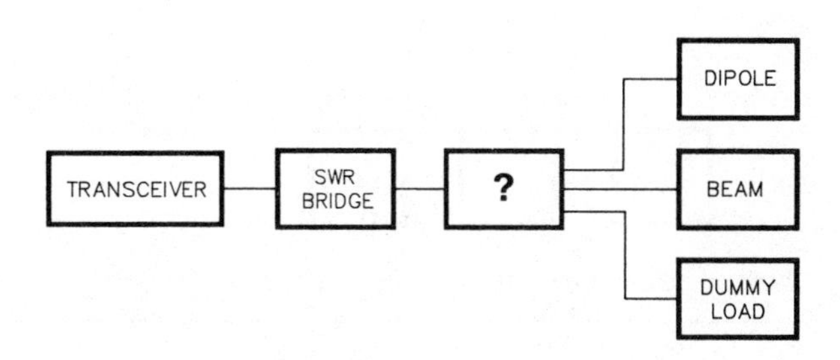

A. A pi network
B. An antenna switch
C. A key click filter
D. A mixer

2G-1-2.2 What is the unlabeled block (?) in this diagram?

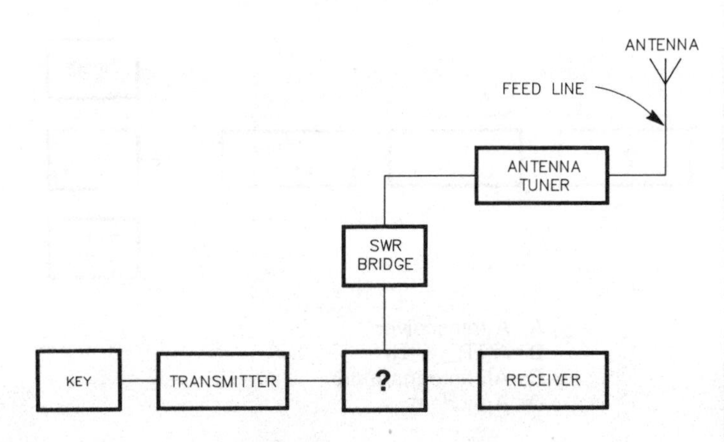

A. A TR switch
B. A variable frequency oscillator
C. A linear amplifier
D. A microphone

2G-1-2.3 What is the unlabeled block (?) in this diagram?

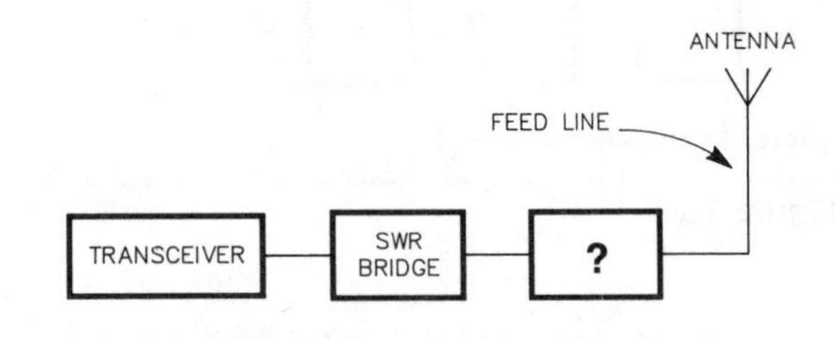

A. An antenna switch
B. An impedance-matching network
C. A key click filter
D. A terminal-node controller

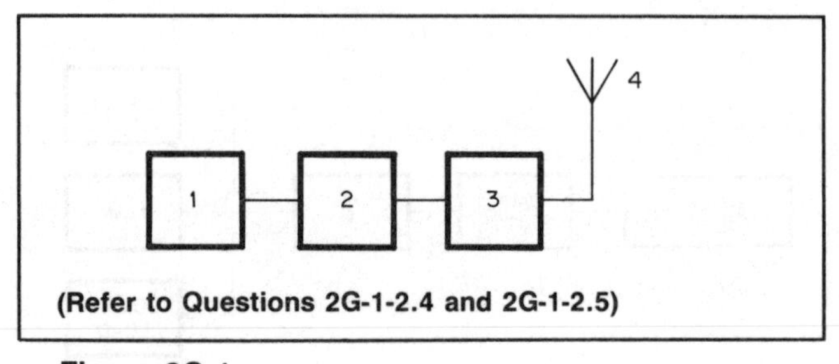

(Refer to Questions 2G-1-2.4 and 2G-1-2.5)

Figure 2G-1

2G-1-2.4 In block diagram 2G-1, if component 1 is a transceiver and component 2 is an SWR meter, what is component 3?
A. A power supply
B. A receiver
C. A microphone
D. An impedance-matching device

2G-1-2.5 In block diagram 2G-1, if component 2 is an SWR meter and component 3 is an impedance-matching device, what is component 1?
A. A power supply
B. An antenna
C. An antenna switch
D. A transceiver

One question must be from the following:

2G-2.1 In an Amateur Radio station designed for Morse radiotelegraph operation, what station accessory will you need to go with your transmitter?
A. A terminal-node controller
B. A telegraph key
C. An SWR meter
D. An antenna switch

2G-2.2 What is the unlabeled block (?) in this diagram of a Morse telegraphy station?

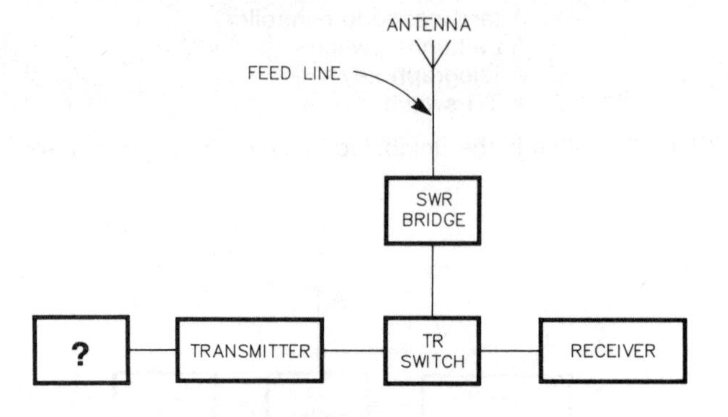

A. A sidetone oscillator
B. A microphone
C. A telegraph key
D. A DTMF keypad

2G-2.3 What station accessory do many amateurs use to help form good Morse code characters?
A. A sidetone oscillator
B. A key-click filter
C. An electronic keyer
D. A DTMF keypad

2G-3.1 In an Amateur Radio station designed for radiotelephone operation, what station accessory will you need to go with your transmitter?
A. A splatter filter
B. A terminal-voice controller
C. A receiver audio filter
D. A microphone

2G-3.2 What is the unlabeled block (?) in this diagram of a radiotelephone station?

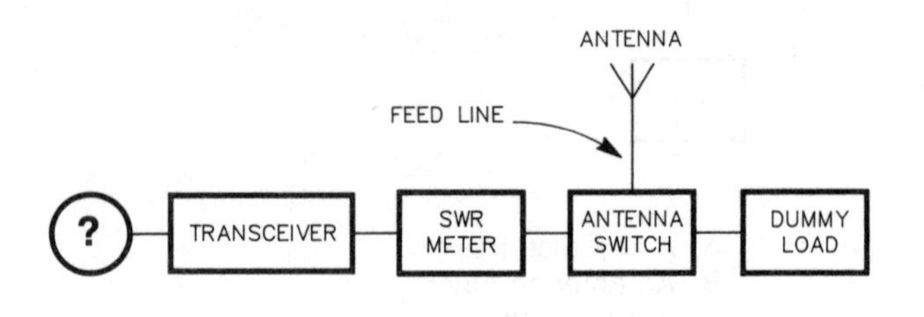

A. A splatter filter
B. A terminal-voice controller
C. A receiver audio filter
D. A microphone

2G-4.1 In an Amateur Radio station designed for radioteletype operation, what station accessories will you need to go with your transmitter?
A. A modem and a teleprinter or computer system
B. A computer, a printer and a RTTY refresh unit
C. A terminal-node controller
D. A modem, a monitor and a DTMF keypad

2G-4.2 What is the unlabeled block (?) in this diagram?

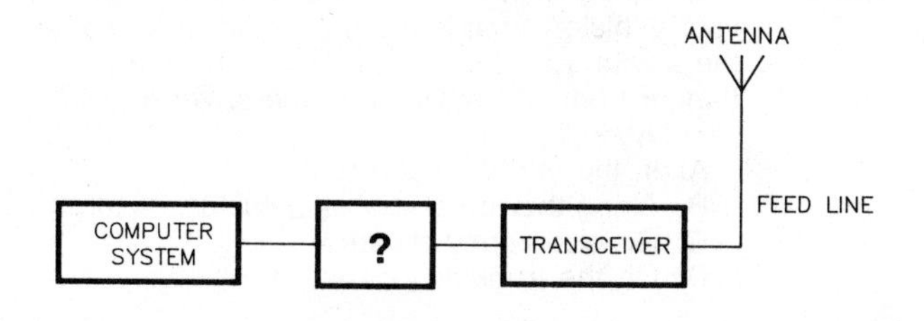

A. An RS-232 interface
B. SWR bridge
C. Modem
D. Terminal-network controller

2G-5.1 In a packet-radio station, what device connects between the radio transceiver and the computer terminal?
A. A terminal-node controller
B. An RS-232 interface
C. A terminal refresh unit
D. A tactical network control system

2G-5.2 What is the unlabeled block (?) in this diagram of a packet-radio station?

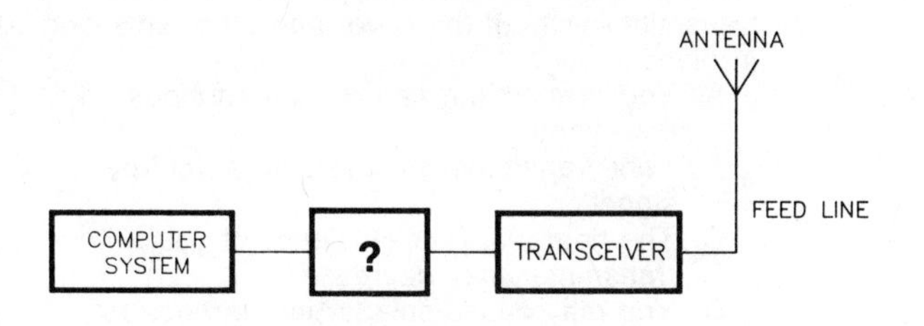

A. A terminal-node controller
B. An RS-232 interface
C. A terminal refresh unit
D. A tactical network control system

2G-5.3 Where does a terminal-node controller connect in an amateur packet-radio station?
A. Between the antenna and the radio
B. Between the computer and the monitor
C. Between the computer or terminal and the radio
D. Between the keyboard and the computer

Subelement 2H—Signals and Emissions (2 Questions)

One question must be from the following:

2H-1-1.1 What keying method is used to transmit A1A radiotelegraph messages?
A. Frequency-shift keying of a radio-frequency signal
B. On/off keying of a radio-frequency signal
C. Audio-frequency-shift keying of an oscillator tone
D. On/off keying of an audio-frequency signal

2H-1-1.2 What emission designator describes the on/off keying of a radio-frequency signal to produce Morse code messages?
A. F1B
B. F2B
C. A1A
D. J3E

2H-1-2.1 What emission designator describes the use of frequency-shift keying to transmit radioteletype messages?
A. F1B
B. F2B
C. A1A
D. J3E

2H-1-2.2 What keying method is used to transmit F1B radioteletype messages?
A. Frequency-shift keying of a radio-frequency signal
B. On/off keying of a radio-frequency signal
C. Audio-frequency-shift keying of an oscillator tone
D. On/off keying of an audio-frequency signal

2H-1-3.1 What emission designator describes frequency-modulated voice transmissions?
A. F3E
B. F2B
C. A1A
D. J3E

2H-1-4.1 What emission designator describes single-sideband suppressed-carrier (SSB) voice transmissions?
A. F3E
B. F2B
C. A1A
D. J3E

2H-2.1 What does the term *key click* mean?
A. The mechanical noise caused by closing a straight key too hard
B. The clicking noise from an excessively square CW keyed waveform
C. The sound produced in a receiver from a CW signal faster than 20 WPM
D. The sound of a CW signal being copied on an AM receiver

2H-2.2 How can key clicks be eliminated?
A. By reducing your keying speed to less than 20 WPM
B. By increasing power to the maximum allowable level
C. By using a power supply with better regulation
D. By using a key-click filter

2H-3.1 What does the term *chirp* mean?
A. A distortion in the receiver audio circuits
B. A high-pitched audio tone transmitted with a CW signal
C. A slight shift in oscillator frequency each time a CW transmitter is keyed
D. A slow change in transmitter frequency as the circuit warms up

2H-3.2 What can be done to the power supply of a CW transmitter to avoid chirp?
A. Resonate the power supply filters
B. Regulate the power supply output voltages
C. Use a buffer amplifier between the transmitter output and the feed line
D. Hold the power supply current to a fixed value

2H-4.1 What is a common cause of superimposed hum?
A. Using a nonresonant random-wire antenna
B. Sympathetic vibrations from a nearby transmitter
C. Improper neutralization of the transmitter output stage
D. A defective filter capacitor in the power supply

2H-4.2 What type of problem can a bad power-supply filter capacitor cause in a transmitter or receiver?
A. Sympathetic vibrations in nearby receivers
B. A superimposed hum or buzzing sound
C. Extreme changes in antenna resonance
D. Imbalance in the mixers

One question must be from the following:

2H-5.1 What is the 4th harmonic of a 7160-kHz signal?
A. 28,640 kHz
B. 35,800 kHz
C. 28,160 kHz
D. 1790 kHz

2H-5.2 You receive an FCC Notice of Violation stating that your station was heard on 21,375 kHz. At the time listed on the notice, you were operating on 7125 kHz. What is a possible cause of this violation?
A. Your transmitter has a defective power-supply filter capacitor
B. Your CW keying speed was excessively fast
C. Your transmitter was radiating excess harmonic signals
D. Your transmitter has a defective power-supply filter choke

2H-6.1 What may happen to body tissues that are exposed to large amounts of UHF or microwave RF energy?
A. The tissue may be damaged because of the heat produced
B. The tissue may suddenly be frozen
C. The tissue may be immediately destroyed because of the Maxwell Effect
D. The tissue may become less resistant to cosmic radiation

2H-6.2 What precaution should you take before working near a high-gain UHF or microwave antenna (such as a parabolic, or dish antenna)?
A. Be certain the antenna is FCC type accepted
B. Be certain the antenna and transmitter are properly grounded
C. Be certain the transmitter cannot be operated
D. Be certain the antenna safety interlocks are in place

2H-6.3 You are installing a VHF or UHF mobile radio in your vehicle. What is the best location to mount the antenna on the vehicle to minimize any danger from RF exposure to the driver or passengers?
A. In the middle of the roof
B. Along the top of the windshield
C. On either front fender
D. On the trunk lid

2H-7.1 You discover that your tube-type transmitter power amplifier is radiating spurious emissions. What is the most likely cause of this problem?
A. Excessively fast keying speed
B. Undermodulation
C. Improper neutralization
D. Tank-circuit current dip at resonance

2H-7.2 Your transmitter radiates signals outside the amateur band where you are transmitting. What term describes this radiation?
A. Off-frequency emissions
B. Transmitter chirp
C. Incidental radiation
D. Spurious emissions

2H-7.3 What problem can occur if you operate your transmitter without the cover and other shielding in place?
A. Your transmitter can radiate spurious emissions
B. Your transmitter may radiate a "chirpy" signal
C. The final amplifier efficiency of your transmitter may decrease
D. You may cause splatter interference to other stations operating on nearby frequencies

2H-7.4 What type of interference will you cause if you operate your SSB transmitter with the microphone gain adjusted too high?
A. You may cause digital interference to computer equipment in your neighborhood
B. You may cause splatter interference to other stations operating on nearby frequencies
C. You may cause atmospheric interference in the air around your antenna
D. You may cause processor interference to the microprocessor in your rig

2H-7.5 What may happen if you adjust the microphone gain or deviation control on your FM transmitter too high?

A. You may cause digital interference to computer equipment in your neighborhood
B. You may cause interference to other stations operating on nearby frequencies
C. You may cause atmospheric interference in the air around your antenna
D. You may cause processor interference to the microprocessor in your rig

2H-7.6 What type of interference can excessive amounts of speech processing in your SSB transmitter cause?

A. You may cause digital interference to computer equipment in your neighborhood
B. You may cause splatter interference to other stations operating on nearby frequencies
C. You may cause atmospheric interference in the air around your antenna
D. You may cause processor interference to the microprocessor in your rig

Subelement 2I—Antennas and Feed Lines (3 Questions)

One question must be from the following:

2I-1.1 What is the approximate length (in feet) of a half-wavelength dipole antenna for 3725 kHz?

A. 126 ft
B. 81 ft
C. 63 ft
D. 40 ft

2I-1.2 What is the approximate length (in feet) of a half-wavelength dipole antenna for 7125 kHz?

A. 84 ft
B. 42 ft
C. 33 ft
D. 66 ft

2I-1.3 What is the approximate length (in feet) of a half-wavelength dipole antenna for 21,125 kHz?

A. 44 ft
B. 28 ft
C. 22 ft
D. 14 ft

2I-1.4 What is the approximate length (in feet) of a half-wavelength dipole antenna for 28,150 kHz?

A. 22 ft
B. 11 ft
C. 17 ft
D. 34 ft

2I-1.5 How is the approximate length (in feet) of a half-wavelength dipole antenna calculated?

A. By substituting the desired operating frequency for f in the formula:

$$\frac{150}{f \text{ (in MHz)}}$$

B. By substituting the desired operating frequency for f in the formula:

$$\frac{234}{f \text{ (in MHz)}}$$

C. By substituting the desired operating frequency for f in the formula:

$$\frac{300}{f \text{ (in MHz)}}$$

D. By substituting the desired operating frequency for f in the formula:

$$\frac{468}{f \text{ (in MHz)}}$$

2I-2.1 What is the approximate length (in feet) of a quarter-wavelength vertical antenna for 3725 kHz?

A. 20 ft
B. 32 ft
C. 40 ft
D. 63 ft

2I-2.2 What is the approximate length (in feet) of a quarter-wavelength vertical antenna for 7125 kHz?

A. 11 ft
B. 16 ft
C. 21 ft
D. 33 ft

2I-2.3 What is the approximate length (in feet) of a quarter-wavelength vertical antenna for 21,125 kHz?
A. 7 ft
B. 11 ft
C. 14 ft
D. 22 ft

2I-2.4 What is the approximate length (in feet) of a quarter-wavelength vertical antenna for 28,150 kHz?
A. 5 ft
B. 8 ft
C. 11 ft
D. 17 ft

2I-2.5 When a vertical antenna is lengthened, what happens to its resonant frequency?
A. It decreases
B. It increases
C. It stays the same
D. It doubles

2I-3.1 Why do many amateurs use a 5/8-wavelength vertical antenna rather than a 1/4-wavelength vertical antenna for their VHF or UHF mobile stations?
A. A 5/8-wavelength antenna can handle more power than a 1/4-wavelength antenna
B. A 5/8-wavelength antenna has more gain than a 1/4-wavelength antenna
C. A 5/8-wavelength antenna exhibits less corona loss than a 1/4-wavelength antenna
D. A 5/8-wavelength antenna looks more like a CB antenna, so it does not attract as much attention as a 1/4-wavelength antenna

2I-3.2 What type of radiation pattern is produced by a 5/8-wavelength vertical antenna?
A. A pattern with most of the transmitted signal concentrated in two opposite directions
B. A pattern with the transmitted signal going equally in all compass directions, with most of the radiation going high above the horizon
C. A pattern with the transmitted signal going equally in all compass directions, with most of the radiation going close to the horizon
D. A pattern with more of the transmitted signal concentrated in one direction than in other directions

One question must be from the following:

2I-4-1.1 What type of antenna produces a radiation pattern with more of the transmitted signal concentrated in a particular direction than in other directions?
A. A dipole antenna
B. A vertical antenna
C. An isotropic antenna
D. A beam antenna

2I-4-1.2 What type of radiation pattern is produced by a Yagi antenna?
A. A pattern with the transmitted signal spread out equally in all compass directions
B. A pattern with more of the transmitted signal concentrated in one direction than in other directions
C. A pattern with most of the transmitted signal concentrated in two opposite directions
D. A pattern with most of the transmitted signal concentrated at high radiation angles

2I-4-1.3 Approximately how long (in wavelengths) is the driven element of a Yagi antenna?
A. 1/4 wavelength
B. 1/3 wavelength
C. 1/2 wavelength
D. 1 wavelength

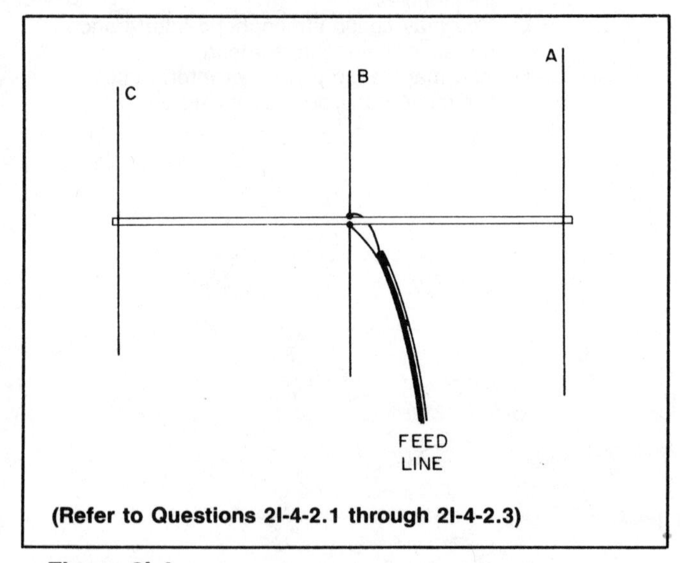

(Refer to Questions 2I-4-2.1 through 2I-4-2.3)

Figure 2I-4

2I-4-2.1 On the Yagi antenna shown in Figure 2I-4, what is the name of section B?
A. Director
B. Reflector
C. Boom
D. Driven element

2I-4-2.2 On the Yagi antenna shown in Figure 2I-4, what is the name of section C?
A. Director
B. Reflector
C. Boom
D. Driven element

2I-4-2.3 On the Yagi antenna shown in Figure 2I-4, what is the name of section A?
A. Director
B. Reflector
C. Boom
D. Driven element

2I-4-2.4 What are the names of the elements in a 3-element Yagi antenna?
A. Reflector, driven element and director
B. Boom, mast and reflector
C. Reflector, base and radiator
D. Driven element, trap and feed line

2I-5.1 How should the antenna on a hand-held transceiver be positioned while you are transmitting?
A. Away from your head and away from others standing nearby
B. Pointed in the general direction of the repeater or other station you are transmitting to
C. Pointed in a general direction 90 degrees away from the repeater or other station you are transmitting to
D. With the top of the antenna angled down slightly to take the most advantage of ground reflections

2I-5.2 Why should you always locate your antennas so that no one can come in contact with them while you are transmitting?
A. Such contact can detune the antenna, causing television interference
B. To prevent RF burns and excessive exposure to RF energy
C. The antenna is more likely to radiate harmonics when it is touched
D. Such contact may reflect the transmitted signal back to the transmitter, damaging the final amplifier

2I-5.3 You are going to purchase a new antenna for your VHF or UHF hand-held radio. Which type of antenna is the best choice to produce a radiation pattern that will be least hazardous to your face and eyes?
A. A 1/8-wavelength whip
B. A 7/8-wavelength whip
C. A 1/2-wavelength whip
D. A short, helically wound, flexible antenna

One question must be from the following:

2I-6.1 What is a *coaxial cable*?
A. Two parallel conductors encased along the edges of a flat plastic ribbon
B. Two parallel conductors held at a fixed distance from each other by insulating rods
C. Two conductors twisted around each other in a double spiral
D. A center conductor encased in insulating material which is covered by a conducting sleeve or shield

2I-6.2 What kind of antenna feed line is constructed of a center conductor encased in insulation which is then covered by an outer conducting shield and weatherproof jacket?
A. Twin lead
B. Coaxial cable
C. Open-wire feed line
D. Wave guide

2I-6.3 What are some advantages of using coaxial cable as an antenna feed line?
A. It is easy to make at home, and it has a characteristic impedance in the range of most common amateur antennas
B. It is weatherproof, and it has a characteristic impedance in the range of most common amateur antennas
C. It can be operated at a higher SWR than twin lead, and it is weatherproof
D. It is unaffected by nearby metallic objects, and has a characteristic impedance that is higher than twin lead

2I-6.4 What commonly-available antenna feed line can be buried directly in the ground for some distance without adverse effects?
A. Twin lead
B. Coaxial cable
C. Parallel conductor
D. Twisted pair

2I-6.5 When an antenna feed line must be located near grounded metal objects, which commonly-available feed line should be used?
A. Twisted pair
B. Twin lead
C. Coaxial cable
D. Ladder-line

2I-7.1 What is parallel-conductor feed line?
A. Two conductors twisted around each other in a double spiral
B. Two parallel conductors held a uniform distance apart by insulating material
C. A conductor encased in insulating material which is then covered by a conducting shield and a weatherproof jacket
D. A metallic pipe whose diameter is equal to or slightly greater than the wavelength of the signal being carried

2I-7.2 How can *TV-type twin lead* be used as a feed line?
A. By carefully running the feed line parallel to a metal post to ensure self resonance
B. TV-type twin lead cannot be used in an Amateur Radio station
C. By installing an impedance-matching network between the transmitter and feed line
D. By using a high-power amplifier and installing a power attenuator between the transmitter and feed line

2I-7.3 What are some advantages of using parallel-conductor feed line?
A. It has a lower characteristic impedance than coaxial cable, and will operate at a higher SWR than coaxial cable
B. It will operate at a higher SWR than coaxial cable, and it is unaffected by nearby metal objects
C. It has a lower characteristic impedance than coaxial cable, and has less loss than coaxial cable
D. It will operate at higher SWR than coaxial cable and it has less loss than coaxial cable

2I-7.4 What are some disadvantages of using parallel-conductor feed line?
A. It is affected by nearby metallic objects, and it has a characteristic impedance that is too high for direct connection to most amateur transmitters
B. It is more difficult to make at home than coaxial cable and it cannot be operated at a high SWR
C. It is affected by nearby metallic objects, and it cannot handle the power output of a typical amateur transmitter
D. It has a characteristic impedance that is too high for direct connection to most amateur transmitters, and it will operate at a high SWR

2I-7.5 What kind of antenna feed line is constructed of two conductors maintained a uniform distance apart by insulated spreaders?
A. Coaxial cable
B. Ladder-line open conductor line
C. Twin lead in a plastic ribbon
D. Twisted pair

2I-8.1 A certain antenna has a feed-point impedance of 35 ohms. You want to use a 50-ohm-impedance coaxial cable to feed this antenna. What type of device will you need to connect between the antenna and the feed line?
A. A balun
B. An SWR bridge
C. An impedance-matching device
D. A low-pass filter

2I-8.2 A certain antenna system has an impedance of 1000 ohms on one band. What must you use to connect this antenna system to the 50-ohm output on your transmitter?
A. A balun
B. An SWR bridge
C. An impedance-matching device
D. A low-pass filter

2I-9.1 The word *balun* is a contraction for what phrase?
A. Balanced-antenna-lobe use network
B. Broadband-amplifier linearly unregulated
C. Balanced unmodulator
D. Balanced to unbalanced

2I-9.2 Where would you install a balun if you wanted to feed your dipole antenna with 450-ohm parallel-conductor feed line?
A. At the transmitter end of the feed line
B. At the antenna feed point
C. In only one conductor of the feed line
D. From one conductor of the feed line to
D. ground

2I-9.3 Where might you install a balun if you wanted to feed your dipole antenna with 50-ohm coaxial cable?
A. You might install a balun at the antenna feedpoint
B. You might install a balun at the transmitter output
C. You might install a balun 1/2 wavelength from the transmitter
D. You might install baluns in the middle of each side of the dipole

2I-10-1.1 A four-element Yagi antenna is mounted with its elements parallel to the ground. A signal produced by this antenna will have what type of polarization?
A. Broadside polarization
B. Circular polarization
C. Horizontal polarization
D. Vertical polarization

2I-11-1.1 A four-element Yagi antenna is mounted with its elements perpendicular to the ground. A signal produced by this antenna will have what type of polarization?
A. Broadside polarization
B. Circular polarization
C. Horizontal polarization
D. Vertical polarization

ELEMENT 2 ANSWER KEY

SUBELEMENT 2A

Numbers in this section refer to the new FCC Rules, Part 97. These new rewritten FCC Rules went into effect September 1, 1989. The VECs have released a supplement to update the questions affected by the new Rules. The supplement questions, which take effect July 1, 1990, are printed at the end of this answer key.

Question	Answer	Reference
2A-1.1	A	{97.1}
2A-1.2	D	{97.1}
2A-1.3	D	{97.1}
2A-1.4	B	{97.1}
2A-2.1	C	{97.3(a)(4)}
2A-2.2	A	{97.3(a)(4)}
2A-3.1	A	{97.3(a)(4)}
2A-3.2	D	{97.3(a)(4)}
2A-4.1	B	{97.3(a)(1)}
2A-4.2	D	{97.3(a)(1)}
2A-5.1	C	{97.5(d)(1)}
2A-5.2	B	{97.5(d)(1)}
2A-6.1	B	{97.5(a), (d)(1)}
2A-6.2	C	{97.5(a), (d)(1)}
2A-7.1	D	{97.3(a)(5)}
2A-8.1	A	{97.3(a)(11)}
2A-8.2	D	{97.3(a)(11)}
2A-9.1	D	{97.9(a)}
2A-9.2	A	{97.9(a)}
2A-9.3	B	{97.9(a)}
2A-10.1	B	{97.301(e)}
2A-10.2	C	{97.301(e)}
2A-10.3	A	{97.301(e)}
2A-10.4	C	{97.301(e)}
2A-10.5	B	{97.301(f)}
2A-10.6	C	{97.301(f)}
2A-10.7	A	{97.301(e)}
2A-10.8	B	{97.301(e)}
2A-10.9	C	{97.301(e)}
2A-10.10	D	{97.301(e)}
2A-11.1	A	{97.5(d)(1)}
2A-11.2	B	{97.5(d)(1)}
2A-12.1	C	{97.501(e)}
2A-12.2	A	{97.503(a)(1)}
2A-12.3	D	{97.503(b)(1)}
2A-13.1	A	{97.5(d)(1)}
2A-14.1	B	{97.21}
2A-15.1	C	{97.17(f)}
2A-15.2	D	{97.17(f)}
2A-15.3	A	{97.17(f)}
2A-15.4	B	{97.17(f)}
2A-15.5	D	{97.17(f)}
2A-16.1	A	{97.23(a)}
2A-17.1	A	{97.3(c)}
2A-17.2	A	{97.305(c); 97.307(f)(9)}
2A-17.3	A	{97.305(c); 97.307(f)(9)}
2A-17.4	A	{97.305(c); 97.307(f)(9)}
2A-17.5	D	{97.305(c); 97.307(f)(9)}
2A-17.6	D	{97.305(c); 97.307(f)9)}
2A-17.7	D	{97.305(c); 97.307(f)(9)}
2A-17.8	C	{97.305(c)}
2A-17.9	C	{97.305(c)}
2A-17.10	D	{97.305(c)}
2A-17.11	D	{97.305(c)}
2A-17.12	D	{97.301(e); 97.305(c)}
2A-17.13	C	{97.301(e); 97.305(c)}
2A-18.1	D	{97.313(a)}
2A-18.2	C	{97.3(b)(6); 97.313(c)(1)}
2A-18.3	C	{97.3(b)(6); 97.313(c)(1)}
2A-18.4	C	{97.3(b)(6); 97.313(c)(1)}
2A-18.5	C	{97.3(b)(6); 97.313(c)(1)}
2A-19.1	C	{97.313(c)(2)}
2A-19.2	B	{97.313(c)(2)}
2A-19.3	C	{97.313(d)}
2A-19.4	D	{97.313(e)}
2A-19.5	B	{97.313(d)}
2A-20.1	C	{97.3(c)(7)}
2A-20.2	C	{97.3(c)(2)}
2A-20.3	D	{97.305(c)}
2A-21.1	C	{97.103(a)}
2A-21.2	A	{97.103(a)}
2A-21.3	D	{97.103(a)}
2A-21.4	B	{97.103(a)}
2A-22.1	C	{97.7}
2A-22.2	A	{97.103(a)}
2A-23.1	B	{97.7}
2A-24.1	B	{97.109(b)}
2A-25.1	C	{97.9(a)}
2A-26.1	D	{97.5(e)}
2A-27.1	C	{97.119(a)}
2A-27.2	B	{97.119(a)}
2A-27.3	B	{97.119(a)}
2A-27.4	A	{97.119(a)}
2A-27.5	B	{97.119(a)}
2A-27.6	C	{97.119(a)}
2A-27.7	B	{97.119(a)}
2A-28.1	D	{97.111(a)(1)}
2A-28.2	C	{97.111(a)(2), (3), (4)}
2A-29.1	B	{97.5(e)}
2A-29.2	D	{97.5(e)}
2A-30.1	A	{97.113(a)}
2A-30.2	B	{97.113(a)}
2A-30.3	D	{97.113(a)}
2A-31.1	D	{97.111(a)(1)}
2A-32.1	A	{97.113(b)}
2A-32.2	D	{97.113(b)}
2A-33.1	A	{97.3(a)(10); 97.113(b)}
2A-34.1	A	{97.3(a)(38)}
2A-34.2	B	{97.3(a)(38)}
2A-34.3	D	{97.115(a)(2)}
2A-35.1	A	{97.113(d)}
2A-36.1	C	{97.113(d)}
2A-36.2	D	{97.113(d)}
2A-37.1	A	{97.113(d)}
2A-37.2	C	{97.113(d)}
2A-38.1	C	{97.119(a)}
2A-38.2	D	{97.119(a)}
2A-38.3	A	{97.119(a)}
2A-39.1	C	{97.101(d)}
2A-39.2	B	{97.101(d)}
2A-39.3	B	{97.101(d)}
2A-40.1	C	{Not covered in the new Part 97 Rules.}
2A-40.2	D	{Not covered in the new Part 97 Rules.}
2A-40.3	D	{Not covered in the new Part 97 Rules.}

SUBELEMENT 2B

Numbers in this section refer to page numbers in *Tune in the World.*

Question	Answer	Page
2B-1-1.1	A	p 10-10
2B-1-1.2	C	p 10-5
2B-1-1.3	C	p 10-14
2B-1-2.1	D	p 10-10
2B-2-1.1	A	p 10-10
2B-2-1.2	B	p 10-11
2B-2-2.1	C	p 10-10
2B-2-3.1	C	p 10-12
2B-2-3.2	D	p 10-12
2B-2-3.3	A	p 10-12
2B-2-3.4	B	p 10-12
2B-2-3.5	C	p 10-12

2B-2-4.1	D	p 10-10
2B-2-4.2	A	p 10-11
2B-2-4.3	A	p 10-12
2B-2-5.1	D	p 10-12
2B-2-5.2	B	p 10-12
2B-2-5.3	A	p 10-12
2B-2-6.1	B	p 10-9
2B-2-6.2	C	p 10-9
2B-2-6.3	D	p 10-9
2B-2-6.4	A	p 10-9
2B-2-6.5	B	p 10-9
2B-3-1.1	C	p 10-15
2B-3-1.2	D	p 10-16
2B-3-2.1	A	p 10-15
2B-3-2.2	C	p 10-15
2B-3-2.3	A	p 10-15
2B-3-2.4	D	p 10-15
2B-3-2.5	A	p 10-15
2B-3-2.6	C	p 10-15
2B-3-2.7	A	p 10-15
2B-3-2.8	D	p 10-15
2B-3-2.9	B	p 10-15
2B-3-2.10	A	p 10-15
2B-3-2.11	C	p 10-15
2B-4-1.1	B	p 10-26
2B-4-2.1	B	p 10-23
2B-5-1.1	C	p 10-32
2B-5-1.2	D	p 10-30
2B-5-2.1	A	p 10-33
2B-5-2.2	B	p 10-35
2B-6-1.1	C	p 10-20
2B-6-2.1	D	p 10-18
2B-6-3.1	A	p 10-19
2B-6-4.1	B	p 10-21
2B-6-5.1	C	p 10-21
2B-6-5.2	D	p 10-20

SUBELEMENT 2C

2C-1.1	A	p 10-3
2C-1.2	B	p 10-8
2C-2.1	D	p 10-3
2C-2.2	B	p 10-3
2C-2.3	B	p 10-3
2C-2.4	A	p 10-3
2C-3.1	A	p 10-3
2C-3.2	B	p 10-3
2C-3.3	D	p 10-3
2C-3.4	A	p 10-4
2C-3.5	C	p 10-4
2C-3.6	D	p 10-3
2C-4.1	C	p 10-3
2C-4.2	D	p 10-3
2C-5.1	A	p 10-4
2C-5.2	B	p 10-4
2C-6.1	C	p 10-8
2C-6.2	C	p 10-8

SUBELEMENT 2D

2D-1.1	B	p 9-2
2D-1.2	A	p 9-9
2D-2.1	D	p 9-10
2D-2.2	C	p 9-10
2D-2.3	D	p 9-10
2D-2.4	B	p 9-10
2D-3.1	A	p 9-3
2D-3.2	C	p 9-10
2D-3.3	B	p 9-3
2D-3.4	C	p 9-3
2D-4.1	B	p 9-11
2D-4.2	A	p 9-11
2D-4.3	C	p 9-11
2D-4.4	A	p 9-11
2D-4.5	B	p 9-11
2D-4.6	B	p 9-11
2D-5.1	D	p 9-10
2D-5.2	A	p 9-10
2D-6.1	D	p 8-20
2D-6.2	D	p 8-21
2D-6.3	A	p 8-20
2D-6.4	D	p 8-21
2D-6.5	A	p 8-21
2D-7-1.1	C	p 8-6
2D-7-1.2	D	p 8-6
2D-7-2.1	B	p 8-6
2D-7-2.2	C	p 8-6
2D-7-2.3	D	p 8-6
2D-7-2.4	C	p 8-6
2D-7-3.1	A	p 8-7
2D-7-3.2	A	p 8-7
2D-7-3.3	B	p 8-6
2D-7-3.4	C	p 8-6
2D-8-1.1	C	p 11-3
2D-8-1.2	B	p 11-3
2D-8-1.3	C	p 11-3
2D-8-1.4	B	p 11-4
2D-8-2.1	A	p 11-4
2D-8-2.2	A	p 11-4
2D-8-2.3	A	p 11-5
2D-8-2.4	C	p 11-5
2D-8-2.5	D	p 11-4
2D-8-2.6	B	p 11-4
2D-8-3.1	A	p 11-3
2D-8-3.2	B	p 11-3

SUBELEMENT 2E

2E-1-1.1	C	p 4-2
2E-1-2.1	B	p 4-2
2E-1-2.2	D	p 4-2
2E-1-3.1	B	p 4-2
2E-1-3.2	D	p 4-2
2E-1-4.1	B	p 4-2
2E-1-5.1	C	p 4-2
2E-1-5.2	C	p 4-2
2E-1-6.1	B	p 4-2
2E-1-7.1	B	p 4-2
2E-2-1.1	D	p 4-5
2E-2-2.1	C	p 4-6
2E-3-1.1	A	p 4-5
2E-3-1.2	D	p 4-5
2E-3-1.3	C	p 4-5
2E-3-2.1	A	p 4-5
2E-4.1	C	p 4-6
2E-5.1	A	p 4-6
2E-6-1.1	D	p 4-7
2E-6-1.2	A	p 4-7
2E-6-2.1	D	p 4-7
2E-7.1	C	p 4-7
2E-7.2	C	p 4-7
2E-7.3	B	p 4-7
2E-7.4	C	p 4-7
2E-8.1	D	p 4-11
2E-8.2	A	p 4-11
2E-9-1.1	C	p 4-11
2E-9-1.2	C	p 4-11
2E-9-2.1	B	p 4-11
2E-10.1	C	p 4-10
2E-11.1	D	p 4-10
2E-12-1.1	B	p 4-12
2E-12-2.1	A	p 4-12
2E-12-3.1	D	p 4-12
2E-12-3.2	A	p 4-12
2E-12-3.3	A	p 4-12
2E-12-4.1	B	p 4-14
2E-12-4.2	A	p 4-14
2E-12-4.3	B	p 4-14
2E-12-5.1	C	p 4-14
2E-12-5.2	B	p 4-14
2E-13.1	C	p 4-14
2E-13.2	D	p 4-14

SUBELEMENT 2F

2F-1.1	B	p 5-2
2F-1.2	C	p 5-2
2F-1.3	C	p 5-2
2F-2.1	A	p 5-2
2F-2.2	A	p 5-2
2F-2.3	B	p 5-2
2F-2.4	D	p 5-3
2F-2.5	A	p 5-2
2F-3.1	C	p 5-3
2F-4.1	C	p 5-4
2F-4.2	B	p 5-4
2F-5.1	D	p 5-5

WEEK 3 QUIZ

1. What are the requirements for properly identifying your amateur radio station during amateur communications? 1 each 10min
   Station call letters each 10min and @ end

2. Explain the difference between direct current and alternating current.
   DC Flows electrons flow one direction
   AC = positive to negative

3. What is an electrical short circuit? What happens if one occurs?
   wires cross fuse blows –

4. What is the approximate wavelength in meters of a radio wave that has a frequency of 7.150 MHz? 40 miles $\lambda = c/f$ $\frac{3 \times 10^8}{7.15 \quad 10^6}$ $\frac{3 \times 10^2}{7.15}$ $\frac{300}{7.15}$ 41

5. What station identification, if any, is required at the end of a QSO?
   originators station call sign's

6. When are business communications permitted in the Amateur Radio Service?
   NONE / emergency

7. What is the name of the characteristic of materials that impedes the flow of electrical current? What unit of measurement is used for this characteristic?
   insolater, resistor - ohms

8. What is the term for the distance an AC signal travels during one complete cycle?
   wave length

9. What is the term for an electrical circuit in which there can be no current?
   open

10. What is the basic unit of electrical power? watt

Two resistors are said to be connected in parallel. What does this mean? (draw a diagram ...u wish).

...ted across this combination, will the current which flows be:

... current which would flow through one resistor alone NO

...hich would flow through either resistor alone YES

... would flow through either resistor alone NO

2F-5.2	B	p 5-5
2F-5.3	A	p 5-5
2F-5.4	D	p 5-5
2F-6.1	D	p 5-5
2F-7.1	C	p 5-6
2F-7.2	A	p 5-6
2F-7.3	C	p 5-6
2F-7.4	B	p 5-6
2F-8.1	A	p 5-7

SUBELEMENT 2G

2G-1-1.1	B	p 6-5
2G-1-1.2	C	p 6-2
2G-1-1.3	D	p 6-2
2G-1-1.4	A	p 6-3
2G-1-1.5	D	p 6-2
2G-1-2.1	B	p 6-3
2G-1-2.2	A	p 6-3
2G-1-2.3	B	p 6-3
2G-1-2.4	D	p 6-3
2G-1-2.5	D	p 6-3
2G-2.1	B	p 6-4
2G-2.2	C	p 6-4
2G-2.3	C	p 6-4
2G-3.1	D	p 6-5
2G-3.2	D	p 6-5
2G-4.1	A	p 6-5
2G-4.2	C	p 6-5
2G-5.1	A	p 6-5
2G-5.2	A	p 6-6
2G-5.3	C	p 6-6

SUBELEMENT 2H

2H-1-1.1	B	p 2-7
2H-1-1.2	C	p 2-7
2H-1-2.1	A	p 2-7
2H-1-2.2	A	p 2-7
2H-1-3.1	A	p 2-8
2H-1-4.1	D	p 2-8
2H-2.1	B	p 11-6
2H-2.2	D	p 11-6
2H-3.1	C	p 11-7
2H-3.2	B	p 11-7
2H-4.1	D	p 11-7
2H-4.2	B	p 11-7
2H-5.1	A	p 11-1
2H-5.2	C	p 11-2
2H-6.1	A	p 11-11
2H-6.2	C	p 11-11
2H-6.3	A	p 11-11
2H-7.1	C	p 11-2
2H-7.2	D	p 11-1
2H-7.3	A	p 11-5
2H-7.4	B	p 11-22
2H-7.5	B	p 11-22
2H-7.6	B	p 11-22

SUBELEMENT 2I

2I-1.1	A	p 8-7
2I-1.2	D	p 8-7
2I-1.3	C	p 8-7
2I-1.4	C	p 8-7
2I-1.5	D	p 8-7
2I-2.1	D	p 8-14
2I-2.2	D	p 8-14
2I-2.3	B	p 8-14
2I-2.4	B	p 8-14
2I-2.5	A	p 8-14
2I-3.1	B	p 8-20
2I-3.2	C	p 8-20
2I-4-1.1	D	p 8-17
2I-4-1.2	B	p 8-19
2I-4-1.3	C	p 8-18
2I-4-2.1	D	p 8-19
2I-4-2.2	A	p 8-19
2I-4-2.3	B	p 8-19
2I-4-2.4	A	p 8-18
2I-5.1	A	p 8-21
2I-5.2	B	p 8-20
2I-5.3	C	p 8-21
2I-6.1	D	p 8-2
2I-6.2	B	p 8-2
2I-6.3	B	p 8-3
2I-6.4	B	p 8-3
2I-6.5	C	p 8-3
2I-7.1	B	p 8-3
2I-7.2	C	p 8-3
2I-7.3	D	p 8-3
2I-7.4	A	p 8-3
2I-7.5	B	p 8-3
2I-8.1	C	p 8-4
2I-8.2	C	p 8-4
2I-9.1	D	p 8-5
2I-9.2	A	p 8-6
2I-9.3	A	p 8-5
2I-10-1.1	C	p 8-20
2I-11-1.1	D	p 8-20

ELEMENT 2 QUESTION POOL SUPPLEMENT

issued by the Volunteer Examiner Coordinators' Question Pool Committee

October 11, 1989

This supplement is scheduled to take effect on July 1, 1990. All Element 2 exams given on or after this date must use the revised question pool that includes the new and revised questions, answers and distractors contained in this supplement. These are only the changes *provided by the Question Pool Committee—questions, answers and distractors left unchanged are contained in the question pool beginning on page 12-2.*

For exams given on or after July 1, 1990, the following changes should be made throughout the *entire* Element 2 question pool by administering VEs. These modifications are based on the new Part 97.

Present wording	*Change to*
(A) Amateur Radio	amateur radio
(B) Amateur Radio Service	amateur service
(C) Amateur Radio Operator	amateur operator
(D) Amateur Radio Station	amateur station
(E) Amateur Radio License	amateur operator/ primary station license
(F) Amateur Radio Station License	primary station license
(G) Amateur Radio Operator License	amateur operator license
(H) Amateur Radio Communications	amateur communications
(I) General state of communication emergency	state of communication emergency
(J) XX-meter band	XX-meter wavelength band
(K) radiocommunication	radio communication

2A-1.2 Which of the following is *not* one of the basic principles for which the *amateur service* rules are designed?

2A-1.3 The amateur service rules were designed to provide a radio communications service that meets five fundamental purposes. Which of the following is *not* one of those principles?

2A-1.4 The amateur service rules were designed to provide a radio communications service that meets five fundamental purposes. What are those principles?

2A-2.1 What is the definition of the *amateur service*?
- C. A radio communication service for the purpose of self-training, intercommunication and technical investigations

2A-2.2 What name is given to the radio communication service that is designed for self-training, intercommunication, and technical investigation?
- A. The amateur service

2A-3.1 What document contains the specific rules and regulations governing the amateur service in the United States?
- A. Part 97 of title 47 CFR (Code of Federal Regulations)
- B. The communications act of 1934 (as amended)
- C. The Radio Amateur's Handbook
- D. The minutes of the International Telecommunication Union meetings

2A-3.2 Which one of the following topics is *not* addressed in the rules and regulations of the amateur service?
- A. Station operation standards
- B. Technical standards
- C. Providing emergency communications
- D. Station construction standards

2A-4.1 What is the definition of an amateur operator?
- B. A person holding a written authorization to be the control operator of an amateur station

2A-4.2 What term describes a person holding a written authorization to be the control operator of an amateur station?
- D. An amateur operator

2A-5.2 What authority is derived from an operator/ primary station license?
- B. The authority to be the control operator of an amateur station

2A-6.1 What authority is derived from a written authorization for an amateur station?
- B. The authority to operate an amateur station

2A-6.2 What part of your amateur license gives you authority to operate an amateur station?
- C. The station license

2A-7.1 What is an *amateur station*?
- D. A station in an amateur service consisting of the apparatus necessary for carrying on radio communications

2A-8.1 Who is a *control operator*?

A. An amateur operator designated by the licensee of a station to be responsible for the transmissions from that station to assure compliance with the FCC rules

2A-8.2 If you designate another amateur operator to be responsible for the transmissions from your station, what is the other operator called?

2A-10.1 What frequencies are available in the amateur 80-meter wavelength band for a control operator holding a Novice class operator license?

2A-10.2 What frequencies are available in the amateur 40-meter wavelength band for a control operator holding a Novice class operator license in ITU Region 2?

2A-10.3 What frequencies are available in the amateur 15-meter wavelength band for a control operator holding a Novice class operator license?

2A-10.4 What frequencies are available in the amateur 10-meter wavelength band for a control operator holding a Novice class operator license?

2A-10.5 What frequencies are available in the amateur 220-MHz band for a control operator holding a Novice class operator license in ITU Region 2?

2A-10.6 What frequencies are available in the amateur 1270-MHz band for a control operator holding a Novice class operator license?

2A-11.1 Who is eligible to obtain a US amateur *operator/primary station* license?

2A-11.2 Who is *not* eligible to obtain a US amateur *operator/primary station* license?

2A-12.1

B. Elements 1(A) and 3(A)
D. Elements 2 and 4

2A-12.2

A. The applicant's ability to send and receive texts in the international Morse code at not less than 5 words per minute
B. The applicant's ability to send and receive texts in the international Morse code at not less than 13 words per minute

2A-12.3

B. The written examination concerning the privileges of a Technician class operator license
D. The written examination concerning the privileges of a Novice class operator license

2A-13.1 Who is eligible to obtain an FCC-issued written authorization for an amateur station?

A. A licensed amateur operator

2A-14.1 Why is an amateur operator required to furnish the FCC with a current mailing address served by the US Postal service?

B. In order to comply with the Commission's rules and so the FCC can correspond with the licensee

2A-16.1

A. On the date specified on the license

2A-17.2 What emission types are Novice control operators permitted to use on the 80-meter wavelength band?

A. CW only
B. Data only
C. RTTY only
D. Phone only

2A-17.3

A. CW only
B. Data only
C. RTTY only
D. Phone only

2A-17.4

A. CW only
B. Data only
C. RTTY only
D. Phone only

2A-17.5

A. Phone only
B. CW and phone
C. All amateur emission privileges authorized for use on those frequencies
D. CW only

2A-17.6 What emission types are Novice control operators permitted to use from 7100 to 7150 kHz in ITU Region 2?

A. CW and Data
B. Phone
C. All amateur emission privileges authorized for use on those frequencies
D. CW only

2A-17.7

A. CW and data only
B. CW and phone only
C. All amateur emission privileges authorized for use on those frequencies
D. CW only

2A-17.8

A. All authorized amateur emission privileges
B. Data or phone only
C. CW, RTTY and data
D. CW and phone only

2A-17.9

A. All authorized emission privileges
B. CW and data only
C. CW and single-sideband phone only
D. Data and phone only

2A-17.10 What emission types are Novice control operators permitted to use on the amateur 220-MHz band in ITU Region 2?
A. CW and phone only
B. CW and data only
C. Data and phone only

2A-17.11

A. Data and phone only
B. CW and data only
C. CW and phone only

2A-17.12 On what frequencies in the 10-meter wavelength band may a Novice control operator use single-sideband phone?

2A-17.13 On what frequencies in the 1.25-meter wavelength band in ITU Region 2 may a Novice control operator use FM phone emission?

2A-18.1

D. The minimum legal power necessary to carry out the desired communications

2A-20.1 What term is used to describe narrow-band direct-printing telegraphy emissions?
C. RTTY communications

2A-20.2 What term is used to describe telemetry, telecommand and computer communications emissions?
C. Data communications

2A-20.3 On what frequencies in the 10-meter wavelength band are Novice control operators permitted to transmit RTTY?

2A-21.2 You allow another amateur operator to use your amateur station. What are your responsibilities, as the station licensee?
A. You and the other amateur operator are equally responsible for the proper operation of your station

2A-21.4 If you are the licensee of an amateur station when are you *not* responsible for its proper operation?

2A-22.1

C. Whenever the station is transmitting

2A-22.2 Another amateur gives you permission to use her amateur station. What are your responsibilities, as the control operator?
A. Both you and she are equally responsible for the proper operation of her station
B. Only the station licensee is responsible for the proper operation of the station, not you the control operator

2A-23.1

B. Any properly licensed amateur operator that is designated by the station licensee

2A-24.1 Where must an amateur operator be when he or she is performing the duties of control operator?
B. At the control point of the amateur station

2A-26.1 Where must you keep your written authorization for an amateur station?
D. The original or a photocopy of the written authorization for an amateur station must be retained at the station

2A-27.2 As an amateur operator, how should you correctly identify your station?

2A-27.3 What station identification, if any, is required at the beginning of communication?

2A-27.4 What station identification, if any, is required at the end of a communication?
A. Both stations must transmit their own call sign, assuming they are FCC-licensed
B. No identification is required at the end of the contact
C. The station originating the contact must always transmit both call signs
D. Both stations must transmit their own call sign followed by a two-letter designator for the nearest FCC field office

2A-27.5 What do the FCC rules for amateur station identification generally require?

2A-30.1 When are communications pertaining to business or commercial affairs of any party permitted in the amateur service?
A. Only when the immediate safety of human life or immediate protection of property is threatened

2A-30.2

B. Never. This would facilitate the commercial affairs of the ARRL

2A-30.3

D. Never. This would facilitate the commercial affairs of a business

2A-31.1 When may an FCC-licensed amateur operator communicate with an amateur operator in a foreign country?

2A-32.2

D. The control operator may accept compensation if he or she works for a club station during the period in which the station is transmitting telegraphy practice or information bulletins if certain exacting conditions are met

2A-33.1 When is an amateur operator permitted to broadcast information intended for the general public?

A. Amateur operators are not permitted to broadcast information intended for the general public

2A-34.1 What is *third-party communications*?

A. A message passed from the control operator of an amateur station to another control operator on behalf of another person

2A-34.2 Who is a *third party* in amateur communications?

B. Any person for whom a message is passed through amateur communication channels other than the control operators of the two stations handling the message

2A-34.3 When is an amateur operator permitted to transmit a message to a foreign country for a third party?

D. When there is a third-party traffic agreement between the US and the foreign government, or when the third party is eligible to be the control operator of the station

2A-35.1 Is an amateur station permitted to transmit music?

2A-36.1 Is the use of codes or ciphers where the intent is to obscure the meaning permitted during a two-way communication in the amateur service?

C. The transmission of codes and ciphers where the intent is to obscure the meaning is not permitted in the amateur service

2A-36.2 When is an operator in the amateur service permitted to use abbreviations that are intended to obscure the meaning of the message?

D. Abbreviations that are intended to obscure the meaning of the message may never be used in the amateur service

2A-38.1 When may an amateur station transmit unidentified communications?

2A-38.2

D. Radio communications in which the station identification is not transmitted

2A-38.3 What is the term used to describe a transmission from an amateur station that does not transmit the required station identification?

A. Unidentified communications or signals

2A-39.1 When may an amateur operator willfully or maliciously interfere with a radio communication or signal

C. You may never willfully or maliciously interfere with another station's transmissions

2A-40.1

C. As specified in the Notice

2A-40.2 If you were to receive a voice distress signal from a station on a frequency outside your operator privileges, what restrictions would apply to assisting the station in distress?

A. You would not be allowed to assist the station because the frequency of its signals were outside your operator privileges

B. You would be allowed to assist the station only if your signals were restricted to the nearest frequency band of your privileges

C. You would be allowed to assist the station on a frequency outside of your operator privileges only if you used international Morse code

D. You would be allowed to assist the station on a frequency outside of your operator privileges using any means of radio communications at your disposal

2A-40.3 If you were in a situation where normal communication systems were disrupted due to a disaster, what restrictions would apply to essential communications you might provide in connection with the immediate safety of human life?

A. You would not be allowed to communicate at all except to the FCC Engineer in Charge of the area concerned

B. You would be restricted to communications using only the emissions and frequencies authorized to your operator privileges

C. You would be allowed to communicate on frequencies outside your operator privileges only if you used international Morse code

D. You would be allowed to use any means of radio communication at your disposal

2H-1-1.1 What keying method is used to transmit CW?

2H-1-1.2 What emission type describes international Morse code telegraphy messages?

A. RTTY

B. Image

C. CW

D. Phone

2H-1-2.1 What emission type describes narrow-band direct-printing telegraphy emissions?

A. RTTY

B. Image

C. CW

D. Phone

2H-1-2.2 What keying method is used to transmit RTTY messages?

C. Digital pulse-code keying of an unmodulated carrier

2H-1-3.1 What emission type describes frequency-modulated voice transmissions?

A. FM phone

B. Image

C. CW

D. Single-sideband phone

2H-1-4.1

A. FM phone

B. Image

C. CW

D. Sideband phone

US Amateur Frequency and Mode Allocations

160 METERS

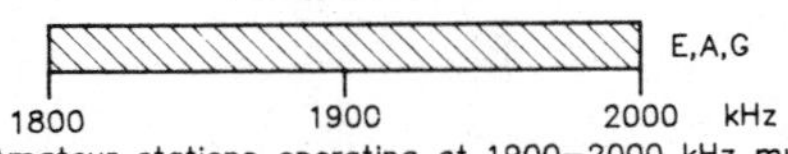

Amateur stations operating at 1900–2000 kHz must not cause harmful interference to the radiolocation service and are afforded no protection from radiolocation operations.

80 METERS

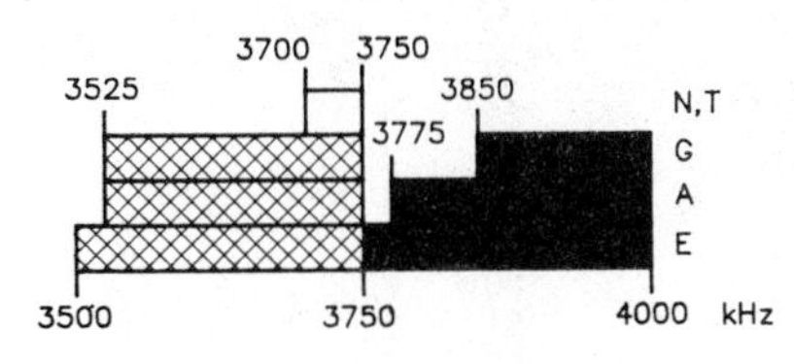

5167.5 kHz (SSB only): Alaska emergency use only.

40 METERS

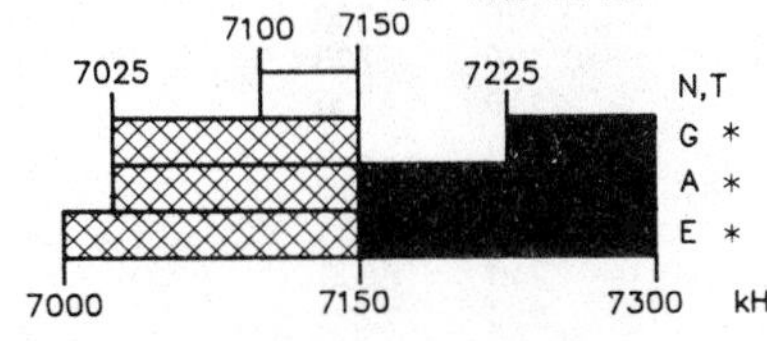

* Phone operation is allowed on 7075–7100 kHz in Puerto Rico, US Virgin islands and areas of the Caribbean south of 20 degrees north latitude; and in Hawaii and areas near ITU Region 3, including Alaska.

30 METERS

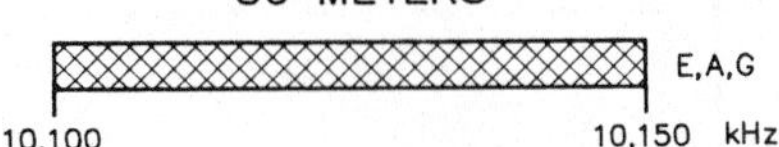

Maximum power on 30 meters is 200 watts PEP output. Amateurs must avoid interference to the fixed service outside the US.

20 METERS

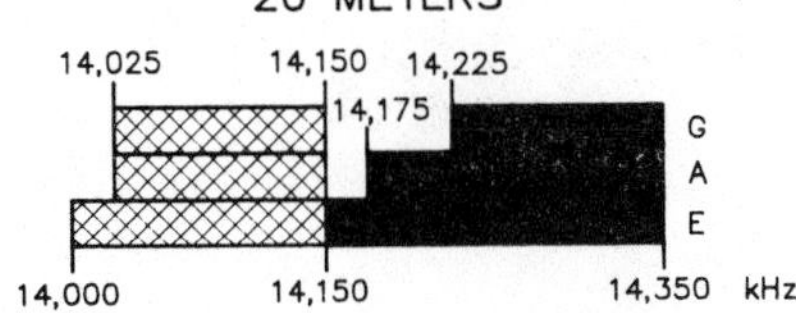

17 METERS

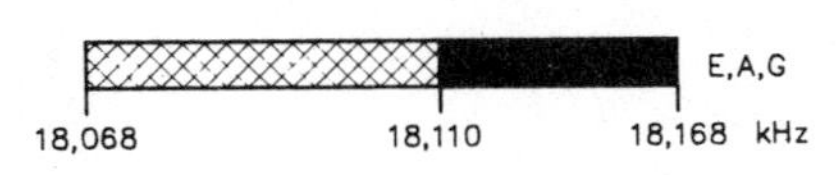

15 METERS

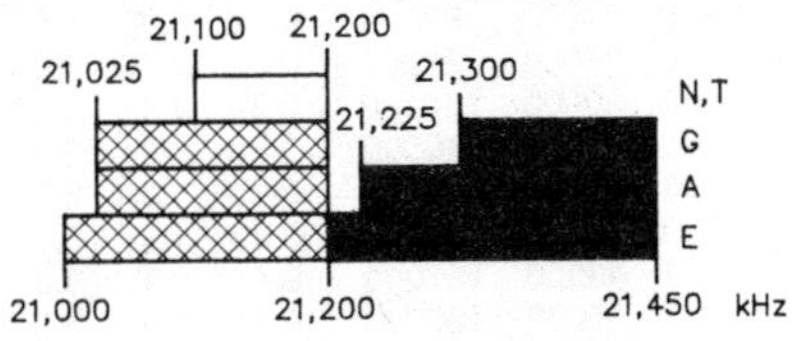

12 METERS

10 METERS

Novices and Technicians are limited to 200 watts PEP output on 10 meters.

6 METERS

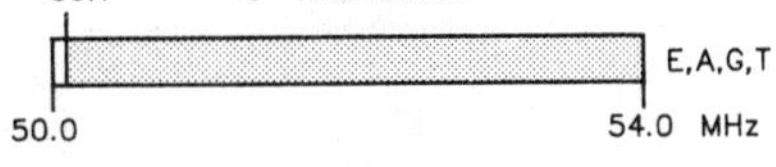

2 METERS

144.1

E,A,G,T

144.0 148.0 MHz

1.25 METERS

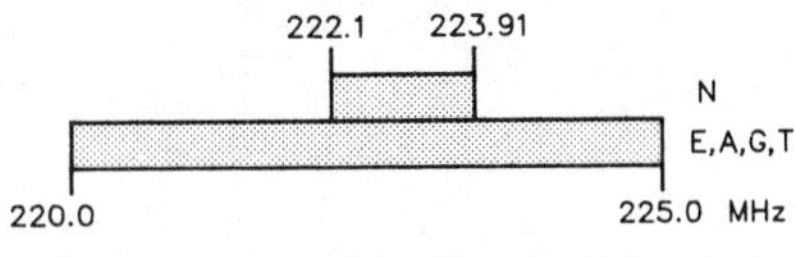

Novices are limited to 25 watts PEP output from 222.1 to 223.91 MHz.

70 CENTIMETERS

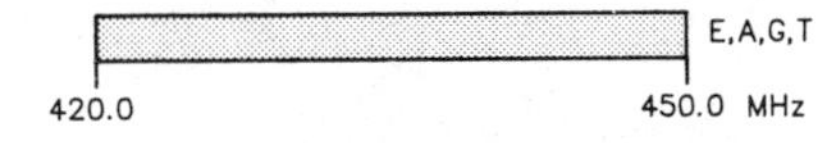

33 CENTIMETERS

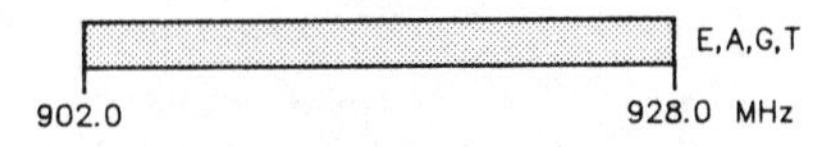

23 CENTIMETERS

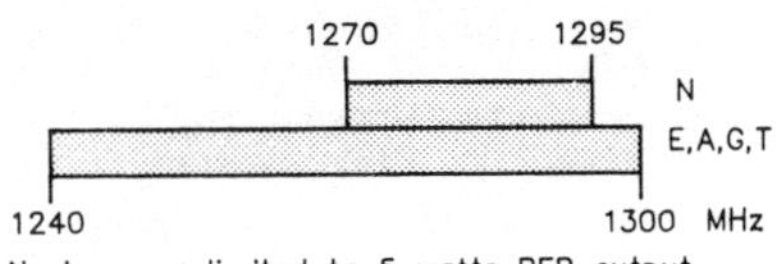

Novices are limited to 5 watts PEP output from 1270 to 1295 MHz.

US AMATEUR BANDS

Revised November 15, 1989

US AMATEUR POWER LIMITS

At all times, transmitter power should be kept down to that necessary to carry out the desired communications. Power is rated in watts PEP output. Unless otherwise stated, the maximum power output is 1500 W. Power for all license classes is limited to 200 W in the 10,100–10,150 kHz band and in all Novice subbands below 28,100 kHz. Novices and Technicians are restricted to 200 W in the 28,100–28,500 kHz subbands In addition, Novices are restricted to 25 W in the 222.1–223.91 MHz subband and 5 W in the 1270–1295 MHz subband.

Operators with Technician class licenses and above may operate on all bands above 50 MHz. For more detailed information see The FCC Rule Book.

KEY

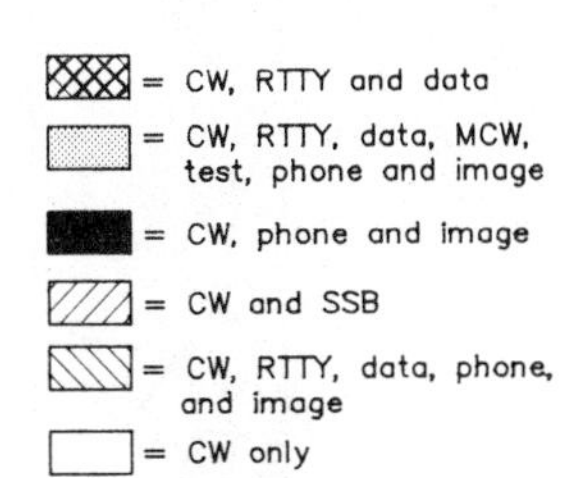

E =AMATEUR EXTRA
A =ADVANCED
G =GENERAL
T =TECHNICIAN
N =NOVICE

Helpful Data Tables

Standard Resistance Values

Numbers in **bold** type are ± 10% values. Others are 5% values.

Ohms										*Megohms*				
1.0	3.6	**12**	43	**150**	510	**1800**	6200	**22000**	75000	0.24	0.62	1.6	4.3	11.0
1.1	**3.9**	13	**47**	160	**560**	2000	**6800**	24000	**82000**	**0.27**	**0.68**	**1.8**	**4.7**	**12.0**
1.2	4.3	**15**	51	**180**	620	**2200**	7500	**27000**	91000	0.30	0.75	2.0	5.1	13.0
1.3	**4.7**	16	**56**	200	**680**	2400	**8200**	30000	**100000**	**0.33**	**0.82**	**2.2**	**5.6**	**15.0**
1.5	5.1	**18**	62	**220**	750	**2700**	9100	**33000**	110000	0.36	0.91	2.4	6.2	16.0
1.6	**5.6**	20	**68**	240	**820**	3000	**10000**	36000	**120000**	**0.39**	**1.0**	**2.7**	**6.8**	**18.0**
1.8	6.2	**22**	75	**270**	910	**3300**	11000	**39000**	130000	0.43	1.1	3.0	7.5	20.0
2.0	**6.8**	24	**82**	300	**1000**	3600	**12000**	43000	**150000**	**0.47**	**1.2**	**3.3**	**8.2**	**22.0**
2.2	7.5	**27**	91	**330**	1100	**3900**	13000	**47000**	160000	0.51	1.3	3.6	9.1	
2.4	**8.2**	30	**100**	360	**1200**	4300	**15000**	51000	**180000**	**0.56**	**1.5**	**3.9**	**10.0**	
2.7	9.1	**33**	110	**390**	1300	**4700**	16000	**56000**	200000					
3.0	**10.0**	36	**120**	430	**1500**	5100	**18000**	62000	**220000**					
3.3	11.0	**39**	130	**470**	1600	**5600**	20000	**68000**						

Resistor Color Code

Color	*Sig. Figure*	*Decimal Multiplier*	*Tolerance (%)*
Black	0	1	
Brown	1	10	
Red	2	100	
Orange	3	1,000	
Yellow	4	10,000	
Green	5	100,000	
Blue	6	1,000,000	
Violet	7	10,000,000	
Gray	8	100,000,000	
White	9	1,000,000,000	
Gold	—	0.1	5
Silver	—	0.01	10
No color	—		20

Standard Capacitance Values

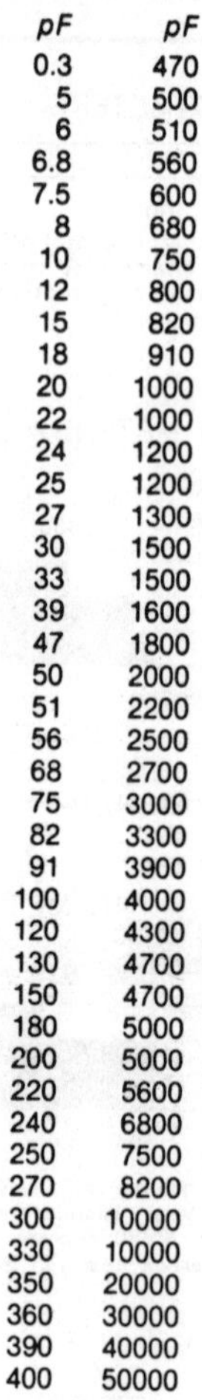

pF	*pF*
0.3	470
5	500
6	510
6.8	560
7.5	600
8	680
10	750
12	800
15	820
18	910
20	1000
22	1000
24	1200
25	1200
27	1300
30	1500
33	1500
39	1600
47	1800
50	2000
51	2200
56	2500
68	2700
75	3000
82	3300
91	3900
100	4000
120	4300
130	4700
150	4700
180	5000
200	5000
220	5600
240	6800
250	7500
270	8200
300	10000
330	10000
350	20000
360	30000
390	40000
400	50000

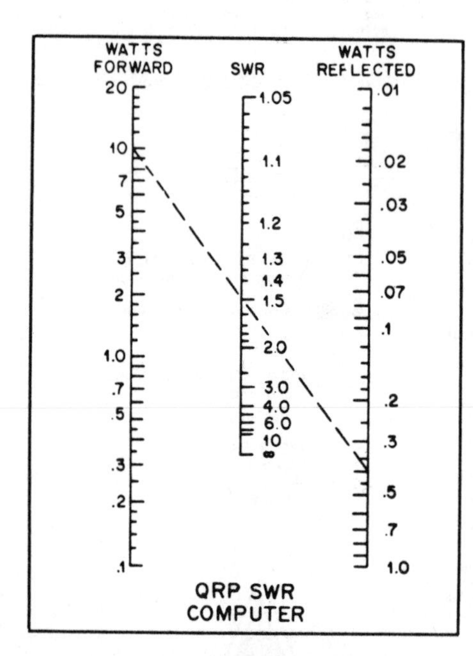

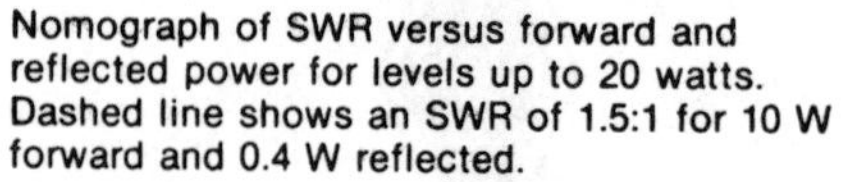

Nomograph of SWR versus forward and reflected power for levels up to 20 watts. Dashed line shows an SWR of 1.5:1 for 10 W forward and 0.4 W reflected.

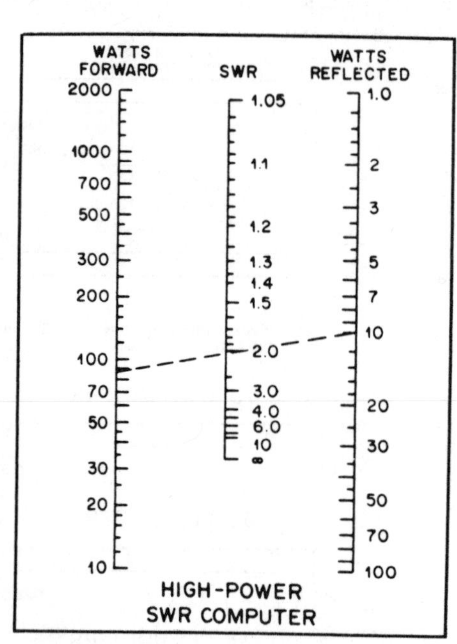

Nomograph of SWR versus forward and reflected power for levels up to 2000 watts. Dashed line shows an SWR of 2:1 for 90 W forward and 10 W reflected.

Fractions of an Inch with Metric Equivalents

Fractions Of An Inch	Decimals Of An Inch	Millimeters	Fractions Of An Inch	Decimals Of An Inch	Millimeters
1/64	0.0156	0.397	33/64	0.5156	13.097
1/32	0.0313	0.794	17/32	0.5313	13.494
3/64	0.0469	1.191	35/64	0.5469	13.891
1/16	0.0625	1.588	9/16	0.5625	14.288
5/64	0.0781	1.984	37/64	0.5781	14.684
3/32	0.0938	2.381	19/32	0.5938	15.081
7/64	0.1094	2.778	39/64	0.6094	15.478
1/8	0.1250	3.175	5/8	0.6250	15.875
9/64	0.1406	3.572	41/64	0.6406	16.272
5/32	0.1563	3.969	21/32	0.6563	16.669
11/64	0.1719	4.366	43/64	0.6719	17.066
3/16	0.1875	4.763	11/16	0.6875	17.463
13/64	0.2031	5.159	45/64	0.7031	17.859
7/32	0.2188	5.556	23/32	0.7188	18.256
15/64	0.2344	5.953	47/64	0.7344	18.653
1/4	0.2500	6.350	3/4	0.7500	19.050
17/64	0.2656	6.747	49/64	0.7656	19.447
9/32	0.2813	7.144	25/32	0.7813	19.844
19/64	0.2969	7.541	51/64	0.7969	20.241
5/16	0.3125	7.938	13/16	0.8125	20.638
21/64	0.3281	8.334	53/64	0.8281	21.034
11/32	0.3438	8.731	27/32	0.8438	21.431
23/64	0.3594	9.128	55/64	0.8594	21.828
3/8	0.3750	9.525	7/8	0.8750	22.225
25/64	0.3906	9.922	57/64	0.8906	22.622
13/32	0.4063	10.319	29/32	0.9063	23.019
27/64	0.4219	10.716	59/64	0.9219	23.416
7/16	0.4375	11.113	15/16	0.9375	23.813
29/64	0.4531	11.509	61/64	0.9531	24.209
15/32	0.4688	11.906	31/32	0.9688	24.606
31/64	0.4844	12.303	63/64	0.9844	25.003
1/2	0.50000	12.700	1	1.0000	25.400

Equations Used in This Book

$$\text{current} = \frac{\text{voltage}}{\text{resistance}} \quad \text{(Eq 4-1)}$$

$$\text{resistance} = \frac{\text{voltage}}{\text{current}} \quad \text{(Eq 4-2)}$$

$$\text{voltage} = \text{current} \times \text{resistance} \quad \text{(Eq 4-3)}$$

$$E = IR \text{ (volts = amperes} \times \text{ohms)} \quad \text{(Eq 4-4)}$$

$$I = \frac{E}{R} \text{ (amperes = volts divided by ohms)} \quad \text{(Eq 4-5)}$$

$$R = \frac{E}{I} \text{ (ohms = volts divided by amperes)} \quad \text{(Eq 4-6)}$$

$$P = IE \quad \text{(Eq 4-7)}$$

$$I = \frac{P}{E} \quad \text{(Eq 4-8)}$$

$$E = \frac{P}{I} \quad \text{(Eq 4-9)}$$

$$c = f\lambda \quad \text{(Eq 4-10)}$$

$$f = \frac{c}{\lambda} \quad \text{(Eq 4-11)}$$

$$\lambda = \frac{c}{f} \quad \text{(Eq 4-12)}$$

$$\lambda \text{ (in feet)} = \frac{984}{f \text{ (in MHz)}} \quad \text{(Eq 8-1)}$$

$$\text{Length (in feet)} = \frac{468}{f \text{ (in MHz)}} \quad \text{(Eq 8-2)}$$

$$\text{Length (in feet)} = \frac{234}{f \text{ (in MHz)}} \quad \text{(Eq 8-3)}$$

US Customary to Metric Conversions

US Customary — Metric Conversion Factors

International System of Units (SI) — Metric Units

Prefix	Symbol			Multiplication Factor
exa	E	10^{18}	=	1,000,000,000,000,000,000
peta	P	10^{15}	=	1,000,000,000,000,000
tera	T	10^{12}	=	1,000,000,000,000
giga	G	10^{9}	=	1,000,000,000
mega	M	10^{6}	=	1,000,000
kilo	k	10^{3}	=	1,000
hecto	h	10^{2}	=	100
deca	da	10^{1}	=	10
(unit)		10^{0}	=	1
deci	d	10^{-1}	=	0.1
centi	c	10^{-2}	=	0.01
milli	m	10^{-3}	=	0.001
micro	μ	10^{-6}	=	0.000001
nano	n	10^{-9}	=	0.000000001
pico	p	10^{-12}	=	0.000000000001
femto	f	10^{-15}	=	0.000000000000001
atto	a	10^{-18}	=	0.000000000000000001

Linear

1 meter (m) = 100 centimeters (cm) = 1000 millimeters (mm)

Area

1 m^2 = 1 × 10^4 cm^2 = 1 × 10^6 mm^2

Volume

1 m^3 = 1 × 10^6 cm^3 = 1 × 10^9 mm^3
1 liter (l) = 1000 cm^3 = 1 × 10^6 mm^3

Mass

1 kilogram (kg) = 1000 grams (g)
(Approximately the mass of 1 liter of water)
1 metric ton (or tonne) = 1000 kg

US Customary Units

Linear Units

12 inches (in) = 1 foot (ft)
36 inches = 3 feet = 1 yard (yd)
1 rod = 5½ yards = 16½ feet
1 statute mile = 1760 yards = 5280 feet
1 nautical mile = 6076.11549 feet

Area

1 ft^2 = 144 in^2
1 yd^2 = 9 ft^2 = 1296 in^2
1 rod^2 = 30¼ yd^2
1 acre = 4840 yd^2 = 43,560 ft^2
1 acre = 160 rod^2
1 $mile^2$ = 640 acres

Volume

1 ft^3 = 1728 in^3
1 yd^3 = 27 ft^3

Liquid Volume Measure

1 fluid ounce (fl oz) = 8 fluidrams = 1.804 in^3
1 pint (pt) = 16 fl oz
1 quart (qt) = 2 pt = 32 fl oz = 57¾ in^3
1 gallon (gal) = 4 qt = 231 in^3
1 barrel = 31½ gal

Dry Volume Measure

1 quart (qt) = 2 pints (pt) = 67.2 in^3
1 peck = 8 qt
1 bushel = 4 pecks = 2150.42 in^3

Avoirdupois Weight

1 dram (dr) = 27.343 grains (gr) or (gr a)
1 ounce (oz) = 437.5 gr
1 pound (lb) = 16 oz = 7000 gr
1 short ton = 2000 lb, 1 long ton = 2240 lb

Troy Weight

1 grain troy (gr t) = 1 grain avoirdupois
1 pennyweight (dwt) or (pwt) = 24 gr t
1 ounce troy (oz t) = 480 grains
1 lb t = 12 oz t = 5760 grains

Apothecaries' Weight

1 grain apothecaries' (gr ap) = 1 gr t = 1 gr a
1 dram ap (dr ap) = 60 gr
1 oz ap = 1 oz t = 8 dr ap = 480 gr
1 lb ap = 1 lb t = 12 oz ap = 5760 gr

Multiply ⟶

Metric Unit = Conversion Factor × US Customary Unit

⟵ **Divide**

Metric Unit ÷ Conversion Factor = US Customary Unit

Metric Unit	= Conversion Factor	× US Unit
(Length)		
mm	25.4	inch
cm	2.54	inch
cm	30.48	foot
m	0.3048	foot
m	0.9144	yard
km	1.609	mile
km	1.852	nautical mile
(Area)		
mm^2	645.16	$inch^2$
cm^2	6.4516	in^2
cm^2	929.03	ft^2
m^2	0.0929	ft^2
cm^2	8361.3	yd^2
m^2	0.83613	yd^2
m^2	4047	acre
km^2	2.59	mi^2
(Mass)	(Avoirdupois Weight)	
grams	0.0648	grains
g	28.349	oz
g	453.59	lb
kg	0.45359	lb
tonne	0.907	short ton
tonne	1.016	long ton

Metric Unit	= Conversion Factor	× US Unit
(Volume)		
mm^3	16387.064	in^3
cm^3	16.387	in^3
m^3	0.028316	ft^3
m^3	0.764555	yd^3
ml	16.387	in^3
ml	29.57	fl oz
ml	473	pint
ml	946.333	quart
l	28.32	ft^3
l	0.9463	quart
l	3.785	gallon
l	1.101	dry quart
l	8.809	peck
l	35.238	bushel
(Mass)	(Troy Weight)	
g	31.103	oz t
g	373.248	lb t
(Mass)	(Apothecaries' Weight)	
g	3.387	dr ap
g	31.103	oz ap
g	373.248	lb ap

Schematic Symbols

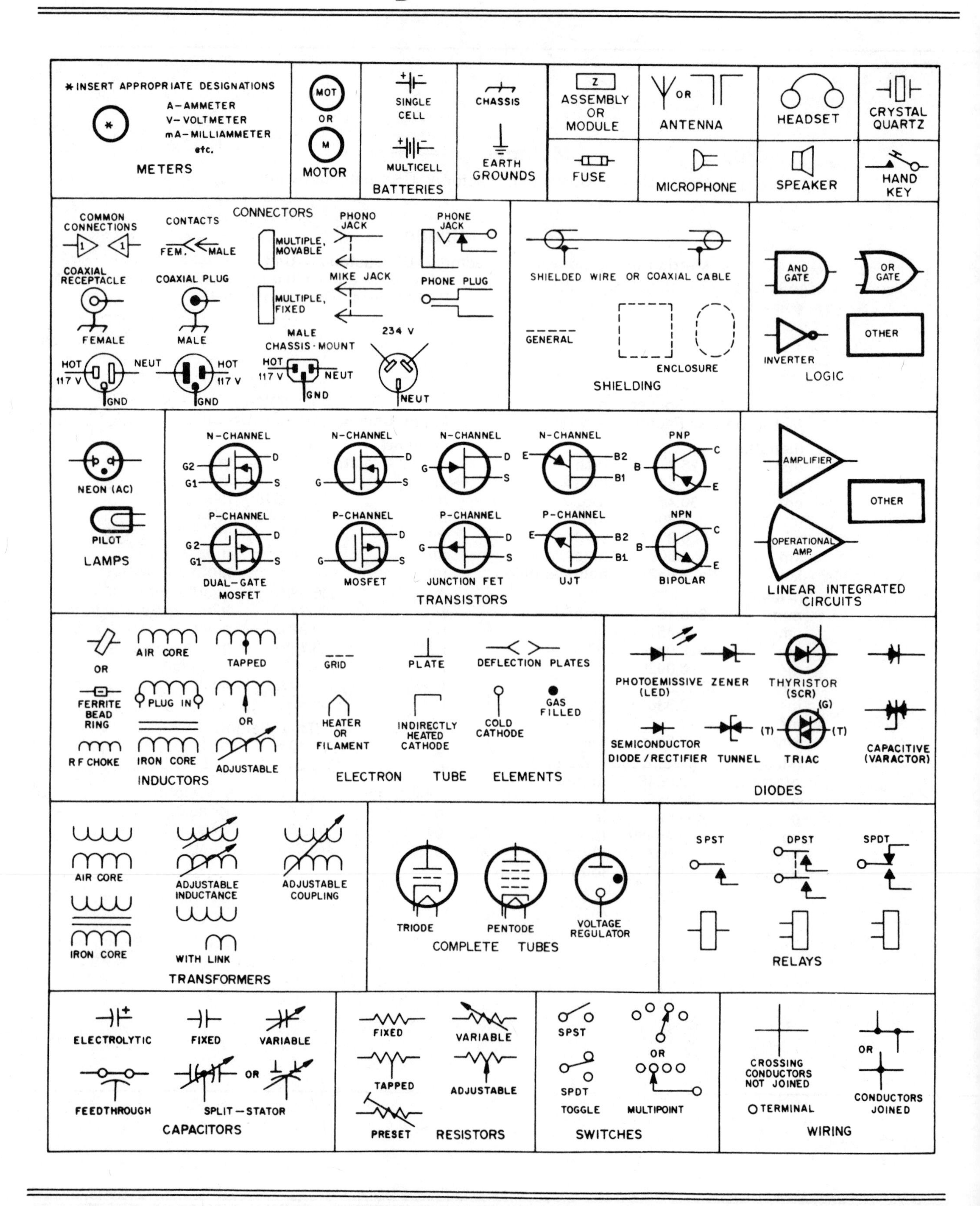

Glossary Of Key Words

A1A emission—The FCC emission designator used to describe Morse code telegraphy (CW) by on/off keying of a radio-frequency signal.

AMTOR—Amateur Teleprinting Over Radio. AMTOR provides error-detecting capabilities. See Automatic Repeat Request and Forward Error Correction.

ASCII—American National Standard Code for Information Interchange. This is a seven-bit digital code used in computer and radioteleprinter applications.

Alternating current (ac)—Electrical current that flows first in one direction in a wire and then in the other. The applied voltage is also changing polarity. This direction reversal continues at a rate that depends on the frequency of the ac.

Alternator—A machine used to generate alternating-current electricity.

Amateur Radio communication—Noncommercial radio communication by or among Amateur Radio stations solely with a personal aim and without pecuniary or business interest. (*Pecuniary* means payment of any type, whether money or other goods.)

Amateur Radio operator—A person holding a valid license to operate an Amateur Radio station. In the US this license is issued by the Federal Communications Commission.

Amateur Radio Service—A radio communication service of self-training, intercommunication, and technical investigation carried on by radio amateurs.

Amateur Radio station—A station licensed in the Amateur Radio Service, including necessary equipment at a particular location, used for Amateur Radio communication.

Ampere (A)—The basic unit of electrical current, equal to 6.24×10^{18} electrons moving past a point in one second.[1] We abbreviate amperes as *amps*.

Amplitude modulation (AM)—A method of combining an information signal and an RF carrier. In double-sideband voice AM transmission, we use the voice information to vary (modulate) the amplitude of a radio-frequency signal. Shortwave broadcast stations use this type of AM, as do stations in the Standard Broadcast Band (510-1600 kHz). Amateurs seldom use double-sideband voice AM, but a variation, known as single sideband, is very popular.

Antenna—A device made from wire or metal tubing. It picks up or sends out radio waves.

Antenna switch—A switch used to connect one transmitter, receiver or transceiver to several different antennas.

[1]Numbers written as a multiple of some power are expressed in *exponential notation*. This notation is explained in detail on page 4-2.

Atom—A basic building block of all matter. Inside an atom, there is a positively charged, dense central core, surrounded by a "cloud" of negatively charged electrons. There are the same number of negative charges as there are positive charges, so the atom is electrically neutral.

Audio frequency (AF)—The range of frequencies that the human ear can detect. Audio frequencies are usually listed as 20 Hz to 20,000 Hz.

Audio-frequency shift keying (AFSK)—A method of transmitting radioteletype information. Two switched audio tones are fed into the microphone input. AFSK RTTY is most often used on VHF.

Automatic Repeat Request (ARQ)—An AMTOR communication mode. In ARQ, also called Mode A, the two stations constantly confirm each other's transmissions. If information is lost, it is repeated until the receiving station confirms correct reception.

Autopatch—A device that allows repeater users to make telephone calls through a repeater.

Balun—Contraction for *bal*anced to *un*balanced. A device to couple a balanced load to an unbalanced source, or vice versa.

Bandwidth—The range of frequencies in the radio spectrum that a radio transmission occupies.

Band spread—A receiver quality used to describe how far apart stations on different nearby frequencies will seem to be. We usually express band spread as the number of kilohertz that the frequency changes per tuning-knob rotation. Note that band spread affects frequency resolution.

Battery—A device that converts chemical energy into electrical energy. It provides excess electrons to produce a current and the voltage or EMF to push those electrons through a circuit.

Baudot—A five-bit digital code used in teleprinter applications.

Baud—The unit used to describe the transmission speed of a digital signal. For a single-channel signal, one baud is equal to one digital bit per second.

Beam antenna—A directional antenna. A beam antenna must be rotated to provide coverage in different directions.

Beat-frequency oscillator (BFO)—A circuit in a receiver that provides a signal to the detector. The BFO signal mixes with the incoming signal to produce an audio tone for CW reception.

Bleeder resistor—A large resistor connected across the output of a power supply. The bleeder discharges the filter capacitors when the supply is turned off.

Block diagram—A picture using boxes to represent sections of a complicated device or process. The block diagram shows the connections between sections.

Breakdown voltage—The voltage that will cause a current in an insulator. Different insulating materials have different breakdown voltages. Breakdown voltage is also related to the thickness of the insulating material.

Calling frequencies—Frequencies set aside for establishing contact. Once two stations are in contact, they should move their QSO to an unoccupied frequency.

Centi—The metric prefix for 10^{-2}, or divide by 100.

Characteristic impedance—The opposition to electric current that an antenna feed line presents. Impedance includes factors other than resistance, and applies to alternating currents. Ideally, the characteristic impedance of a feed line will be the same as the transmitter output impedance and the antenna input impedance.

Chassis ground—The common connection for all parts of a circuit that connect to the negative side of the power supply.

Coaxial cable—coax (pronounced kō'aks). This is a type of feed line with one conductor inside the other. Insulation surrounds the inner conductor, and in turn, the insulation is surrounded by a braided shielding conductor. A plastic covering protects the shield. Sometimes the shielding conductor is solid.

Code-practice oscillator—A device that produces an audio tone, used for learning the code.

Code key—A device used as a switch to generate Morse code.

Communications terminal—A computer-controlled device that demodulates RTTY and CW for display by a computer or ASCII terminal. The communications terminal also accepts information from a computer or terminal and modulates a transmitted signal.

Computer-Based Message System (CBMS)—A system in which a computer is used to store messages for later retrieval. Also called a RTTY mailbox.

Conductor—A material that has a loose grip on its electrons, so that an electrical current can pass through it.

Connected—The condition in which two packet-radio stations are sending information to each other. Each is acknowledging when the data has been received correctly.

Contests—On-the-air operating events. Different contests have different objectives: contacting as many other amateurs as possible in a given amount of time, contacting amateurs in as many different countries as possible or contacting an amateur in each county in one particular state, to name only a few.

Continuous wave (CW)—A term used by amateurs as a synonym for Morse code communication. Hams usually produce Morse code signals by interrupting the continuous-wave signal from a transmitter to form the dots and dashes.

Control operator—A licensed amateur designated to be responsible for the transmissions of an Amateur Radio station.

Coordinated Universal Time (UTC)—A system of time referenced to time at the prime meridian, which passes through Greenwich, England.

CQ—The general call when requesting a conversation with anyone.

Current—A flow of electrons in an electrical circuit.

Dash—The long sound used in Morse code. Pronounce this as "dah" when verbally sounding Morse code characters.

Digipeater—A packet-radio station used to retransmit signals that are specifically addressed to be retransmitted by that station.

Digital communications—The term used to describe Amateur Radio transmissions that are designed to be received and printed automatically. The term also describes transmissions used for the direct transfer of information from one computer to another.

Dipole antenna—See **Half-wave dipole**. A dipole may have lengths other than 1/2 wavelength.

Direct current (dc)—Electrical current that flows in one direction only.

Directivity—The ability of an antenna to focus transmitter power into a beam. Also its ability to enhance received signals from specific directions.

Director—An element in "front" of the driven element in a Yagi antenna.

Direct waves—Radio waves that travel directly from a transmitting antenna to a receiving antenna. Also called "line-of-sight" communications.

Dot—The short sound used in Morse code. Pronounce this as "dit" when verbally sounding Morse code characters if the dot comes at the end of the character. If the dot comes at the beginning or in the middle of the character, pronounce it as "di."

Double-pole, double throw (DPDT) switch—A switch that has six contacts. The DPDT switch has two center contacts. The two center contacts can each be connected to one of two other contacts.

Driven element—The element of an antenna that connects directly to the feed line.

Dummy load (dummy antenna)—A resistor that provides a transmitter with a proper load. The resistor gets rid of the transmitter output power without radiating a signal.

DX—Distance, foreign countries.

DX Century Club (DXCC)—A prestigious award given to amateurs who can prove contact with amateurs in at least 100 DXCC countries.

Earth ground—A circuit connection to a cold-water pipe or to a ground rod driven into the earth.

Electromotive force (EMF)—The force or pressure that pushes a current through a circuit.

Electronic keyer—A device used to generate Morse code dots and dashes electronically. One input generates dots, the other, dashes. Character speed is usually adjusted by turning a control knob. Speeds range from 5 or 10 words per minute up to 60 or more.

Electron—A tiny, negatively charged particle, normally found in an area surrounding the nucleus of an atom. Moving electrons make up an electrical current.

Emission—The transmitted signal from an Amateur Radio station.

Emission privilege—Permission to use a particular emission type (such as Morse code or voice).

Energy—The ability to do work; the ability to exert a force to move some object.

False or deceptive signals—Transmissions that are intended to mislead or confuse those who may receive the transmissions. For example, distress calls transmitted when there is no actual emergency are false or deceptive signals.

Feed line (feeder)—The wires or cable used to connect your transmitter and receiver to an antenna. Also see **Transmission line**.

Field Day—An annual event in which amateurs set up stations in outdoor locations. Emergency power is also encouraged.

Field-effect transistor volt-ohm-milliammeter (FET VOM)—A multiple-range meter used to measure voltage, current and resistance. The meter circuit uses an FET amplifier to provide a high input impedance. This leads to more accurate readings than can be obtained with a VOM. This is the solid-state equivalent of a VTVM.

Fills—Repeats of parts of a previous transmission—usually requested because of interference.

Fist—The unique rhythm of an individual amateur's Morse code sending.

Forward Error Correction (FEC)—A mode of AMTOR communication. In FEC mode, also called Mode B, each character is sent twice. The receiving station checks the mark/space ratio of the received characters. If an error is detected, the receiving station prints a space to show that an incorrect character was received.

Frequency—The number of complete cycles of an alternating current that occur per second.

Frequency bands—A group of frequencies where amateur communications are authorized.

Frequency privilege—Permission to use a particular group of frequencies.

Frequency resolution—The space between markings on a receiver dial. The greater the frequency resolution, the easier it is to separate signals that are close together. Note that frequency resolution affects band spread.

Frequency-shift keying (FSK)—A method of transmitting radioteletype information by switching an RF carrier between two separate frequencies. FSK RTTY is most often used on HF.

Fuse—A thin strip of metal mounted in a holder. When too much current passes through the fuse, the metal strip melts and opens the circuit.

F1B emission—The FCC emission designator used to describe frequency-shift keyed (FSK) digital communications.

F2B emission—The FCC emission designator used to describe audio-frequency shift keyed (AFSK) digital communications.

F3E emission—The FCC emission designator used to describe FM voice communications.

Gain—A comparison of how much signal two different antennas will pick up. Also a comparison of the transmitter signal enhancement from two different antennas.

General-coverage receiver—A receiver used to listen to both the shortwave-broadcast frequencies and the amateur bands.

Giga—The metric prefix for 10^9, or times 1,000,000,000.

Grid—The control element (or elements) in a vacuum tube.

Ground connection—A connection made to the earth for electrical safety.

Ground waves—Radio waves that travel along the surface of the earth.

Half-wave dipole—A basic antenna used by radio amateurs. It consists of a length of wire or tubing, opened and fed at the center. The entire antenna is 1/2 wavelength long at the desired operating frequency.

Hand key—A simple switch used to send Morse code.

Ham-bands-only receiver—A receiver designed to cover only the bands used by amateurs.

Hertz (Hz)—An alternating-current frequency of one cycle per second. The basic unit of frequency.

Impedance-matching network—A device that matches the impedance of an antenna system to the impedance of a transmitter or receiver. Also called an antenna-matching network or Transmatch.

Input frequency—A repeater's receiving frequency.

Insulator—A material that maintains a tight grip on its electrons, so that an electrical current cannot pass through it.

Inverted-V dipole—A half-wave dipole antenna with its center elevated and the ends drooping toward the ground. Amateurs sometimes call this antenna an "inverted V."

Ionosphere—A region of charged particles high above the earth. The ionosphere bends radio waves as they travel through it, returning them to earth.

Ion—An electrically charged particle. An electron is an ion. Another example of an ion is the nucleus of an atom that is surrounded by too few or too many electrons. An atom like this has a net positive or negative charge.

J3E emission—The FCC emission designator used to describe single-sideband suppressed-carrier voice communications.

Kilo—The metric prefix for 10^3, or times 1000.

Ladder line—Parallel-conductor feeder with insulating spacer rods every few inches.

Line of sight—The term used to describe VHF and UHF propagation in a straight line directly from one station to another.

Lower sideband (LSB)—The common mode of single-sideband transmission used on the 40, 80 and 160-meter amateur bands.

Malicious interference—Intentional, deliberate obstruction of radio transmissions.

Matching network—A device that matches one impedance level to another. For example, it may match the impedance of an antenna system to the impedance of a transmitter or receiver. Amateurs also call such devices a Transmatch, impedance-matching network, or match box.

Maximum usable frequency (MUF)—The greatest frequency at which radio signals will return to a particular location from the ionosphere. The MUF may vary for radio signals sent to different destinations.

Mega—The metric prefix for 10^6, or times 1,000,000.

Metric prefixes—A series of terms used in the metric system of measurement. We use metric prefixes to describe a quantity as compared to a basic unit. The metric prefixes indicate multiples of 10.

Metric system—A system of measurement developed by scientists and used in most countries of the world. This system uses a set of prefixes that are multiples of 10 to indicate quantities larger or smaller than the basic unit.

Micro—The metric prefix for 10^{-6}, or divide by 1,000,000.

Microphone—A device that converts sound waves into electrical energy.

Milli—The metric prefix for 10^{-3}, or divide by 1000.

Mobile operation—Amateur Radio operation conducted while in motion or at temporary stops at different locations.

Modem—Short for modulator/demodulator. A modem modulates a radio signal to transmit data and demodulates a received signal to recover transmitted data.

Monitor mode—One type of packet-radio receiving mode. In monitor mode, everything transmitted on a packet frequency is displayed by the monitoring TNC. This occurs whether the transmissions are addressed to the monitoring station or not.

Multiband antenna—An antenna that will operate well on more than one frequency band.

Multimode transceiver—A VHF or UHF transceiver capable of SSB, CW and FM operation.

Negative Charge—One of two types of electrical charge. The electrical charge of a single electron.

Nets—Groups of amateurs who meet on the air to pass traffic or communicate about a specific subject. One station (called the *net control station*) usually directs the net.

Network—A term used to describe several packet stations linked together to transmit data over long distances.

Neutral—Having no electrical charge, or having an equal number of positive and negative charges.

Nucleus—The dense central portion of an atom. The nucleus contains positively charged particles.

Offset—The slight difference in transmitting and receiving frequencies in a transceiver.

Ohm's Law—A basic law of electronics. Ohm's Law gives a relationship between voltage, resistance and current ($E = IR$).

Ohm—The basic unit of electrical resistance, used to describe the amount of opposition to current.

Omnidirectional—Antenna characteristic meaning it radiates equal power in all compass directions.

Operator license—The portion of an Amateur Radio license that gives permission to operate an Amateur Radio station.

Open circuit—An electrical circuit that does not have a complete path, so current can't flow through the circuit.

Open-wire feed line—Parallel-conductor feeder with air as its primary insulation material.

Output frequency—A repeater's transmitting frequency.

Packet Bulletin-Board System (PBBS)—A computer system used to store packet-radio messages for later retrieval by other amateurs.

Packet radio—A system of digital communication whereby information is broken into short bursts. The bursts ("packets") also contain addressing and error-detection information.

Parallel circuit—An electrical circuit where the electrons follow more than one path.

Parallel-conductor feed line—Feed line with two conductors spaced uniformly.

Peak envelope power (PEP)—The average power of a signal at its largest amplitude peak.

Phone—Voice communications.

Pico—The metric prefix for 10^{-12}, or divide by 1,000,000,000,000.

Polarization—Describes the characteristic of a radio wave. An antenna that is parallel to the surface of the earth, such as a dipole, produces horizontal polarization. One that is perpendicular to the earth's surface, such as a quarter-wave vertical, produces vertical polarization.

Portable operation—Amateur Radio operation conducted away from the location shown on the station license.

Positive charge—One of two types of electrical charge. A positive charge is the opposite of a negative charge. Electrons have a negative charge. The nucleus of an atom has a positive charge.

Potentiometer—Another name for a variable resistor. The value of a potentiometer can be changed without removing it from a circuit.

Power—The rate of energy consumption. We calculate power in an electrical circuit by multiplying the voltage applied to the circuit times the current through the circuit.

Power supply—That part of an electrical circuit that provides excess electrons to flow into a circuit. The power supply also supplies the voltage or EMF to push the electrons along. Power supplies convert a power source (such as the ac mains) to a useful form. (A circuit that provides a direct-current output at some desired voltage from an ac input voltage.)

Procedural signal (prosign)—One or two letters sent as a single character. Amateurs use prosigns in CW QSOs as a short way to indicate the operator's intention. Some examples are "K" for "Go Ahead," or "$\overline{AR}$" for "End of Message." (The bar over the letters indicates that we send the prosign as one character.)

Propagation—The study of how radio waves travel from one place to another.

Q signals—Three-letter symbols beginning with "Q." Q signals are used in amateur CW work to save time and for better communication.

QSL card—A postcard sent to another radio amateur to confirm a contact.

QSO—A conversation between two radio amateurs.

Radiate—To convert electric energy into electromagnetic (radio) waves. Radio waves radiate from an antenna.

Radioteletype (RTTY)—Radio signals sent from one teleprinter machine to another machine. Anything that one operator types on his teleprinter will be printed on the other machine. (A type of digital communications.)

Radio frequency (RF)—The range of frequencies that can be radiated through space in the form of electromagnetic radiation. We usually consider RF to be those frequencies higher than the audio frequencies, or above 20 kilohertz.

Ragchew—A lengthy conversation (or QSO) between two radio amateurs.

Random-length wire antenna—An antenna having a length that is not necessarily related to the wavelength of a desired signal.

Receiver—A device that converts radio signals into audio signals.

Receiver incremental tuning (RIT)—A transceiver control that allows for a slight change in the receiver frequency without changing the transmitter frequency. Some manufacturers call this a Clarifier (CLAR) control.

Reflector—An element in "back" of the driven element in a Yagi antenna.

Repeater—An amateur station that receives a signal and retransmits it for greater range.

Resistance—The ability to oppose an electric current.

Resistor—Any material that opposes a current in an electrical circuit. An electronic component especially designed to oppose current.

Resonant frequency—The desired operating frequency of a tuned circuit. In an antenna, the resonant frequency is one where the feed-point impedance contains only resistance.

RF burn—A flesh burn caused by exposure to a strong field of RF energy.

Rig—The radio amateur's term for a transmitter, receiver or transceiver.

Rotary switch—A switch that connects one center contact to several individual contacts. An antenna switch is one common use for a rotary switch.

RST—A system of numbers used for signal reports: R is readability, S is strength and T is tone.

Schematic symbol—A drawing used to represent a circuit component on a wiring diagram.

Secondary station identifier (SSID)—A number added to a packet-radio station's call sign so that one amateur call sign can be used for several packet stations.

Selective-call identifier—A four-character AMTOR station identifier.

Selectivity—The ability of a receiver to separate two closely spaced signals.

Semiconductor—Material that has some properties of a conductor and some properties of an insulator.

Sensitivity—The ability of a receiver to detect weak signals.

Series circuit—An electrical circuit where the electrons must all flow through every part of the circuit. There is only one path for the current to follow.

Shack—The room where an Amateur Radio operator keeps his or her station equipment.

Short circuit—An electrical circuit where the current does not take the desired path, but finds a shortcut instead. Often the current goes directly from the negative power-supply terminal to the positive one, bypassing the rest of the circuit.

Simplex operation—A term normally used in relation to VHF and UHF operation. Simplex means you are receiving and transmitting on the same frequency.

Single sideband (SSB)—A common mode for voice operation on the amateur high-frequency bands. This is a variation of amplitude modulation.

Single-pole, double-throw (SPDT) switch—A switch that connects one center contact to one of two other contacts.

Single-pole, single-throw (SPST) switch—A switch that only connects one center contact to another contact.

Sine wave—A smooth curve, usually drawn to represent the variation in voltage or current over time for an ac signal.

Skip—Radio waves that are bent back to earth by the ionosphere. Skip is also called *sky-wave propagation*.

Skip zone—An area past the maximum range of ground waves and before the range of waves returned from the ionosphere. An area where radio communications between stations is not possible on a certain frequency.

Sky waves—Radio waves that travel through the ionosphere and back to earth. Sky-wave propagation is sometimes called skip.

Sloper—A 1/2-wave dipole antenna that has one end elevated and one end nearer the ground.

Solid-state devices—Circuit components that use semiconductor materials. Semiconductor diodes, transistors and integrated circuits are all solid-state devices.

Speech processor—A device that increases the average power of a sideband signal, making the voice easier to understand under weak signal conditions.

Splatter—The term used to describe a very wide-bandwidth signal. Splatter is usually caused by an improperly adjusted sideband transmitter.

Standing-wave ratio (SWR)—Sometimes denoted as VSWR. A measure of the impedance match between the feed line and the antenna. Also, with a Transmatch in use, a measure of the match between the feeder from the transmitter and the antenna *system*. The system includes the Transmatch and the line to the antenna.

Standing-wave-ratio (SWR) meter—A device used for measuring SWR. SWR is a relative measure of the impedance match between an antenna, feed line and transmitter.

Station license—The portion of an Amateur Radio license that authorizes an amateur station at a specific location. The station license also lists the call sign of that station.

Subatomic particles—The building blocks of atoms. Electrons, protons and neutrons are the most common subatomic particles.

Sunspots—Dark spots on the surface of the sun. When there are few sunspots, long-distance radio propagation is poor on the higher-frequency bands.

Switch—A device used to connect or disconnect electrical contacts.

SWR meter—A device used to measure SWR. A measuring instrument that can indicate when an antenna system is working well.

Teleprinter—A machine that can convert keystrokes (typing) into electrical impulses. The teleprinter can also convert the proper electrical impulses back into text. Hams use teleprinters for radioteletype work.

Terminal node controller (TNC)—A TNC accepts information from a computer or terminal and converts the information into packets by including address and error-checking information. The TNC also receives packet signals and extracts transmitted information for display by a computer.

Third-party participation (or communication)—The way an unlicensed person can participate in Amateur Radio communications. A control operator must ensure compliance with FCC rules.

Third-party traffic—Messages passed from one amateur to another on behalf of a third person.

Ticket—The radio amateur's term for an Amateur Radio license.

Traffic—Messages passed from one amateur to another in a relay system; the amateur version of a telegram.

Traffic net—An on-the-air meeting of amateurs, for the purpose of relaying messages.

Transceiver—A radio transmitter and receiver combined in one unit.

Transformer—A device that changes ac voltage levels.

Transmatch—See **Matching network**.

Transmission line—The wires or cable used to connect a transmitter or receiver to an antenna.

Transmitter—A device that produces radio-frequency signals.

Transmit-receive (TR) switch—A device used for switching between transmit and receive operation. The TR switch allows you to connect one antenna to a receiver and a transmitter. As you operate the switch, it connects the antenna to the correct unit.

Triode—A vacuum tube with three active elements: cathode, plate and control grid.

Tropospheric enhancement—A weather-related phenomenon. Tropo can produce unusually long propagation on the VHF and UHF bands.

Twin lead—Parallel-conductor feeder with wires encased in insulation.

Unidentified communications or signals—Signals or radio communications in which the transmitting station's call sign is not transmitted.

Upper sideband (USB)—The common single-sideband operating mode on the 20, 17, 15, 12 and 10-meter HF amateur bands. Hams also use upper sideband on all the VHF and UHF bands.

Vacuum-tube voltmeter (VTVM)—A multiple-range meter used to measure voltage, current and resistance. The meter circuit includes a vacuum-tube amplifier to provide a high input impedance. This leads to more accurate readings than can be obtained with a VOM.

Variable-frequency oscillator (VFO)—A circuit used to control the frequency of an amateur transmitter.

Vertical antenna—A common amateur antenna, usually made of metal tubing. The radiating element is vertical. There are usually four or more radial conductors parallel to or on the ground.

Voltage—The EMF or pressure that causes electrons to move through an electrical circuit.

Voltage source—Any source of excess electrons. A voltage source produces a current and the force to push the electrons through an electrical circuit.

Volt (V)—The basic unit of electrical pressure or EMF.

Volt-ohm-milliammeter (VOM)—A multiple-range meter used to measure voltage, current and resistance. This is the least expensive (and least accurate) type of meter.

W1AW—The headquarters station of the American Radio Relay League. This station is a memorial to the League's cofounder, Hiram Percy Maxim. The station provides daily on-the-air code practice and bulletins of interest to hams.

Watt (W)—The unit of power in the metric system. The watt describes how fast a circuit uses electrical energy.

Wavelength—Often abbreviated λ. The distance a radio wave travels in one RF cycle. The wavelength relates to frequency. Higher frequencies have shorter wavelengths.

Yagi antenna—The most popular type of amateur directional (beam) antenna. It has one driven element and one or more additional elements.

Zero beat—When two operators in a QSO are transmitting on the same frequency.

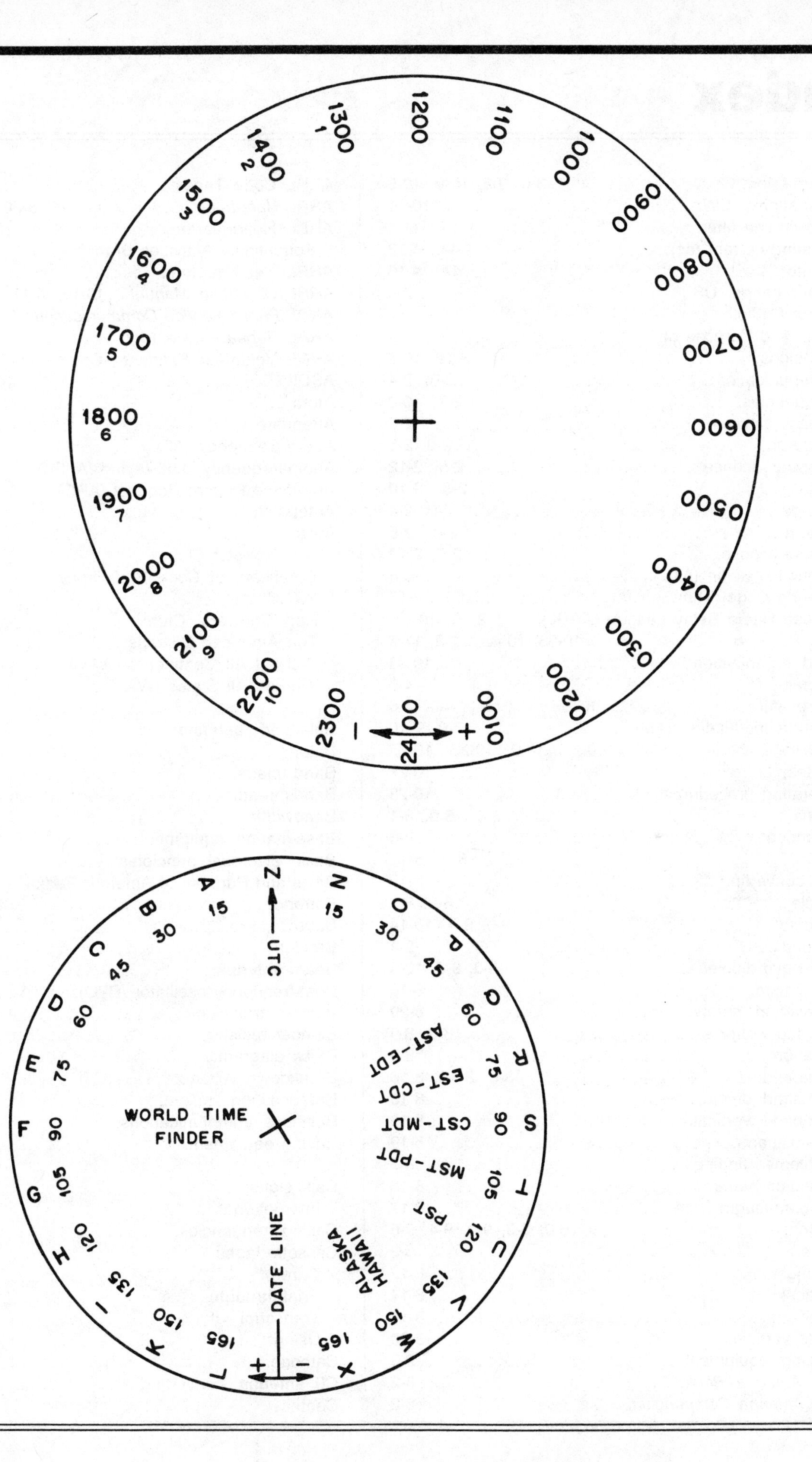

1200
1100
1000
0900
0800
0700
0600
0500
0400
0300
0200
0100
2400
2300
2200
2100
2000
1900
1800
1700
1600
1500
1400
1300
WORLD TIME
FINDER
UTC
DATE LINE
ALASKA
HAWAII
AST-EDT
EST-CDT
CST-MDT
MST-PDT
PST

Index

TUNE IN THE WORLD

PROOF OF PURCHASE

Advertising Index

BEST OF MFJ

MFJ, Bencher and Curtis team up to bring you America's most popular keyer in a compact package for smooth easy CW

MFJ-422B **$129.95**

The best of all CW world's -- a deluxe MFJ Keyer using a Curtis 8044ABM chip in a compact package that fits right on the Bencher iambic paddle!

This MFJ Keyer is small in size but big in features. you get iambic keying, adjustable weight and tone and front panel volume and speed controls (8-50 WPM), dot-dash memories, speaker, sidetone and push button selection of automatic or semi-automatic/ tune modes. It's also totally RF proof and has ultra-reliable solid state outputs that key both tube and solid state rigs. Use 9 volt battery or 110 VAC with MFJ-1305, $12.95.

The keyer mounts on a Bencher paddle to form a small (4-1/8 x 2-5/8 x 5½ inches) attractive combination that is a pleasure to look at and use.

The Bencher paddle has adjustable gold plated silver contacts, lucite paddles, chrome plated brass and a heavy steel base with non-skid feet.

You can buy just the keyer assembly, MFJ-422BX, for only $79.95 to mount on **your** Bencher paddle.

Deluxe 300 W Tuner

MFJ-949D **$149.95**

MFJ-949D is the world's most popular 300 watt PEP tuner. It covers 1.8-30 MHz, gives you a new peak and average reading Cross-Needle SWR/Wattmeter, built-in dummy load, 6 position antenna switch and 4:1 balun -- in a compact 10 x 3 x 7 inch cabinet. Meter lamp uses 12 VDC or 110 VAC with MFJ-1312, $12.95.

Antenna Bridge

MFJ-204B **$79.95**

Now you can quickly optimize your antenna for peak performance with this portable, totally self-contained antenna bridge.

No other equipment needed -- take it to your antenna site. Determine if your antenna is too long or too short, measure its resonate frequency and antenna resistance to 500 ohms. It's the easiest, most convenient way to determine antenna performance. Built in resistance bridge, null meter, tunable oscillator-driver (1.8-30 MHz). Use 9 V battery or 110 VAC with AC adapter, $12.95.

Super Active Antenna

"World Radio TV Handbook" says MFJ-1024 is a "first rate easy-to-operate active antenna ... quiet ... excellent dynamic range ... good gain ... very low noise ... broad frequency coverage ... excellent choice."

Mount it outdoors away from electrical noise for maximum signal, minimum noise. Covers 50 KHz to 30 MHz.

Receives strong, clear signals from all over the world. 20 dB attenuator, gain control, ON LED. Swtich two receivers and aux. or active antenna. 6x3x5 in. Remote unit has 54 inch whip, 50 ft. coax and connector. 3x2x4 in. Use 12 VDC or 110 VAC with MFJ-1312, $12.95.

MFJ-1024 **$129.95**

VHF SWR/Wattmeter

MFJ-812B **$29.95**

Covers 2 Meters and 220 MHz. 30 or 300 Watt scales. Also reads relative field strength 1-170 MHz and SWR above 14 MHz. 4½x2¼x3 in.

MFJ Coax Antenna Switches

$34.95 MFJ-1701

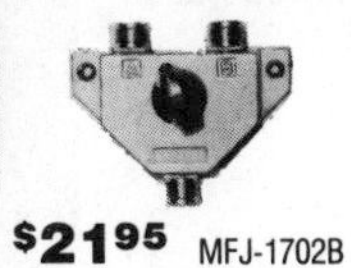

$21.95 MFJ-1702B

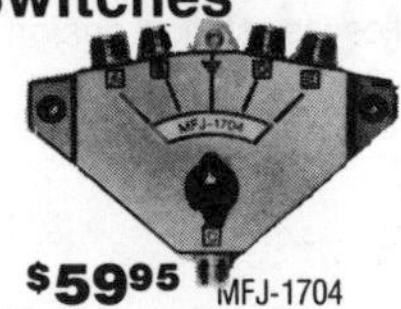

$59.95 MFJ-1704

Select any of several antennas from your operating desk with these MFJ Coax Switches. They feature mounting holes and automatic grounding of unused terminals. They come with MFJ's one year **unconditional** guarantee.

MFJ-1701, $34.95. Six position antenna switch. SO-239 connectors. 50-75 ohm loads. 2 KW PEP, 1 KW CW. Black alum. cabinet. 10x3x1½ inches.

MFJ-1702B, $21.95. 2 positions plus new Center Ground. 2.5 KW PEP, 1 KW CW. Insertion loss below .2 dB. 50 dB isolation at 450 MHz. 50 ohm. 3x2x2 in.

MFJ-1704, $59.95. 4 position cavity switch with lightening/surge protection device. Center ground. 2.5 KW PEP, 1 KW CW. Low SWR. Isolation better than 50 dB at 500 MHz. Negligible loss. 50 ohm. 6¼x4¼x1¼ in.

"Dry" Dummy Loads for HF/VHF/UHF

MFJ-260B **$28.95**

MFJ-262 **$69.95**

MFJ-264 **$109.95**

MFJ has a full line of dummy loads to suit your needs. Use a dummy load for tuning to reduce needless (and illegal) QRM and save your finals.

MFJ-260B, $28.95. VHF/HF. Air cooled, non-inductive 50 ohm resistor. SO-239 connector. Handles 300 Watts. Run full load for 30 seconds, derating curve to 5 minutes. SWR less than 1.3:1 to 30 MHz, 1.5:1 to 150 MHz. 2½x2½x7 in.

MFJ-262, $69.95. HF.1 KW. SWR less than 1.5:1 to 30 MHz. 3x3x13 in.

MFJ-264, $109.95. Versatile UHF/VHF/HF 1.5 KW load. Low SWR to 650 MHz, usable to 750 MHz. Run 100 watts for 10 minutes, 1500 watts for 10 seconds. SWR is 1.1:1 to 30 MHz, below 1.3:1 to 650 MHz. 3x3x7 inches.

MFJ Ham License Upgrade Theory Tutor

MFJ Theory Tutor practically guarantees you'll pass the theory part of any FCC ham license exam. Versatile MFJ software is the best computer tutor ever tailor-made for ham radio. You can study the entire FCC question pool, selected areas and take (or print) sample tests. Auto. saves each study session (ex. sample tests), gives you all FCC test graphics (ex. mono.), explanations of hard questions, pop-up calculator, weighted scoring analysis, color change option and more. **Order** MFJ-1610-**Novice**; MFJ-1611-**Tech.**; MFJ-1612-**Gen.**; MFJ-1613-**Adv.**; MFJ-1614-**Ex.** for **IBM compatible.** For **Macintosh:** MFJ-1630-N; MFJ-1631-T; MFJ-1632-G; MFJ-1633-A; MFJ-1634-E, **$29.95** per license class.

MFJ Speaker Mics

MFJ-284 or MFJ-286 **$24.95**

MFJ's compact Speaker/Mics let you carry your HT on your belt and never have to remove it to monitor calls or talk.

You get a wide range speaker and first-rate electret mic element for superb audio on both transmit and receive.

Earphone jack, handy lapel/pocket clip, PTT, lightweight retractable cord. Gray. One year **unconditional** guarantee.

MFJ-284 fits ICOM, Yaesu, Santec. MFJ-286 fits Kenwood.

MFJ-1278 Multi-Mode Data Controller

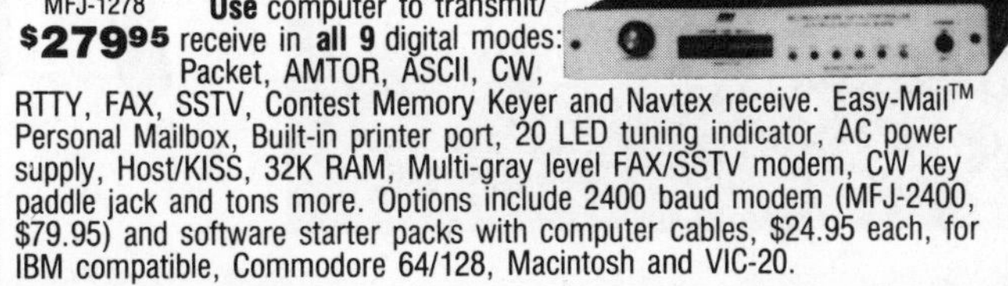

MFJ-1278 **$279.95**

Use computer to transmit/ receive in **all 9** digital modes: Packet, AMTOR, ASCII, CW, RTTY, FAX, SSTV, Contest Memory Keyer and Navtex receive. Easy-Mail™ Personal Mailbox, Built-in printer port, 20 LED tuning indicator, AC power supply, Host/KISS, 32K RAM, Multi-gray level FAX/SSTV modem, CW key paddle jack and tons more. Options include 2400 baud modem (MFJ-2400, $79.95) and software starter packs with computer cables, $24.95 each, for IBM compatible, Commodore 64/128, Macintosh and VIC-20.

12/24 Hour LCD Clocks

$19.95 MFJ-108B

$9.95 MFJ-107B

Huge 5/8 inch bold LCD digits let you see the time from anywhere in your shack. Choose from the dual clock that has separate UTC/local time display or the single 24 hour ham clock.

Mounted in a brushed aluminum frame. Easy to set. The world's most popular ham clocks for accurate logs. MFJ-108B 4½x1x2; MFJ-107B 2¼x1x2 in.

Cross-Needle SWR/Wattmeter

MFJ-815B **$69.95**

MFJ Cross-Needle SWR/ Wattmeter has a new **peak** reading function! It shows you SWR, forward and reflected power in 2000/500 and 200/50 watt ranges. Covers 1.8-30 MHz.

Mechanical zero adjusts for movement. SO-239 connectors. Lamp uses 12 VDC or 110 VAC with MFJ-1312, $12.95.

Deluxe Code Practice Oscillator

MFJ-557 **$24.95**

MFJ-557 Deluxe Code Practice Oscillator has a Morse key and oscillator unit mounted together on a heavy steel base so it stays put on your table. Portable because it runs on a 9-volt battery (not included) or an AC adapter ($12.95) that plugs into a jack on the side.

Earphone jack for private practice, Tone and Volume controls for a wide range of sound. Speaker. Key has adjustable contacts and can be hooked to your transmitter. Sturdy. 8½x2¼x3¾ in.

AC Volt Monitor

MFJ-850 **$19.95**

Prevent damage to rig, computer or other gear. Monitor AC line voltage for potentially damaging power surge or brown out conditions. Expanded 95-135 volt 2-color scale. Plugs into any AC outlet. 2% accuracy. 2¼x2¼x1½ in.

Nearest Dealer/Orders: 800-647-1800

MFJ ENTERPRISES, INC.
Box 494, Miss. State, MS 39762
(601) 323-5869; **FAX:** (601) 323-6551
TELEX: 53 4590 MFJ STKV

• One year **unconditional** guarantee • 30 day **money back** guarantee (less s/h) on orders from MFJ • Add $5.00 each s/h • **FREE** catalog

MFJ . . . making quality affordable

Upgrade

After you receive your Novice license and get on the air, you'll most likely wish for more operating privileges. That's where ARRL instructional materials come in. Our critically acclaimed *License Manual Series* will take you from Technician class through the top-of-the-line Extra Class license which allows you the most privileges available to Radio Amateurs. The *FCC Rulebook* is invaluable as a study guide for the regulatory material found on the exams and as a handy reference. *Every* Amateur needs an up-to-date copy. *First Steps in Radio* presents electronic principles for the beginner.

Morse Code the Essential Language has tips on learning the code, high speed operation and history. If you have a Commodore 64™ or C 128 computer, *Morse University* provides hours of fun and competition in improving your code proficiency. *GGTE Morse Tutor Software* for the IBM PC and compatibles teaches the code, gives you plenty of practice for exams, and helps keep your code skills sharp. Besides the cassettes in *Tune In The World*, we have four sets of tapes that give you practice from 5 to 22 words per minute.

the easy way!

***Tune In The World With Ham Radio**
With book and cassettes ...... # **$19**
Book only .... # **$14**

License Manual Series
*Technician Class .... **#237-5 $ 6**
*General Class .... **#238-3 $ 6**
Advanced Class .... **#016-X $ 5**
Extra Class .... **#239-1 $ 8**
*FCC Rulebook .... # **$ 9**

First Steps in Radio .... #228-6 $ 5

Code Proficiency
Morse Code the Essential Language .... **#035-6 $ 5**

Morse University TM
(includes *Tune In* book) ..... **#025-9 $40**
Morse Tutor .... **#208-1 $20**

Code Practice Cassettes—Each set of two C-90 tapes gives 3 hours of instruction.
Set 1: 5 to 10 WPM (18 WPM Char Speed) ...... **#222-7 $10**
Set 2: 10 to 15 WPM (18 WPM Char Speed) ...... **#223-5 $10**
Set 3: 15 to 22 WPM (See ** below) .... **#224-3 $10**
Set 4: 13 to 14 WPM (Actual Speed) .... **#225-1 $10**

***New Edition**
****Up to and including 18 WPM, the Character Speed is 18 WPM. Above 18 WPM, the Character Speed is the actual speed.**

ARRL
225 MAIN ST.
NEWINGTON, CT 06111

Please use this form to give us your comments on this book and what you'd like to see in future editions.

Name ______________________________ Age __________ Call Sign ______________

Address ______________________________ Daytime Phone (   ) ______________

City ____________________ State/Province ______________ ZIP/Postal Code ____________

From ______________________________

____

____

Please affix postage. Post office will not deliver without sufficient postage

Editor, Tune in the World
Eighth Edition
American Radio Relay League
225 Main Street
Newington, CT USA 06111

please fold and tape